Environmental and Occupational Exposure to Chemical Agents and Health Challenges

Environmental and Occupational Exposure to Chemical Agents and Health Challenges

Guest Editor

Carina Ladeira

MDPI

Basel • Beijing • Wuhan • Barcelona • Belgrade • Novi Sad • Cluj • Manchester

Guest Editor
Carina Ladeira
Escola Superior de
Tecnologia da Saúde
Polytechnic University of Lisbon
Lisbon
Portugal

Editorial Office
MDPI AG
Grosspeteranlage 5
4052 Basel, Switzerland

This is a reprint of the Special Issue, published open access by the journal *Toxics* (ISSN 2305-6304), freely accessible at: https://www.mdpi.com/journal/toxics/special_issues/DAECZZA0C1.

For citation purposes, cite each article independently as indicated on the article page online and as indicated below:

Lastname, A.A.; Lastname, B.B. Article Title. *Journal Name* **Year**, *Volume Number*, Page Range.

ISBN 978-3-7258-2703-9 (Hbk)
ISBN 978-3-7258-2704-6 (PDF)
https://doi.org/10.3390/books978-3-7258-2704-6

Contents

About the Editor

Carina Ladeira

Carina Ladeira is PhD in Biology with specialization in Genetics and Master in Human Molecular Biology from the Faculty of Sciences of the University of Lisbon and has a 4-year degree in Biomedical Laboratorial Sciences. She is Full Professor at the Lisbon School of Health Technology in the Biomedical Laboratory Sciences degree and in the Master's Degree in Clinical-Laboratory Technologies, collaborating also in other postgraduate programs. She has participated in several international and national scientific projects, author of 1 book, 8 book chapters, around 40 scientific articles in journals with an impact factor, 50 oral communications and 100 poster communications. She was member of scientific and organizing committees for several national and international events, and served as moderator of roundtables at scientific events. She is European Registered Toxicologist (ERT) and has been expert evaluator of scientific projects at international level. She is the Chair of the International Comet Assay Working Group (ICAWG), a satellite group of the European Environmental Mutagenesis and Genomics Society (EEMGS). Her research interests are focused in human biomonitoring, genetic toxicology, biomarkers, chemicals mixtures, exposome, risk assessment, carcinogenesis and prevention.

Preface

Human beings live in constant contact with chemical agents. Environmental exposure is ubiquitous, originating from air, water, and soil. While some chemicals are short-lived in the environment and may elicit no harmful effects in humans, others bioaccumulate and/or persist for a long time in the environment or the human body. It is also necessary to highlight that environmental exposure refers to lifestyle, including diet and hobbies, which can lead to exposure to other substances such as drugs, food additives, pesticides, and nanomaterials, among other daily products. Another important source of chemical exposure is professional activities. Occupational exposure concerns the potentially harmful exposure of humans to hazardous chemicals in the workplace, often resulting in more constant and substancial contact with specific substances. Possible health effects can arise from these types of exposure, which can be measured, controlled and prevented using biomonitoring.

This reprint aimed to gather studies on scientific environmental and occupational exposure that employ biomonitoring tools to enhnance knowledge by providing a better understanding of human biomonitoring outcomes, and studies that demonstrate the role of biomonitoring in ensuring better regulatory decision-making.

Finally, I would like to express my personal gratitute to all the authors involved in this Special Issue and to Ms. Selena Li, the Editorial Assistant for *Toxics*.

Carina Ladeira
Guest Editor

toxics

MDPI

Editorial

Environmental and Occupational Exposure to Chemical Agents and Health Challenges I—What Message Can Bring to Regulatory Science?

Carina Ladeira [1,2]

1 H&TRC-Health & Technology Research Center, ESTeSL-Escola Superior de Tecnologia da Saúde, Instituto Politécnico de Lisboa, 1990-096 Lisbon, Portugal; carina.ladeira@estesl.ipl.pt

2 NOVA National School of Public Health, Public Health Research Centre, Universidade NOVA de Lisboa, Comprehensive Health Research Centre (CHRC), 1600-560 Lisbon, Portugal

Human beings live in constant contact with chemical agents—mainly through environmental exposure—and also derived from the occupational settings. Environmental exposure is ubiquitous, originating from air, water, and soil [1]. While some chemicals are short-lived in the environment and may elicit no harmful effects in humans, other chemicals bioaccumulate and/or persist for a long time in the environment or the human body due to frequent exposure. However, it is valuable to refer that environmental exposure also includes diet, lifestyle, hobbies, and exposure to other substances such as drugs, food additives, pesticides, and nanomaterials, among other daily products, which are currently significant areas of research, such as bisphenols, per- and polyfluoroalkyl substances (PFAS), micro-nanoplastics (MP-NPs), among others [2–8]. Occupational exposure concerns the potentially harmful exposure to hazardous chemicals in the workplace; however, more specifically, it involves a substantial contact with hazardous substances [9]. Possible health effects can arise from these types of exposure, which can be measured and prevented by biomonitoring, and the outcome should be integrated to ensure better regulatory decision making.

This editorial provides a brief overview of the major findings of each of the research published, and taken together which information can give to surveillance health and regulatory science on the future research that should be considered.

This Special Issue is focused on 9 selected topics, one perspective paper (contribution 1), 2 reviews (contributions 2 and 3), and six original research papers (contributions 4–9), being three regarding human biomonitoring and occupational exposure (contributions 4-6), one in human biomonitoring and environmental exposure (contribution 7), other of environmental exposure (contribution 8), and an in vitro study (contribution 9).

The perspective paper from Myriam Borgatta and Florian Breider (contribution 1) addresses important health concerns related to inhalation of microplastics (MPs), taking in consideration three aspects—particles characteristics, additives present in plastics, and exogenous substances absorbed onto them, aiming to present a comprehensive toxicological profile of deposited MPs in the lungs, encompassing local and systemic effects.

The systematic scoping review from Ladeira et al. (contribution 2) gathered around 300 original studies of human biomonitoring of environmental and occupational exposures to chemicals using comet assay as biomarkers of DNA damage. The studies are divided in six groups of chemicals, namely: air pollutants, anaesthetics, antineoplastic drugs, heavy metals, pesticides and solvents; covering a wide range of chemicals which have been target of human biomonitoring activities.

The review from Ma et al. (contribution 3) provides an extensive overview of the application scenarios, classification, and challenges associated with hand sanitizer gels (HSGs)—due to SARS-CoV-2 their use highly increased—while emphasizing the emergence of novel components with biological functions, aiming to contribute to the advancement of

Citation: Ladeira, C. Environmental and Occupational Exposure to Chemical Agents and Health Challenges I—What Message Can Bring to Regulatory Science? *Toxics* **2024**, *12*, 778. https://doi.org/10.3390/toxics12110778

Received: 10 September 2024
Revised: 21 October 2024
Accepted: 23 October 2024
Published: 25 October 2024

hand hygiene practices and offer novel insights for the development of novel HSGs with outstanding antimicrobial properties with other multiple biological functions and desirable biosafety profiles.

Regarding the studies of human biomonitoring in occupational exposure, the study from Fustinoni et al. (contribution 4) investigated the kinetics of excretion of a new generation perfluoroalkyl surfactant (cC_6O_4) using blood and urine samples from workers occupationally exposed to this substance. The study concluded that cC_6O_4 has a shorter biopersistence in comparison with legacy per- and polyfluoroalkyl substances (PFAS), giving important scientific based evidence information which can directly impact workers health and the regulatory bodies for substances substitution. The study from Cui et al. (contribution 5) used urinary biomarkers to estimate benzene exposure levels in individuals exposed to benzene. In order to achieve that, biomarkers of internal exposure to benzene were quantified by UPLC-MS/MS such as trans, trans-muconic (*t,t*-MA) and S-phenyl mercapturic acid (S-PMA) metabolites, and a simple pharmacokinetic model was used to calculate back-calculated airborne benzene levels (BCABL). The results of this study showed that the *t,t*-MA-based BCABL can reflect the actual airborne benzene level and that the S-PMA-based BCABL is more reliable for non-professional benzene exposure. These results are very important since can offer a new approach, based on modelling, which can be use in occupational and in environmental exposures monitoring. The study from Bunsri et al. (contribution 6) explored hematological and biochemical effects of occupational pesticide exposure of Thai vegetable farmers. It was found a variety of parameters, both hematological and biochemical, to be different between pesticides users and non-users in general, and in female pesticides users in particular, concluding that the pesticide use among Thai vegetable farmers may cause hematological alterations and increase the risk of hepatic and renal dysfunction. These results showed that some hematological and biochemical parameters may be used for monitoring and surveillance in occupational exposure to pesticides.

The work from Salamanca-Fernández et al. (contribution 7) present results from phthalate metabolites and protein of brain-derived neurotrophic factor (BDNF) levels environmental exposure, among European children from five HBM4EU aligned studies and its relation with neurodevelopmental alterations. For context, the EU Horizon European Human Biomonitoring Initiative (HBM4EU) yielded data on human exposure to chemicals from harmonized human biomonitoring studies across Europe. The data provide a picture of how chemicals burden the body and impact health in Europe. For the study, it was used data from 5 cohorts from Italy, Slovakia, Slovenia, Hungary, and Norway. In brief, the results suggested that phthalates metabolites associated with more externalizing problems in boys, and a higher exposure to a non-phthalate plasticizer may associate with lower systemic BDNF levels also in boys; higher phthalate exposure is associated with higher urinary BDNF concentrations; and higher urinary BDNF concentrations may predict internalizing problems.

The work from Passarelli et al. (contribution 8) showed environmental exposure assessment results of mercury (Hg), from gold mining activity, in six provinces of the Ecuadorian Amazon region in order to assess the resulting risks to environment and human health using deterministic and probabilistic methods.

The work from Cuéllar-Pérez et al. (contribution 9) is an in vivo study, which used male Wistar rats to study the effects of blocking the brain-derived neurotrophic factor receptor TrkB on microglia. The study evidenced that BDNF and its receptor TrkB are involved in sodium cyanide (NaCN) stimulation. The TrkB receptor inhibitor, K252a, significantly inhibited microglial reactivity, indicating a role for the tyrosine-kinase receptor in these changes.

Although all papers have its major focus on Toxicology, there is a wide variety of research approaches to the published studies, from in vitro to One Health concept, and from (human and environmental) biomonitoring original researches and systematic review.

In summary, all the studies have as major objective to achieve more knowledge, at a toxicological point of view, for a better health promotion and surveillance, and disease prevention. There is an effort from the scientific community to gather information regarding the reliable value of the use of biomarkers—exposure and/or effect—in environmental and occupational exposure assessment, and to correlate or incorporate with in silico methods, such as pharmacokinetic models, allowing other the extraction of accurate information in less invasive way. The expectation is that the results taken from these studies can add data and integrate information to encourage national and international bodies to make decisions regarding old used and newly identified chemicals substances from environmental and occupational exposures; and to move forward to continuing the research needed to fill the still existing knowledge gaps.

Funding: This research received no external funding.

Institutional Review Board Statement: Not applicable.

Informed Consent Statement: Not applicable.

Data Availability Statement: Not applicable.

Acknowledgments: As Guest Editor of the Special Issue "Environmental and Occupational Exposure to Chemical Agents and Health Challenges I", I would like to express my deep appreciation to all authors whose valuable work was published under this issue and thus contributed to the success of the edition.

Conflicts of Interest: The author declares no conflict of interest.

List of Contributors:

1. Borgatta, M.; Breider, F. Inhalation of Microplastics—A Toxicological Complexity. *Toxics* **2024**, *12*, 358. https://doi.org/10.3390/toxics12050358.
2. Ladeira, C.; Møller, P.; Giovannelli, L.; Gajski, G.; Haveric, A.; Bankoglu, E.E.; Azqueta, A.; Gerić, M.; Stopper, H.; Cabêda, J.; et al. The Comet Assay as a Tool in Human Biomonitoring Studies of Environmental and Occupational Exposure to Chemicals—A Systematic Scoping Review. *Toxics* **2024**, *12*, 270. https://doi.org/10.3390/toxics12040270.
3. Ma, Y.; Yi, J.; Ma, J.; Yu, H.; Luo, L.; Wu, W.; Jin, L.; Yang, Q.; Lou, T.; Sun, D.; et al. Hand Sanitizer Gels: Classification, Challenges, and the Future of Multipurpose Hand Hygiene Products. *Toxics* **2023**, *11*, 687. https://doi.org/10.3390/toxics11080687.
4. Fustinoni, S.; Mercadante, R.; Lainati, G.; Cafagna, S.; Consonni, D. Kinetics of Excretion of the Perfluoroalkyl Surfactant cC6O4 in Humans. *Toxics* **2023**, *11*, 284. https://doi.org/10.3390/toxics11030284.
5. Cui, S.; Pang, B.; Yan, H.; Wu, B.; Li, M.; Xing, C.; Li, J. Using Urinary Biomarkers to Estimate the Benzene Exposure Levels in Individuals Exposed to Benzene. *Toxics* **2022**, *10*, 636. https://doi.org/10.3390/toxics10110636.
6. Bunsri, S.; Muenchamnan, N.; Naksen, W.; Ong-Artborirak, P. The Hematological and Biochemical Effects from Pesticide Exposure on Thai Vegetable Farmers. *Toxics* **2023**, *11*, 707. https://doi.org/10.3390/toxics11080707.
7. Salamanca-Fernández, E.; Espín-Moreno, L.; Olivas-Martínez, A.; Pérez-Cantero, A.; Martín-Rodríguez, J.L.; Poyatos, R.M.; Barbone, F.; Rosolen, V.; Mariuz, M.; Ronfani, L.; et al. Associations between Urinary Phthalate Metabolites with BDNF and Behavioral Function among European Children from Five HBM4EU Aligned Studies. *Toxics* **2024**, *12*, 642. https://doi.org/10.3390/toxics12090642.
8. Passarelli, I.; Villacis Verdesoto, M.V.; Jiménez-Oyola, S.; Flores Huilcapi, A.G.; Mora-Silva, D.; Anfuso, G.; Esparza Parra, J.F.; Jimenez-Gutierrez, M.; Carrera Almendáriz, L.S.; Avalos Peñafiel, V.G.; et al. Analysis of Mercury in Aquifers in Gold Mining Areas in the Ecuadorian Amazon and Its Associated Risk for Human Health. *Toxics* **2024**, *12*, 162. https://doi.org/10.3390/toxics12020162.
9. Cuéllar-Pérez, R.; Jauregui-Huerta, F.; Ruvalcaba-Delgadillo, Y.; Montero, S.; Lemus, M.; Roces de Álvarez-Buylla, E.; García-Estrada, J.; Luquín, S. K252a Prevents Microglial Activation Induced by Anoxic Stimulation of Carotid Bodies in Rats. *Toxics* **2023**, *11*, 871. https://doi.org/10.3390/toxics11100871.

References

1. Naidu, R.; Biswas, B.; Willett, I.R.; Cribb, J.; Kumar Singh, B.; Paul Nathanail, C.; Coulon, F.; Semple, K.T.; Jones, K.C.; Barclay, A.; et al. Chemical pollution: A growing peril and potential catastrophic risk to humanity. *Environ. Int.* **2021**, *156*, 106616. [CrossRef] [PubMed]
2. Brunn, H.; Arnold, G.; Körner, W.; Rippen, G.; Steinhäuser, K.G.; Valentin, I. PFAS: Forever chemicals—Persistent, bioaccumulative and mobile. Reviewing the status and the need for their phase out and remediation of contaminated sites. *Environ. Sci. Eur.* **2023**, *35*, 20. [CrossRef]
3. Simonds, N.I.; Ghazarian, A.A.; Pimentel, C.B.; Schully, S.D.; Ellison, G.L.; Gillanders, E.M.; Mechanic, L.E. Review of the Gene-Environment Interaction Literature in Cancer: What Do We Know? *Genet. Epidemiol.* **2016**, *40*, 356–365. [CrossRef] [PubMed]
4. Fenton, S.E.; Ducatman, A.; Boobis, A.; DeWitt, J.C.; Lau, C.; Ng, C.; Smith, J.S.; Roberts, S.M. Per- and Polyfluoroalkyl Substance Toxicity and Human Health Review: Current State of Knowledge and Strategies for Informing Future Research. *Environ. Toxicol. Chem.* **2021**, *40*, 606–630. [CrossRef] [PubMed]
5. Pathak, V.M.; Verma, V.K.; Rawat, B.S.; Kaur, B.; Babu, N.; Sharma, A.; Dewali, S.; Yadav, M.; Kumari, R.; Singh, S.; et al. Current status of pesticide effects on environment, human health and it's eco-friendly management as bioremediation: A comprehensive review. *Front. Microbiol.* **2022**, *13*, 962619. [CrossRef] [PubMed]
6. Klaine, S.J.; Alvarez, P.J.J.; Batley, G.E.; Fernandes, T.F.; Handy, R.D.; Lyon, D.Y.; Mahendra, S.; McLaughlin, M.J.; Lead, J.R. Nanomaterials in the environment: Behavior, fate, bioavailability, and effects. *Environ. Toxicol. Chem.* **2008**, *27*, 1825–1851. [CrossRef] [PubMed]
7. Lee, I.; Ji, K. Identification of combinations of endocrine disrupting chemicals in household chemical products that require mixture toxicity testing. *Ecotoxicol. Environ. Saf.* **2022**, *240*, 113677. [CrossRef] [PubMed]
8. Kahn, L.G.; Philippat, C.; Nakayama, S.F.; Slama, R.; Trasande, L. Endocrine-disrupting chemicals: Implications for human health. *Lancet Diabetes Endocrinol.* **2020**, *8*, 703–718. [CrossRef] [PubMed]
9. International Labour Organization. *International Labour Organization X Exposure to Hazardous Chemicals at Work and Resulting Health Impacts: A Global Review*; International Labour Office: Geneva, Switzerland, 2021.

toxics

MDPI

Perspective

Inhalation of Microplastics—A Toxicological Complexity

Myriam Borgatta [1,*] and Florian Breider [2]

1 Center for Primary Care and Public Health (Unisanté-Lausanne), University of Lausanne, 1015 Lausanne, Switzerland
2 Central Environmental Laboratory, Ecole Polytechnique Fédérale de Lausanne (EPFL), 1015 Lausanne, Switzerland; florian.breider@epfl.ch
* Correspondence: myriam.borgatta@unisante.ch

Abstract: Humans are chronically exposed to airborne microplastics (MPs) by inhalation. Various types of polymer particles have been detected in lung samples, which could pose a threat to human health. Inhalation toxicological studies are crucial for assessing the effects of airborne MPs and for exposure-reduction measures. This communication paper addresses important health concerns related to MPs, taking into consideration three levels of complexity, i.e., the particles themselves, the additives present in the plastics, and the exogenous substances adsorbed onto them. This approach aims to obtain a comprehensive toxicological profile of deposited MPs in the lungs, encompassing local and systemic effects. The physicochemical characteristics of MPs may play a pivotal role in lung toxicity. Although evidence suggests toxic effects of MPs in animal and cell models, no established causal link with pulmonary or systemic diseases in humans has been established. The transfer of MPs and associated chemicals from the lungs into the bloodstream and/or pulmonary circulation remains to be confirmed in humans. Understanding the toxicity of MPs requires a multidisciplinary investigation using a One Health approach.

Keywords: microplastics; inhalation; lung; human; toxicity

Citation: Borgatta, M.; Breider, F. Inhalation of Microplastics—A Toxicological Complexity. *Toxics* **2024**, *12*, 358. https://doi.org/10.3390/toxics12050358

Academic Editor: Carina Ladeira

Received: 27 March 2024
Revised: 27 April 2024
Accepted: 7 May 2024
Published: 11 May 2024

1. Introduction

The presence and toxicity of plastic in the environment have been studied for decades now, and plastic is recognized as a persistent organic pollutant that causes harmful health effects on living organisms [1]. Nevertheless, studies on human exposure to and the toxicity of plastic are scarce and focused predominantly on the oral route within contaminated food and beverages. Recent publications have reported the presence of plastic fragments in the human lungs, revealing inhalation as an additional route of absorption [2–6]. Plastic is a generic term including numerous polymers and rubbers that are used in the production of countless single-use and reusable products intended for human activities, such as toys, furniture, tires, vehicle interiors, medical and sport equipment, and carpets [7–9]. During the manufacturing process of these products, a plethora of chemicals, such as plasticizers, solvents, antioxidants, biocides, colorants, fillers, flame retardants, light stabilizers, nucleating agents, and fragrances, can be incorporated into the polymer matrix [10]. These additives impart additional properties to the polymers, such as functionality, longevity, color, brightness, and homogenous blending of the final product [11,12]. Plastics may also contain non-intentionally added chemicals, like by-products and breakdown products [13]. Therefore, one plastic item is not a composite blend of a polymer but a cocktail of different chemicals. With use and aging, plastic items wear out, releasing countless fragments into the soil, water, and the atmosphere [14,15]. MPs are plastic fragments with a size ranging from 5 mm to 0.001 mm, the smallest (<0.001 mm) being nanoplastics [16,17]. MPs are predominant as dust [18–20] and airborne particles [16,20–23] in urban areas [24] and indoor environments [23,25,26]. Dust is easily resuspended in the air with the action of air flow (i.e., wind, ventilation systems, passing vehicles, doors opening/closing), making

MPs available for inhalation [8,27,28]. Human inhalation exposure to airborne MPs may vary during the day, depending on human activity and location. Workers may be exposed to indoor and outdoor airborne MPs during specific activities such as the production of plastic items, the treatment of plastic waste, road maintenance, and 3D printing [29–36]. Children crawling on the floor can lead to resuspending MPs in the air at short distance to their nose and mouth. Knowing that children and adults spend on average 19 h and 21 h per day indoors (i.e., at home, in the workplace, and in a vehicle) [37], respectively, inhalation can therefore be considered as a chronic exposure route to airborne MPs, with concentration peaks depending on the activities.

Little is known about the inhalation exposure of humans to MPs and the characteristics of these particles in the lungs. The interactions of MPs with the pulmonary environment (i.e., local effects) and their potential distribution throughout the human body need to be investigated. In the environmental field, studies have predominantly focused on relatively large MPs using analytical methods reliable for water, soil, and air samples. Analytical method improvements such as Raman and infrared microscopy (e.g., μFTIR), and pyrolysis–gas chromatography–mass spectrometry (Py-GC/MS) can nowadays provide detailed descriptions of the MPs present in human fluids and tissues [2–4]. Airborne MPs display a range of sizes, shapes, and chemical compositions that may influence both their lung deposition and pulmonary (local) and systemic toxicity [7,38,39]. Systemic effects occur when a chemical enters the lymphatic or circulatory system, which is interconnected with various organs and tissues [40]. Moreover, the hydrophobic surface of MPs likely promotes the adsorption of exogenous pollutants with high octanol–water partition coefficients (K_{OW}), such polycyclic aromatic hydrocarbons (PAHs), organochlorinated biphenyls (PCBs), per- and poly-fluoroalkyl substances (PFASs), pesticides, and pharmaceuticals [7,41–43]. These adsorbed pollutants may be released in the lungs depending on the local conditions (e.g., temperature, pH, humidity, macrophages, residence time). MPs can thus serve as reservoirs for additives and as carriers for hazardous pollutants that may threaten human health [44–47].

The toxicological assessment of airborne MPs via inhalation relies on a combination of physical and chemical factors that could potentially act simultaneously to induce local and/or systemic toxic effects (Figure 1). The aim of this communication is to outline three levels of complexity that are important to consider when assessing inhalation exposure to MPs and human health: (1) MPs as deposited particles in the lungs, (2) MPs as reservoirs of additives, and (3) MPs as carriers of adsorbed pollutants. From our perspective, adopting a holistic approach facilitates a comprehensive assessment of the risks associated with MP exposure, thereby providing valuable insights for decision-makers in the public health and environmental protection domains.

1.1. MPs Deposited in the Lungs

As with other airborne particles, MPs with a size range of 1–5 μm are likely deposited in the nasopharyngeal and bronchial sections of the respiratory tract [48]. MPs of 1 μm and nanoplastics can reach the deepest lung regions (i.e., the alveoli), where vital gas exchanges with blood occur. Inhalation exposure to MPs can lead to an accumulation of 26 to 130 MPs per day in the human lungs [22]. Deposition of MPs in the respiratory system depends on their size, shape, and surface properties. Until now, only seven studies have characterized MPs in human lung tissues and bronchoalveolar lung fluid (BALF) using optimized analytical methods [2–6,49,50]. The results of these studies are summarized in Table 1. Among the polymers that were targeted in lung tissues, the predominant one were polypropylene (PP) and polyethylene terephthalate (PET) [2,4,5,49]. Focusing specifically on MP fibers, Chen et al., in 2022 [4], observed that the quantity of these fibers was twice as high in lung tumors than in normal tissues. Three other studies also primarily focused on fibers but in human BALF samples [3,6,50]. The MP size ranged from <5 μm to 34 μm, and the four dominant polymers were PE, PP, PET, and PS. A single study reported a concentration of 9.18 ± 2.45 MP fibers/100 mL of BALF [3]. In the BALF of children, up to 332 MP particles were found, and the concentration was even higher than in adult samples [50]. Within these studies, fibers, fragments, and films were detected. The length range reported was from <5.5 μm up to 2475 μm, with fibers usually exhibiting the most elongated shape. One study characterized the surface of the fibers as having rough and porous features with cracks or being significantly damaged [6]. These seven studies provide the first insights into the presence of MPs in human lungs. The detection and characterization methods used in these studies relied on infrared spectro-imaging (i.e., μFTIR, μRaman, LDIR). While these imaging techniques provide particle-based concentrations, their detection is limited by the spatial resolution of the instrument, typically ranging between 5–20 μm. This limitation hinders the detection of small MPs and nanoplastics. The identification of particles using these techniques is heavily influenced by particle aging, as most spectral libraries only contain data on virgin polymers unaffected by solar radiation, atmospheric gases, abrasion, and heat. Additionally, these techniques solely provide information on the nature of the polymer but not the additives present in the particles. Therefore, additives cannot be analyzed using these techniques. To our knowledge, Py-GC/MS has not been employed for analyzing MPs in human lungs up to now, although Py-GC/MS is not based on the spectral characteristics of the MPs but on the analysis of molecular tracers by mass spectrometry. This approach offers several advantages over infrared-based methods, since the detection of MPs is not limited by the size but by the mass. Py-GC/MS can provide information on the additives present in MPs. Detection and characterization methods based on infrared spectroscopy and mass spectrometry complement each other. Combining these approaches could enhance our understanding of the chemical and physical complexities of MPs, leading to a more comprehensive assessment of the nature of airborne MPs and, therefore, the potential health effects.

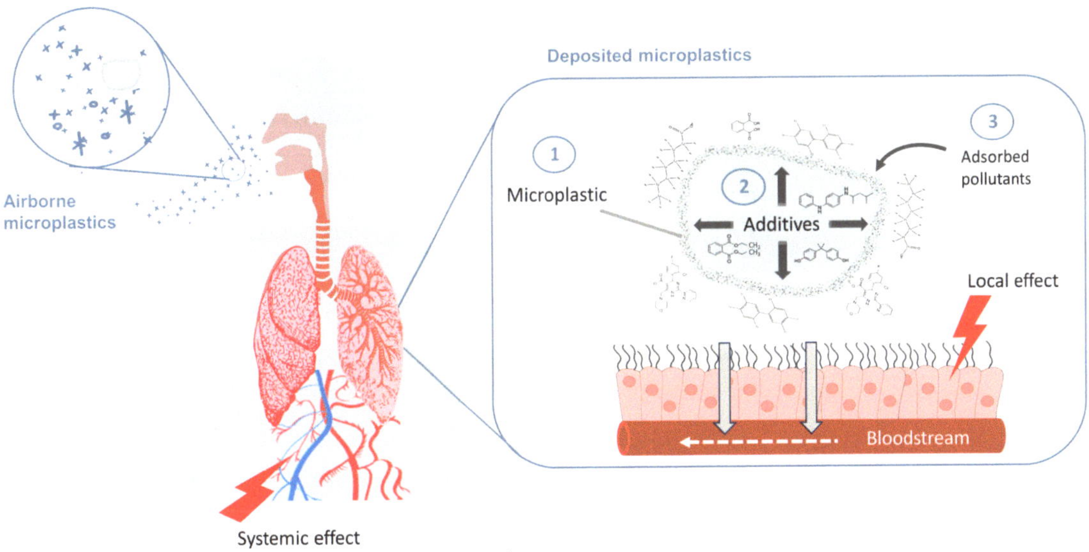

Figure 1. Inhalation exposure to microplastics (MPs) of varying physical and chemical properties. Three situations related to MP deposition in the pulmonary system may harm human health by acting simultaneously or independently. 1. MP particles deposited in the lungs. 2. MPs as reservoirs of additives. 3. MPs as carriers of environmental pollutants. This figure was created with https://www.canva.com/ (Accessed on 15 March 2024).

Table 1. MP characteristics in human lung tissues and bronchoalveolar lavage (BALF). The most abundant MP is highlighted in bold when reported by the studies. The size refers to either the average or the maximal size or the most abundant MP size found in the studies. Acrylates (ACRs), acrylic (AC), butadiene rubber (BR), chlorinated polyethylene (CPE), chlorinated polyisoprene (PICH), ethylenevinyl acetate (EVA), poly(ethylene terephthalate) (PET), poly(methylmethacrylate) (PMMA), poly(tetrafluoroethylene) (PTFE), poly-(vinylchloride) (PVC), polycarbonate (PC), polyester (PES), polyethylene (PE), polyimide (PI), polystyrene (PS), polysulfone (PSU), polytetrafluoroethylene (PTFE), polyurethane (PUR), polyamide (PA), polybutylene terephthalate (PBT), and polypropylene (PP).

Sample Type	Polymer	Shape	Size (μm)	Concentration	Detection Method	Reference
Lung tissue	**PP**, PET	Fibers, fragments, films	12–2475 (length) 4–88 (width)	0.41–3.12 MPs/g tissue	mFTIR	[2]
Lung tissue	**PP**, PET, PS, PVC, PTFE, CPE, PE, ACR, EVA, BR, PUR, silicone	Fibers	20–100 (diameter)	2.19 MPs/g tissue	LDIR	[49]
Lung tissue	**PP**, PE, PVC, cellulose acetate, PE, co-PP, PS, PS-co-PVC, PUR	Fibers, fragments	<5 (fragments) 8.12–16.8 (fibers)	0.56 MPs/g tissue	mRaman	[5]
Lung tissue	**PE**, PET, AC, phenoxy resin, rayon	Fibers (>20 μm)	Up to 1750 (length) Up to 34.29 (width)	-	mFTIR	[4]
BALF	-	Fibers	1730	9.2 ± 2.5 MPs/ 100 mL BALF	mFTIR	[3]
BALF	PP, PE, PES, PET, PVC, PC, PTFE, AC, PA, PBT	Fragment, fibers, pellets, sheet	<5	4.31 ± 2.77 MPs/ 10 mL BALF	mRaman	[50]
BALF	**PE**, PET, PP, PS, PC, PUR, PSU, PP, PVC, PMMA, PI, PTFE, ACR, PICH	Fiber (length-to-diameter ratio ≥3 μm), "Irregular particles" (length-to-diameter ratio <3 μm)	30–34	0.91 MPs/g BALF	LDIR	[6]

1.2. Toxicity of MPs in the Lungs

Alveolar macrophages are efficient in clearing particles with diameters >1 μm, while smaller particles tend to persist for longer period [51]. In the bronchial region, a layer of mucus lines the airway walls, where cilia actively vibrate to remove deposited particles back up to the esophagus within 24 h following particle inhalation [52]. In vitro studies showed that polystyrene (PS) microspheres of a 1 μm diameter can be internalized in less than 24 h by human alveolar A549 cells, altering the proliferation, metabolism, morphology, and cohesion of these cells [53–55]. The size and shape of polyethylene (PE) and PS MPs are related to cytotoxic effects in human cerebral and lung epithelial cells [56,57]. In addition to the size, the surface characteristics may induce pulmonary effects in animal-models exposed to PS microspheres by tracheal instillation [58]. Except in respiratory therapy, where spherical MPs are used as carriers for drug delivery application [59], humans are mainly exposed to airborne MPs with irregular shapes [6,49,50]. Workers exposed to polypropylene (PP) microfibers with an irregular shape and size coupled with lamellar fragmentation at the surface exhibited a 3.6-fold higher risk of respiratory symptoms compared to non-exposed individuals [60]. Toxicological studies on the effects of PP and other polymers with an irregular shape at the pulmonary level are lacking [61]. PS microspheres have been the most extensively studied polymer and shape in vitro and in vivo, demonstrating apoptosis in human alveolar macrophages, alveolar epithelial cells, and lung cancer cells. Alveolar lesions were reported in rodents exposed by inhalation to nylon dust and acrylic fibers [62]. Lung biopsies revealing acute alveolar injury and bronchiolitis were also documented in workers suffering from occupational interstitial lung disease (flock worker's lung) after repeated and prolonged exposures to PE and nylon microfibers [29,63–65]. Chronic occupational exposure to PVC dust was associated with breathlessness symptoms and abnormal pulmonary function responses [31,60]. PVC and vinyl chloride were linked to lung inflammation and diseases, including cancers [66]. While evidence supports the effects of MPs on pulmonary cells in vitro and in animal models as well as on respiratory diseases in plastic industry workers, a causal link has never been established in humans. One reason for the lack of causality may be attributed to advances in methods specifically designed to characterize and quantify MPs, which are only recent. These methods hold promise for demonstrating a causal link in humans and, thus, a better understanding of the health risks of inhaled MPs.

Py-GC/MS detected PET, PE, and styrene polymers with sizes equal to or exceeding 0.7 μm in human blood samples [67], raising questions about the route and mechanism of absorption of MPs. Diffusion through the alveolar wall into the lymphatic system was reported for particles ranging between 0.1 and 1 mm in diameter [68]. After oral exposure, nano- and micro-plastics (polystyrene) penetrated the lymphatic and/or blood system in rodents [69,70]. PS microparticle absorption across membranes was reported to be fast, within minutes in the small intestine of rats orally exposed to PS microspheres [71,72]. PS MPs with a diameter ranging from 500 nm to 5 μm were detected in the blood and further distributed to the brain and the heart, with the highest concentrations found in the liver, kidneys, and lungs. In humans, the likelihood that MPs reach the placenta and the blood–brain barrier is starting to be tackled, along with their distribution into the liver, muscles, and brain [7]. In an in vitro 3D lung cell model, Rothen-Rutishauser et al. [54] showed that 1 μm PS particles either adhered to or were internalized in macrophages present on the epithelial surface. These immune cells aim to remove foreign particles through the bloodstream and the lymphatic system. Subsequently, MPs may also be found in the regional lymph nodes [73]. The relative contribution of lymphatic transport and mucociliary clearance varies among species [74] and is unknown with MPs in humans, nor are the systemic effects of MPs. A broad spectrum of MP toxic effects at the whole-organism level were observed in various species, mainly from marine ecosystems [75–79]. These effects span multiple life stages, encompassing developmental, behavioral, genotoxic, and metabolic aspects, along with increased mortality, immune responses, and organs dysfunction. To understand the potential local and systemic effects in humans chronically exposed to airborne MPs, the characterization and quantification of both airborne MPs in indoor environments and deposited MPs in the lungs would serve as a starting point. Basic toxicological principles assert that the frequency and dose (i.e., the mass of MPs deposited in the lungs) of exposure, as well as the residence time of MPs in the lungs, can determine the toxicity of both the MPs and the released chemicals (i.e., additives and adsorbed pollutants).

1.3. MPs as Reservoirs of Chemicals

MPs provide a durable reservoir for both classified and unclassified hazardous chemical intentionally or unintentionally incorporated in the polymer, which may leak into the pulmonary environment (i.e., lung tissues and fluids). Approximately 10,550 chemicals were identified as likely being used in plastics across various industrial sectors, and 4300 chemicals are likely used in plastic packaging [10,13]. Among these additives, up to half of them lack hazard classifications, with the others having been reported as having endocrine-disrupting effects, neurotoxicity, or reprotoxicity, such as phthalates, bisphenol A (BPA), triclosan, bisphenone, organotins, and brominated flame retardants [80]. Phthalates, used as plasticizers to impart flexibility to polymer plastics, are an example of additives that are not chemically bound to the polymer matrix and that can migrate to the surface of the product [81]. Phthalates are additives known to leak from plastic medical devices and accessories when in contact with solvents such as *n*-hexane and a mixture of ethanol/water in laboratory conditions [82,83]. Phthalates (along with other additives) are prevalent in indoor air and dust, particularly in environments with plastic products and devices such as hospitals and freshly decorated spaces [84,85]. After inhalation exposure, these additives have the potential to traverse the human alveolar–capillary membrane and be distributed throughout the body [86]. The lungs are a large surface of chemical absorption into the blood. This route of absorption bypasses the first-pass metabolism, a primary important "detoxifying" step inherent in the oral route [40]. Therefore, hazardous chemicals can be distributed to organs prior to being detoxified. To the best of our knowledge, no study has specifically investigated the inhalation of MPs containing additives nor the release of these chemicals in vivo or in a simulated pulmonary environment (e.g., pulmonary fluid, T°, pH, and enzymes). It is known that additives can be released by tire particles in the gastric and intestinal fluid [87], and similar studies should be performed with MPs exposed to simulated pulmonary fluid. Toxicological data are needed to understand the diffusion kinetics of plastic additives, their concentrations in the lungs and residence time, and their transfer across membranes. These data are important in assessing the potential effects of additives in the lungs and their absorption rate into the blood following inhalation exposure to MPs.

1.4. Toxicity of Additives

Given the multitude of additives and unintentionally added chemicals that could be present in MPs, we have chosen to summarize some effects related to phthalates and bisphenols as examples. In the lungs, phthalates interfere with the nuclear hormone receptor superfamily, influencing gene transcription [88] and lung maturation [89]. Following inhalation exposure, phthalates alter lung weight, cell proliferation, and the structure of the alveoli, potentially impacting gas exchange in rats [90–92]. Phthalates are associated with lung inflammation, oxidative stress, and a clinical decrease in the pulmonary functions [93–99]. Multiple studies have linked phthalates exposure with asthma and allergic reactions [100–102], including in children [97]. Oral exposure to phthalates also induces numerous lung effects, including abnormal lung histology [90], morphological changes in pneumocytes cells, and an increase in the number of alveolar macrophages in the pulmonary blood vessels [103]. These results demonstrate the affinity of phthalates for lung tissues, even when the exposure is not via inhalation. Bisphenols are other common plastic additives that cause pulmonary damage when in contact with the respiratory system. Similarly to phthalates, bisphenol A is associated with asthma [93,97,104–106]. Bisphenols also cause collapsed alveoli and morphological changes in the mucosa, blood vessels, and interalveolar septum [107,108], as well as lung development disorders [109,110]. These two examples of additives exhibiting lung toxicity are concerning considering the numerous other chemicals that may also leak into the lungs following MP inhalation exposure. This mixture of chemicals may induce individual, additive, or synergistic effects at both the local and systemic levels.

Phthalates and bisphenols (A and B) are endocrine-disrupting chemicals (EDCs) [111,112] with toxic effects on female and male reproduction, development, and various organs, along with the initiation of cancers [102,113–118]. In humans, toxic effects have been associated with phthalates such as retarded male reproductive development [119,120], altered semen quality [121,122], and allergic symptoms in children [97,104]. However, it remains unknown whether plastic additives such as phthalates and bisphenols can be absorbed in the bloodstream after MP inhalation exposure. Some evidence suggests that inhaled phthalates can cross the alveolar–blood membrane before being eliminated through urine [86,123,124]. Therefore, chronic exposure to MPs may be seen as a hidden health threat with significant toxic effects in humans [125].

1.5. MPs as Carriers for Environmental Pollutants

MPs can be a carrier for hazardous chemicals present in the environment [126,127]. MP surfaces are mainly hydrophobic and often have large specific surface areas favoring the adsorption of environmental pollutants (i.e., exogenous chemicals) with medium-to-high log K_{ow} constants, such as PAHs, PCBs, PFASs, and PBDEs [7,41,42]. PFASs, with log K_{ow} values ranging from 3 to over 6, are hydrophobic compounds that easily adsorb on plastics and are mainly associated with particulate matter. In the lungs, MPs and the carried hazardous pollutants come into contact with the epithelial fluid that lines the airways, extending from the larger airways down to the alveoli. This aqueous layer of biological solutes lining the pulmonary epithelium contains surfactants [128]. Composed mainly of phospholipids and proteins, the surfactant reduces surface tension, thus maintaining the alveoli in an open conformation optimal for gas exchange. With their amphiphilic structure, the phospholipids may favor the desorption and solubilization of hydrophobic chemicals. The local conditions (e.g., temperature, pH, humidity, enzymes) in the lungs may also influence the release of adsorbed pollutants. These chemicals become bioavailable in the surfactant and epithelial fluid, acting possibly similarly to the additives described previously. However, studies having specifically assessed the inhalation of MPs containing hazardous pollutants are still lacking. Studies performed with pulmonary fluids exist for particle matter (PM2.5 and PM10) and adsorbed metals and inorganic and organic matter such as PAHs [129]. MPs have been demonstrated to adsorb PAHs [130], but the desorption kinetics is unknown. In the lungs, PAHs easily diffuse through cell membranes and have been quantified in cancer tissues [131,132]. PAHs are reported to alter pulmonary functions and to be responsible for lung cancer [130]. PAHs have mutagenic, carcinogenic, and endocrine-disrupting effects, mainly due to enzymatic transformation leading to the formation of toxic metabolites [131]. After inhalation exposure, PAHs induced cancers, cardiovascular diseases [133], and blood hemolysis [134]. PAHs are an example of exogenous chemicals, among others, that MPs can carry. Therefore, airborne MPs can act as a "Trojan Horse", hiding and carrying toxic exogenous and indigenous (additives) chemicals along the airways and likely throughout the entire organism.

2. Key Messages

In contrast to previous beliefs, MPs should no longer be considered inert and harmless particles. The physical heterogeneity of MPs can induce localized effects on the lungs. Evidence shows the pulmonary and systemic effects of MPs in animal or cell models, with PS microspheres being predominantly studied in terms of polymer and shape. As carriers for additives and environmental chemicals, inhalation exposure to MPs may pose additional risks to human health. Studies investigating the inhalation of MPs containing significant amounts of additives and/or pollutants are lacking. The transfer of MPs and endogenous/exogenous chemicals from the lungs into the bloodstream and/or lymphatic circulation still needs to be confirmed in humans, along with the local conditions favoring the release of these chemicals in the lungs. Similarly to other airborne particles, MPs may cross pulmonary membranes and accumulate in lymphatic nodules, impairing the immune system or being bioavailable for blood transport within the human body [74]. So far, only one article has reported the presence of MPs in human blood [68], leaving a significant gap in the knowledge on the target organs and Trojan Horse potential of these particles. Characterizing human exposure to MPs, monitoring plastic additives in the urine and blood of individuals with high exposure levels (e.g., workers in plastic industries and plastic recycling), and performing inhalation exposure studies using labeled MPs are promising research avenues for understanding the human health risk of MPs.

Chronic inhalation exposure to MPs, from birth to death, requires thorough investigations. Given the increasing prevalence of MPs in the air, a "One Health" approach integrating disciplines such as toxicology, environmental sciences, chemistry, biology, and medicine appears imperative. Only a multidisciplinary approach will enable a comprehensive understanding of the toxicological effects of MPs via inhalation, providing guidance for public health policies. Continuing the effort to characterize airborne MPs and those deposited in the respiratory system is important, considering all particle shapes, polymers, and even overlooked sources like rubber from tires. A non-target screening analysis is likely valuable for analyzing and studying the different chemicals associated with MPs in both human and environment samples, such as indoor buildings, urban centers, and plastic recycling areas. Nowadays, advanced instrumental methods are available for analyzing air and human biological samples. Encouraging the adoption of standardized protocols is important for consistently quantifying and characterizing MPs, considering factors like polymer types, sizes, shapes, and surface characteristics. Obtaining human samples poses ethical challenges, making them valuable and scarce resources in scientific research. Therefore, adopting a uniform approach can enhance the comparability of results across different studies, aiding in a comprehensive understanding of the health risks posed by airborne MPs.

Author Contributions: Conceptualization, M.B. and F.B.; methodology, M.B. and F.B.; investigation, M.B.; resources, M.B. and F.B.; data curation, M.B. and F.B.; writing—original draft preparation, M.B.; writing—review and editing, M.B. and F.B.; visualization, M.B.; project administration, M.B. and F.B. All authors have read and agreed to the published version of the manuscript.

Funding: The APC was funded by University of Lausanne and Unisanté.

Institutional Review Board Statement: Not applicable.

Informed Consent Statement: Not applicable.

Data Availability Statement: Data are contained within the article.

Conflicts of Interest: The authors declare no conflicts of interest.

References

1. Basel Convention Plastic Waste—Overview. Available online: https://www.basel.int/Implementation/Plasticwaste/Overview/tabid/8347/Default.aspx (accessed on 29 November 2023).
2. Jenner, L.C.; Rotchell, J.M.; Bennett, R.T.; Cowen, M.; Tentzeris, V.; Sadofsky, L.R. Detection of Microplastics in Human Lung Tissue Using MFTIR Spectroscopy. *Sci. Total Environ.* **2022**, *831*, 154907. [CrossRef]
3. Baeza-Martínez, C.; Olmos, S.; González-Pleiter, M.; López-Castellanos, J.; García-Pachón, E.; Masiá-Canuto, M.; Hernández-Blasco, L.; Bayo, J. First Evidence of Microplastics Isolated in European Citizens' Lower Airway. *J. Hazard. Mater.* **2022**, *438*, 129439. [CrossRef]
4. Chen, Q.; Gao, J.; Yu, H.; Su, H.; Yang, Y.; Cao, Y.; Zhang, Q.; Ren, Y.; Hollert, H.; Shi, H.; et al. An Emerging Role of Microplastics in the Etiology of Lung Ground Glass Nodules. *Environ. Sci. Eur.* **2022**, *34*, 25. [CrossRef]
5. Amato-Lourenço, L.F.; Carvalho-Oliveira, R.; Júnior, G.R.; Dos Santos Galvão, L.; Ando, R.A.; Mauad, T. Presence of Airborne Microplastics in Human Lung Tissue. *J. Hazard. Mater.* **2021**, *416*, 126124. [CrossRef]
6. Qiu, L.; Lu, W.; Tu, C.; Li, X.; Zhang, H.; Wang, S.; Chen, M.; Zheng, X.; Wang, Z.; Lin, M.; et al. Evidence of Microplastics in Bronchoalveolar Lavage Fluid among Never-Smokers: A Prospective Case Series. *Environ. Sci. Technol.* **2023**, *57*, 2435–2444. [CrossRef]
7. Wright, S.L.; Kelly, F.J. Plastic and Human Health: A Micro Issue? *Environ. Sci. Technol.* **2017**, *51*, 6634–6647. [CrossRef]
8. Amato-Lourenço, L.F.; dos Santos Galvão, L.; de Weger, L.A.; Hiemstra, P.S.; Vijver, M.G.; Mauad, T. An Emerging Class of Air Pollutants: Potential Effects of Microplastics to Respiratory Human Health? *Sci. Total Environ.* **2020**, *749*, 141676. [CrossRef]
9. Zimmermann, L.; Dierkes, G.; Ternes, T.; Völker, C.; Wagner, M. Benchmarking the in Vitro Toxicity and Chemical Composition of Plastic Consumer Products. *Environ. Sci. Technol.* **2019**, *53*, 11467–11477. [CrossRef]
10. Wiesinger, H.; Wang, Z.; Hellweg, S. Deep Dive into Plastic Monomers, Additives, and Processing Aids. *Environ. Sci. Technol.* **2021**, *55*, 9339–9351. [CrossRef] [PubMed]
11. Weschler, C.J.; Nazaroff, W.W. Semivolatile Organic Compounds in Indoor Environments. *Atmos. Environ.* **2008**, *42*, 9018–9040. [CrossRef]
12. Hahladakis, J.N.; Velis, C.A.; Weber, R.; Iacovidou, E.; Purnell, P. An Overview of Chemical Additives Present in Plastics: Migration, Release, Fate and Environmental Impact during Their Use, Disposal and Recycling. *J. Hazard. Mater.* **2018**, *344*, 179–199. [CrossRef] [PubMed]
13. Groh, K.J.; Backhaus, T.; Carney-Almroth, B.; Geueke, B.; Inostroza, P.A.; Lennquist, A.; Leslie, H.A.; Maffini, M.; Slunge, D.; Trasande, L.; et al. Overview of Known Plastic Packaging-Associated Chemicals and Their Hazards. *Sci. Total Environ.* **2019**, *651*, 3253–3268. [CrossRef] [PubMed]

14. Liu, Z.; Little, J.C. Semivolatile Organic Compounds (SVOCs): Phthalates and Flame Retardants. In *Toxicity of Building Materials*; Pacheco-Torgal, F., Jalali, S., Fucic, A., Eds.; Woodhead Publishing Series in Civil and Structural Engineering; Woodhead Publishing: Sawston, UK, 2012; pp. 122–137. ISBN 978-0-85709-122-2.
15. Liu, C.; Zhao, B.; Zhang, Y. The Influence of Aerosol Dynamics on Indoor Exposure to Airborne DEHP. *Atmos. Environ.* **2010**, *44*, 1952–1959. [CrossRef]
16. Zhang, Y.; Kang, S.; Allen, S.; Allen, D.; Gao, T.; Sillanpää, M. Atmospheric Microplastics: A Review on Current Status and Perspectives. *Earth-Sci. Rev.* **2020**, *203*, 103118. [CrossRef]
17. Smith, M.; Love, D.C.; Rochman, C.M.; Neff, R.A. Microplastics in Seafood and the Implications for Human Health. *Curr. Environ. Health Rep.* **2018**, *5*, 375–386. [CrossRef]
18. Kitahara, K.-I.; Nakata, H. Plastic Additives as Tracers of Microplastic Sources in Japanese Road Dusts. *Sci. Total Environ.* **2020**, *736*, 139694. [CrossRef]
19. Patchaiyappan, A.; Dowarah, K.; Zaki Ahmed, S.; Prabakaran, M.; Jayakumar, S.; Thirunavukkarasu, C.; Devipriya, S.P. Prevalence and Characteristics of Microplastics Present in the Street Dust Collected from Chennai Metropolitan City, India. *Chemosphere* **2021**, *269*, 128757. [CrossRef]
20. Dehghani, S.; Moore, F.; Akhbarizadeh, R. Microplastic Pollution in Deposited Urban Dust, Tehran Metropolis, Iran. *Environ. Sci. Pollut. Res. Int.* **2017**, *24*, 20360–20371. [CrossRef]
21. Gasperi, J.; Wright, S.L.; Dris, R.; Collard, F.; Mandin, C.; Guerrouache, M.; Langlois, V.; Kelly, F.J.; Tassin, B. Microplastics in Air: Are We Breathing It In? *Curr. Opin. Environ. Sci. Health* **2018**, *1*, 1–5. [CrossRef]
22. Prata, J.C. Airborne Microplastics: Consequences to Human Health? *Environ. Pollut.* **2018**, *234*, 115–126. [CrossRef]
23. Dris, R.; Gasperi, J.; Saad, M.; Mirande, C.; Tassin, B. Synthetic Fibers in Atmospheric Fallout: A Source of Microplastics in the Environment? *Mar. Pollut. Bull.* **2016**, *104*, 290–293. [CrossRef] [PubMed]
24. Liao, Z.; Ji, X.; Ma, Y.; Lv, B.; Huang, W.; Zhu, X.; Fang, M.; Wang, Q.; Wang, X.; Dahlgren, R.; et al. Airborne Microplastics in Indoor and Outdoor Environments of a Coastal City in Eastern China. *J. Hazard. Mater.* **2021**, *417*, 126007. [CrossRef] [PubMed]
25. Zhang, J.; Wang, L.; Kannan, K. Microplastics in House Dust from 12 Countries and Associated Human Exposure. *Environ. Int.* **2020**, *134*, 105314. [CrossRef] [PubMed]
26. Jenner, L.C.; Sadofsky, L.R.; Danopoulos, E.; Rotchell, J.M. Household Indoor Microplastics within the Humber Region (United Kingdom): Quantification and Chemical Characterisation of Particles Present. *Atmos. Environ.* **2021**, *259*, 118512. [CrossRef]
27. Balasubramanian, R.; Lee, S.S. Characteristics of Indoor Aerosols in Residential Homes in Urban Locations: A Case Study in Singapore. *J. Air Waste Manag. Assoc.* **2007**, *57*, 981–990. [CrossRef] [PubMed]
28. Henry, C.; Minier, J.-P.; Brambilla, S. Particle Resuspension: Challenges and Perspectives for Future Models. *Phys. Rep.* **2023**, *1007*, 1–98. [CrossRef]
29. Kern, D.G.; Crausman, R.S.; Durand, K.T.; Nayer, A.; Kuhn, C. Flock Worker's Lung: Chronic Interstitial Lung Disease in the Nylon Flocking Industry. *Ann. Intern. Med.* **1998**, *129*, 261–272. [CrossRef] [PubMed]
30. Shahsavaripour, M.; Abbasi, S.; Mirzaee, M.; Amiri, H. Human Occupational Exposure to Microplastics: A Cross-Sectional Study in a Plastic Products Manufacturing Plant. *Sci. Total Environ.* **2023**, *882*, 163576. [CrossRef] [PubMed]
31. Soutar, C.A.; Copland, L.H.; Thornley, P.E.; Hurley, J.F.; Ottery, J.; Adams, W.G.; Bennett, B. Epidemiological Study of Respiratory Disease in Workers Exposed to Polyvinylchloride Dust. *Thorax* **1980**, *35*, 644–652. [CrossRef]
32. Murashov, V.; Geraci, C.L.; Schulte, P.A.; Howard, J. Nano- and Microplastics in the Workplace. *J. Occup. Environ. Hyg.* **2021**, *18*, 489–494. [CrossRef]
33. Dris, R.; Gasperi, J.; Mirande, C.; Mandin, C.; Guerrouache, M.; Langlois, V.; Tassin, B. A First Overview of Textile Fibers, Including Microplastics, in Indoor and Outdoor Environments. *Environ. Pollut* **2017**, *221*, 453–458. [CrossRef] [PubMed]
34. Stefaniak, A.B.; Johnson, A.R.; du Preez, S.; Hammond, D.R.; Wells, J.R.; Ham, J.E.; LeBouf, R.F.; Martin, S.B.; Duling, M.G.; Bowers, L.N.; et al. Insights Into Emissions and Exposures From Use of Industrial-Scale Additive Manufacturing Machines. *Saf. Health Work* **2019**, *10*, 229–236. [CrossRef] [PubMed]
35. Wagner, S.; Hüffer, T.; Klöckner, P.; Wehrhahn, M.; Hofmann, T.; Reemtsma, T. Tire Wear Particles in the Aquatic Environment—A Review on Generation, Analysis, Occurrence, Fate and Effects. *Water Res.* **2018**, *139*, 83–100. [CrossRef] [PubMed]
36. Mathissen, M.; Scheer, V.; Vogt, R.; Benter, T. Investigation on the Potential Generation of Ultrafine Particles from the Tire–Road Interface. *Atmos. Environ.* **2011**, *45*, 6172–6179. [CrossRef]
37. Mercier, F.; Glorennec, P.; Thomas, O.; Le Bot, B. Organic Contamination of Settled House Dust, a Review for Exposure Assessment Purposes. *Environ. Sci. Technol.* **2011**, *45*, 6716–6727. [CrossRef] [PubMed]
38. Poerio, T.; Piacentini, E.; Mazzei, R. Membrane Processes for Microplastic Removal. *Molecules* **2019**, *24*, 4148. [CrossRef] [PubMed]
39. Leslie, H.A.; Depledge, M.H. Where Is the Evidence That Human Exposure to Microplastics Is Safe? *Environ. Int.* **2020**, *142*, 105807. [CrossRef] [PubMed]
40. Burcham, P.C. *An Introduction to Toxicology*; Springer: London, UK, 2014; ISBN 978-1-4471-5552-2.
41. González-Pleiter, M.; Pedrouzo-Rodríguez, A.; Verdú, I.; Leganés, F.; Marco, E.; Rosal, R.; Fernández-Piñas, F. Microplastics as Vectors of the Antibiotics Azithromycin and Clarithromycin: Effects towards Freshwater Microalgae. *Chemosphere* **2021**, *268*, 128824. [CrossRef] [PubMed]
42. Verdú, I.; González-Pleiter, M.; Leganés, F.; Rosal, R.; Fernández-Piñas, F. Microplastics Can Act as Vector of the Biocide Triclosan Exerting Damage to Freshwater Microalgae. *Chemosphere* **2021**, *266*, 129193. [CrossRef] [PubMed]

43. Sangkham, S. Global Perspective on the Impact of Plastic Waste as a Source of Microplastics and Per- and Polyfluoroalkyl Substances in the Environment. *ACS EST Water* **2024**, *4*, 1–4. [CrossRef]
44. Hartmann, N.B.; Rist, S.; Bodin, J.; Jensen, L.H.; Schmidt, S.N.; Mayer, P.; Meibom, A.; Baun, A. Microplastics as Vectors for Environmental Contaminants: Exploring Sorption, Desorption, and Transfer to Biota. *Integr. Environ. Assess. Manag.* **2017**, *13*, 488–493. [CrossRef] [PubMed]
45. Menéndez-Pedriza, A.; Jaumot, J. Interaction of Environmental Pollutants with Microplastics: A Critical Review of Sorption Factors, Bioaccumulation and Ecotoxicological Effects. *Toxics* **2020**, *8*, 40. [CrossRef] [PubMed]
46. Heinrich, P.; Braunbeck, T. Bioavailability of Microplastic-Bound Pollutants *in Vitro*: The Role of Adsorbate Lipophilicity and Surfactants. *Comp. Biochem. Physiol. Part C Toxicol. Pharmacol.* **2019**, *221*, 59–67. [CrossRef] [PubMed]
47. Rafa, N.; Ahmed, B.; Zohora, F.; Bakya, J.; Ahmed, S.; Ahmed, S.F.; Mofijur, M.; Chowdhury, A.A.; Almomani, F. Microplastics as Carriers of Toxic Pollutants: Source, Transport, and Toxicological Effects. *Environ. Pollut.* **2024**, *343*, 123190. [CrossRef] [PubMed]
48. Costa, A.; Pinheiro, M.; Magalhães, J.; Ribeiro, R.; Seabra, V.; Reis, S.; Sarmento, B. The Formulation of Nanomedicines for Treating Tuberculosis. *Adv. Drug Deliv. Rev.* **2016**, *102*, 102–115. [CrossRef]
49. Wang, S.; Lu, W.; Cao, Q.; Tu, C.; Zhong, C.; Qiu, L.; Li, S.; Zhang, H.; Lan, M.; Qiu, L.; et al. Microplastics in the Lung Tissues Associated with Blood Test Index. *Toxics* **2023**, *11*, 759. [CrossRef] [PubMed]
50. Chen, C.; Liu, F.; Quan, S.; Chen, L.; Shen, A.; Jiao, A.; Qi, H.; Yu, G. Microplastics in the Bronchoalveolar Lavage Fluid of Chinese Children: Associations with Age, City Development, and Disease Features. *Environ. Sci. Technol.* **2023**, *57*, 12594–12601. [CrossRef] [PubMed]
51. Bakand, S.; Hayes, A.; Dechsakulthorn, F. Nanoparticles: A Review of Particle Toxicology Following Inhalation Exposure. *Inhal. Toxicol.* **2012**, *24*, 125–135. [CrossRef]
52. Matijević, E. (Ed.) *Fine Particles in Medicine and Pharmacy*; Springer: Boston, MA, USA, 2012; ISBN 978-1-4614-0378-4.
53. Goodman, K.E.; Hare, J.T.; Khamis, Z.I.; Hua, T.; Sang, Q.-X.A. Exposure of Human Lung Cells to Polystyrene Microplastics Significantly Retards Cell Proliferation and Triggers Morphological Changes. *Chem Res Toxicol* **2021**, *34*, 1069–1081. [CrossRef]
54. Rothen-Rutishauser, B.M.; Kiama, S.G.; Gehr, P. A Three-Dimensional Cellular Model of the Human Respiratory Tract to Study the Interaction with Particles. *Am. J. Respir. Cell Mol. Biol.* **2005**, *32*, 281–289. [CrossRef]
55. Blank, F.; Rothen-Rutishauser, B.M.; Schurch, S.; Gehr, P. An Optimized in Vitro Model of the Respiratory Tract Wall to Study Particle Cell Interactions. *J. Aerosol. Med.* **2006**, *19*, 392–405. [CrossRef] [PubMed]
56. Dong, C.-D.; Chen, C.-W.; Chen, Y.-C.; Chen, H.-H.; Lee, J.-S.; Lin, C.-H. Polystyrene Microplastic Particles: In Vitro Pulmonary Toxicity Assessment. *J. Hazard. Mater.* **2020**, *385*, 121575. [CrossRef] [PubMed]
57. Schirinzi, G.F.; Pérez-Pomeda, I.; Sanchís, J.; Rossini, C.; Farré, M.; Barceló, D. Cytotoxic Effects of Commonly Used Nanomaterials and Microplastics on Cerebral and Epithelial Human Cells. *Environ. Res* **2017**, *159*, 579–587. [CrossRef] [PubMed]
58. Brown, D.M.; Wilson, M.R.; MacNee, W.; Stone, V.; Donaldson, K. Size-Dependent Proinflammatory Effects of Ultrafine Polystyrene Particles: A Role for Surface Area and Oxidative Stress in the Enhanced Activity of Ultrafines. *Toxicol. Appl. Pharmacol.* **2001**, *175*, 191–199. [CrossRef] [PubMed]
59. Craparo, E.F.; Cabibbo, M.; Emanuele Drago, S.; Casula, L.; Lai, F.; Cavallaro, G. Inhalable Polymeric Microparticles as Pharmaceutical Porous Powder for Drug Administration. *Int. J. Pharm.* **2022**, *628*, 122325. [CrossRef] [PubMed]
60. Atis, S.; Tutluoglu, B.; Levent, E.; Ozturk, C.; Tunaci, A.; Sahin, K.; Saral, A.; Oktay, I.; Kanik, A.; Nemery, B. The Respiratory Effects of Occupational Polypropylene Flock Exposure. *Eur. Respir. J.* **2005**, *25*, 110–117. [CrossRef] [PubMed]
61. Yee, M.S.-L.; Hii, L.-W.; Looi, C.K.; Lim, W.-M.; Wong, S.-F.; Kok, Y.-Y.; Tan, B.-K.; Wong, C.-Y.; Leong, C.-O. Impact of Microplastics and Nanoplastics on Human Health. *Nanomaterials* **2021**, *11*, 496. [CrossRef] [PubMed]
62. Pimentel, J.C.; Avila, R.; Lourenço, A.G. Respiratory Disease Caused by Synthetic Fibres: A New Occupational Disease. *Thorax* **1975**, *30*, 204–219. [CrossRef]
63. Kern, D.G.; Kuhn, C.; Ely, E.W.; Pransky, G.S.; Mello, C.J.; Fraire, A.E.; Müller, J. Flock Worker's Lung: Broadening the Spectrum of Clinicopathology, Narrowing the Spectrum of Suspected Etiologies. *Chest* **2000**, *117*, 251–259. [CrossRef]
64. Eschenbacher, W.L.; Kreiss, K.; Lougheed, M.D.; Pransky, G.S.; Day, B.; Castellan, R.M. Nylon Flock–Associated Interstitial Lung Disease. *Am. J. Respir. Crit. Care Med.* **1999**, *159*, 2003–2008. [CrossRef]
65. Barroso, E.; Ibañez, M.D.; Aranda, F.I.; Romero, S. Polyethylene Flock-Associated Interstitial Lung Disease in a Spanish Female. *Eur. Respir. J.* **2002**, *20*, 1610–1612. [CrossRef] [PubMed]
66. Lilis, R. Review of Pulmonary Effects of Poly(Vinyl Chloride) and Vinyl Chloride Exposure. *Environ. Health Perspect.* **1981**, *41*, 167–169. [CrossRef] [PubMed]
67. Leslie, H.A.; van Velzen, M.J.M.; Brandsma, S.H.; Vethaak, A.D.; Garcia-Vallejo, J.J.; Lamoree, M.H. Discovery and Quantification of Plastic Particle Pollution in Human Blood. *Environ. Int.* **2022**, *163*, 107199. [CrossRef] [PubMed]
68. Saghir, S.A. Absorption. In *Encyclopedia of Toxicology*, 3rd ed.; Wexler, P., Ed.; Academic Press: Oxford, UK, 2014; pp. 1–6. ISBN 978-0-12-386455-0.
69. Hodges, G.M.; Carr, E.A.; Hazzard, R.A.; Carr, K.E. Uptake and Translocation of Microparticles in Small Intestine. *Dig. Dis. Sci.* **1995**, *40*, 967–975. [CrossRef] [PubMed]
70. Eldridge, J.H.; Meulbroek, J.A.; Staas, J.K.; Tice, T.R.; Gilley, R.M. Vaccine-Containing Biodegradable Microspheres Specifically Enter the Gut-Associated Lymphoid Tissue Following Oral Administration and Induce a Disseminated Mucosal Immune Response. *Adv. Exp. Med. Biol.* **1989**, *251*, 191–202. [CrossRef]

71. Jenkins, P.G.; Howard, K.A.; Blackhall, N.W.; Thomas, N.W.; Davis, S.S.; O'Hagan, D.T. The Quantitation of the Absorption of Microparticles into the Intestinal Lymph of Wistar Rats. *Int. J. Pharm.* **1994**, *102*, 261–266. [CrossRef]
72. Reineke, J.J.; Cho, D.Y.; Dingle, Y.-T.; Morello, A.P.; Jacob, J.; Thanos, C.G.; Mathiowitz, E. Unique Insights into the Intestinal Absorption, Transit, and Subsequent Biodistribution of Polymer-Derived Microspheres. *Proc. Natl. Acad. Sci. USA* **2013**, *110*, 13803–13808. [CrossRef]
73. Ural, B.B.; Caron, D.P.; Dogra, P.; Wells, S.B.; Szabo, P.A.; Granot, T.; Senda, T.; Poon, M.M.L.; Lam, N.; Thapa, P.; et al. Inhaled Particulate Accumulation with Age Impairs Immune Function and Architecture in Human Lung Lymph Nodes. *Nat. Med.* **2022**, *28*, 2622–2632. [CrossRef]
74. Bailey, M.R.; Kreyling, W.G.; Andre, S.; Batchelor, A.; Collier, C.G.; Drosselmeyer, E.; Ferron, G.A.; Foster, P.; Haider, B.; Hodgson, A.; et al. An Interspecies Comparison of the Lung Clearance of Inhaled Monodisperse Cobalt Oxide Particles—Part I: Objectives and Summary of Results. *J. Aerosol Sci.* **1989**, *20*, 169–188. [CrossRef]
75. Chang, X.; Xue, Y.; Li, J.; Zou, L.; Tang, M. Potential Health Impact of Environmental Micro- and Nanoplastics Pollution. *J. Appl. Toxicol.* **2020**, *40*, 4–15. [CrossRef]
76. Prüst, M.; Meijer, J.; Westerink, R.H.S. The Plastic Brain: Neurotoxicity of Micro- and Nanoplastics. *Part. Fibre Toxicol.* **2020**, *17*, 24. [CrossRef] [PubMed]
77. Hale, R.C.; Seeley, M.E.; La Guardia, M.J.; Mai, L.; Zeng, E.Y. A Global Perspective on Microplastics. *J. Geophys. Res. Ocean* **2020**, *125*, e2018JC014719. [CrossRef]
78. Li, Z.; Zhu, S.; Liu, Q.; Wei, J.; Jin, Y.; Wang, X.; Zhang, L. Polystyrene Microplastics Cause Cardiac Fibrosis by Activating Wnt/β-Catenin Signaling Pathway and Promoting Cardiomyocyte Apoptosis in Rats. *Environ. Pollut.* **2020**, *265*, 115025. [CrossRef] [PubMed]
79. Park, E.-J.; Han, J.-S.; Park, E.-J.; Seong, E.; Lee, G.-H.; Kim, D.-W.; Son, H.-Y.; Han, H.-Y.; Lee, B.-S. Repeated-Oral Dose Toxicity of Polyethylene Microplastics and the Possible Implications on Reproduction and Development of the next Generation. *Toxicol. Lett.* **2020**, *324*, 75–85. [CrossRef] [PubMed]
80. Galloway, T.S. Micro- and Nano-Plastics and Human Health. In *Marine Anthropogenic Litter*; Bergmann, M., Gutow, L., Klages, M., Eds.; Springer International Publishing: Cham, Switzerland, 2015; pp. 343–366. ISBN 978-3-319-16510-3.
81. Peijnenburg, W.J.G.M. Phthalates. In *Encyclopedia of Ecology*; Jørgensen, S.E., Fath, B.D., Eds.; Academic Press: Oxford, UK, 2008; pp. 2733–2738. ISBN 978-0-08-045405-4.
82. Wang, W.; Kannan, K. Leaching of Phthalates from Medical Supplies and Their Implications for Exposure. *Environ. Sci. Technol.* **2023**, *57*, 7675–7683. [CrossRef] [PubMed]
83. Wang, X.; Okoffo, E.D.; Banks, A.P.; Li, Y.; Thomas, K.V.; Rauert, C.; Aylward, L.L.; Mueller, J.F. Phthalate Esters in Face Masks and Associated Inhalation Exposure Risk. *J. Hazard. Mater.* **2022**, *423*, 127001. [CrossRef] [PubMed]
84. Wang, X.; Song, M.; Guo, M.; Chi, C.; Mo, F.; Shen, X. Pollution Levels and Characteristics of Phthalate Esters in Indoor Air in Hospitals. *J. Environ. Sci.* **2015**, *37*, 67–74. [CrossRef] [PubMed]
85. U.S. Food and Drug Administration; Center for Devices and Radiological Health. *Safety Assessment of Di(2-Ethylhexyl)Phthalate (DEHP) Released from PVC Medical Devices*; Center for Devices and Radiological Health: Rockville, MD, USA. Available online: https://noharm-global.org/documents/safety-assessment-dehp-released-pvc-medical-devices (accessed on 15 March 2024).
86. Reale, E.; Hopf, N.B.; Breider, F.; Grandjean, D.; Pirard, C.; Charlier, C.; Koch, H.M.; Berthet, A.; Suarez, G.; Borgatta, M. Repeated Human Exposure to Semivolatile Organic Compounds by Inhalation: Novel Protocol for a Nonrandomized Study. *JMIR Res. Protoc.* **2023**, *12*, e51020. [CrossRef] [PubMed]
87. Masset, T.; Ferrari, B.J.D.; Dudefoi, W.; Schirmer, K.; Bergmann, A.; Vermeirssen, E.; Grandjean, D.; Harris, L.C.; Breider, F. Bioaccessibility of Organic Compounds Associated with Tire Particles Using a Fish In Vitro Digestive Model: Solubilization Kinetics and Effects of Food Coingestion. *Environ. Sci. Technol.* **2022**, *56*, 15607–15616. [CrossRef]
88. Chang, T.-H.; Szabo, E. Induction of Differentiation and Apoptosis by Ligands of Peroxisome Proliferator-Activated Receptor γ in Non-Small Cell Lung Cancer. *Cancer Res.* **2000**, *60*, 1129–1138.
89. Miller, M.D.; Marty, M.A. Impact of Environmental Chemicals on Lung Development. *Environ. Health Perspect.* **2010**, *118*, 1155–1164. [CrossRef] [PubMed]
90. Magliozzi, R.; Nardacci, R.; Scarsella, G.; Di Carlo, V.; Stefanini, S. Effects of the Plasticiser DEHP on Lung of Newborn Rats: Catalase Immunocytochemistry and Morphometric Analysis. *Histochem. Cell Biol.* **2003**, *120*, 41–49. [CrossRef]
91. Klimisch, H.J.; Gamer, A.O.; Hellwig, J.; Kaufmann, W.; Jäckh, R. Di-(2-Ethylhexyl) Phthalate: A Short-Term Repeated Inhalation Toxicity Study Including Fertility Assessment. *Food Chem. Toxicol.* **1992**, *30*, 915–919. [CrossRef] [PubMed]
92. Rosicarelli, B.; Stefanini, S. DEHP Effects on Histology and Cell Proliferation in Lung of Newborn Rats. *Histochem. Cell Biol.* **2009**, *131*, 491–500. [CrossRef] [PubMed]
93. Kim, Y.-M.; Kim, J.; Cheong, H.-K.; Jeon, B.-H.; Ahn, K. Exposure to Phthalates Aggravates Pulmonary Function and Airway Inflammation in Asthmatic Children. *PLoS ONE* **2018**, *13*, e0208553. [CrossRef]
94. Kondo, K.; Kagi, N.; Namiki, N. Study on the Mechanism of SVOC Adsorption onto Airborne Particles in Indoor Air. *Jpn. Archit. Rev.* **2018**, *1*, 528–537. [CrossRef]
95. Ataei, Y.; Sun, Y.; Liu, W.; Ellie, A.S.; Dong, H.; Ahmad, U.M. Health Effects of Exposure to Indoor Semi-Volatile Organic Compounds in Chinese Building Environment: A Systematic Review. *Int. J. Environ. Res. Public Health* **2023**, *20*, 678. [CrossRef]

96. Kocbach-Bølling, A.; Holme, J.A.; Bornehag, C.G.; Nygaard, U.C.; Bertelsen, R.J.; Nånberg, E.; Bodin, J.; Sakhi, A.K.; Thomsen, C.; Becher, R. Pulmonary Phthalate Exposure and Asthma—Is PPAR a Plausible Mechanistic Link? *EXCLI J.* **2013**, *12*, 733–759.
97. Bornehag, C.-G.; Sundell, J.; Weschler, C.J.; Sigsgaard, T.; Lundgren, B.; Hasselgren, M.; Hägerhed-Engman, L. The Association between Asthma and Allergic Symptoms in Children and Phthalates in House Dust: A Nested Case-Control Study. *Environ. Health Perspect.* **2004**, *112*, 1393–1397. [CrossRef]
98. Brassea-Pérez, E.; Hernández-Camacho, C.J.; Labrada-Martagón, V.; Vázquez-Medina, J.P.; Gaxiola-Robles, R.; Zenteno-Savín, T. Oxidative Stress Induced by Phthalates in Mammals: State of the Art and Potential Biomarkers. *Environ. Res.* **2022**, *206*, 112636. [CrossRef]
99. Sree, C.G.; Buddolla, V.; Lakshmi, B.A.; Kim, Y.-J. Phthalate Toxicity Mechanisms: An Update. *Comp. Biochem. Physiol. C Toxicol. Pharmacol.* **2023**, *263*, 109498. [CrossRef]
100. Gaudin, R.; Marsan, P.; Ndaw, S.; Robert, A. Surveillance Biologique de l'exposition Au Phtalate de Di-(2-Éthylhexyle) (DEHP) Dans Six Entreprises Françaises—INRS. 2013. Available online: https://www.inrs.fr/dms/inrs/CataloguePapier/DMT/TI-TF-205/tf205.pdf (accessed on 24 February 2024).
101. Hauser, R.; Calafat, A.M. Phthalates and Human Health. *Occup. Environ. Med.* **2005**, *62*, 806–818. [CrossRef]
102. Swan, S.H. Environmental Phthalate Exposure in Relation to Reproductive Outcomes and Other Health Endpoints in Humans. *Environ. Res.* **2008**, *108*, 177–184. [CrossRef]
103. Atia, T.; Abdel-Gawad, S. Pulmonary Toxicity Induced by Exposure to Phthalates, an Experimental Study. *Inhal. Toxicol.* **2019**, *31*, 376–383. [CrossRef]
104. Øie, L.; Hersoug, L.G.; Madsen, J.O. Residential Exposure to Plasticizers and Its Possible Role in the Pathogenesis of Asthma. *Environ. Health Perspect.* **1997**, *105*, 972–978. [CrossRef]
105. Spanier, A.J.; Fiorino, E.K.; Trasande, L. Bisphenol A Exposure Is Associated with Decreased Lung Function. *J. Pediatr.* **2014**, *164*, 1403–1408.e1. [CrossRef]
106. Wang, I.-J.; Chen, C.-Y.; Bornehag, C.-G. Bisphenol A Exposure May Increase the Risk of Development of Atopic Disorders in Children. *Int. J. Hydrogen Environ. Health* **2016**, *219*, 311–316. [CrossRef]
107. Soliman, M.A.E.; Noya, D.A.E. Effect of Bisphenol A on the Lung of Adult Male Albino Rats and the Possible Protective Role of Propolis: Light and Electron Microscopic Study. *Egypt. J. Histol.* **2023**, *46*, 290–307. [CrossRef]
108. Wang, S.; Yang, Y.; Luo, D.; Wu, D.; Liu, H.; Li, M.; Sun, Q.; Jia, L. Lung Inflammation Induced by Exposure to Bisphenol-A Is Associated with MTOR-Mediated Autophagy in Adolescent Mice. *Chemosphere* **2020**, *248*, 126035. [CrossRef]
109. Nakajima, Y.; Goldblum, R.M.; Midoro-Horiuti, T. Fetal Exposure to Bisphenol A as a Risk Factor for the Development of Childhood Asthma: An Animal Model Study. *Environ. Health* **2012**, *11*, 8. [CrossRef]
110. Midoro-Horiuti, T.; Tiwari, R.; Watson, C.S.; Goldblum, R.M. Maternal Bisphenol A Exposure Promotes the Development of Experimental Asthma in Mouse Pups. *Environ. Health Perspect.* **2010**, *118*, 273–277. [CrossRef]
111. Saeidnia, S. Phthalates. In *Encyclopedia of Toxicology*, 3rd ed.; Wexler, P., Ed.; Academic Press: Oxford, UK, 2014; pp. 928–933. ISBN 978-0-12-386455-0.
112. Agence Nationale de Sécurité Sanitaire de L'alimentation, de L'environnement et du Travail (Anses) Bisphenol B, an Endocrine Disruptor for Humans and the Environment. Available online: https://www.anses.fr/en/content/bisphenol-b-endocrine-disruptor-humans-and-environment (accessed on 24 February 2024).
113. Chen, Q.; Yang, H.; Zhou, N.; Sun, L.; Bao, H.; Tan, L.; Chen, H.; Ling, X.; Zhang, G.; Huang, L.; et al. Phthalate Exposure, Even below US EPA Reference Doses, Was Associated with Semen Quality and Reproductive Hormones: Prospective MARHCS Study in General Population. *Environ. Int.* **2017**, *104*, 58–68. [CrossRef]
114. Duty, S.M.; Silva, M.J.; Barr, D.B.; Brock, J.W.; Ryan, L.; Chen, Z.; Herrick, R.F.; Christiani, D.C.; Hauser, R. Phthalate Exposure and Human Semen Parameters. *Epidemiology* **2003**, *14*, 269–277. [CrossRef]
115. Buck Louis, G.M.; Gray, L.E., Jr.; Marcus, M.; Ojeda, S.R.; Pescovitz, O.H.; Witchel, S.F.; Sippell, W.; Abbott, D.H.; Soto, A.; Tyl, R.W.; et al. Environmental Factors and Puberty Timing: Expert Panel Research Needs. *Pediatrics* **2008**, *121*, S192–S207. [CrossRef]
116. Colón, I.; Caro, D.; Bourdony, C.J.; Rosario, O. Identification of Phthalate Esters in the Serum of Young Puerto Rican Girls with Premature Breast Development. *Environ. Health Perspect.* **2000**, *108*, 895–900. [CrossRef]
117. Ventrice, P.; Ventrice, D.; Russo, E.; De Sarro, G. Phthalates: European Regulation, Chemistry, Pharmacokinetic and Related Toxicity. *Environ. Toxicol. Pharmacol.* **2013**, *36*, 88–96. [CrossRef]
118. Weaver, J.A.; Beverly, B.E.J.; Keshava, N.; Mudipalli, A.; Arzuaga, X.; Cai, C.; Hotchkiss, A.K.; Makris, S.L.; Yost, E.E. Hazards of Diethyl Phthalate (DEP) Exposure: A Systematic Review of Animal Toxicology Studies. *Environ. Int.* **2020**, *145*, 105848. [CrossRef]
119. Adibi, J.J.; Perera, F.P.; Jedrychowski, W.; Camann, D.E.; Barr, D.; Jacek, R.; Whyatt, R.M. Prenatal Exposures to Phthalates among Women in New York City and Krakow, Poland. *Environ. Health Perspect.* **2003**, *111*, 1719–1722. [CrossRef]
120. Swan, S.H.; Main, K.M.; Liu, F.; Stewart, S.L.; Kruse, R.L.; Calafat, A.M.; Mao, C.S.; Redmon, J.B.; Ternand, C.L.; Sullivan, S.; et al. Decrease in Anogenital Distance among Male Infants with Prenatal Phthalate Exposure. *Environ. Health Perspect.* **2005**, *113*, 1056–1061. [CrossRef]
121. Hauser, R.; Meeker, J.D.; Duty, S.; Silva, M.J.; Calafat, A.M. Altered Semen Quality in Relation to Urinary Concentrations of Phthalate Monoester and Oxidative Metabolites. *Epidemiology* **2006**, *17*, 682–691. [CrossRef]
122. Hauser, R.; Meeker, J.D.; Singh, N.P.; Silva, M.J.; Ryan, L.; Duty, S.; Calafat, A.M. DNA Damage in Human Sperm Is Related to Urinary Levels of Phthalate Monoester and Oxidative Metabolites. *Hum. Reprod.* **2007**, *22*, 688–695. [CrossRef]

123. Green, R.; Hauser, R.; Calafat, A.M.; Weuve, J.; Schettler, T.; Ringer, S.; Huttner, K.; Hu, H. Use of Di(2-Ethylhexyl) Phthalate-Containing Medical Products and Urinary Levels of Mono(2-Ethylhexyl) Phthalate in Neonatal Intensive Care Unit Infants. *Environ. Health Perspect* **2005**, *113*, 1222–1225. [CrossRef]
124. Calafat, A.M.; Needham, L.L.; Silva, M.J.; Lambert, G. Exposure to Di-(2-Ethylhexyl) Phthalate among Premature Neonates in a Neonatal Intensive Care Unit. *Pediatrics* **2004**, *113*, e429–e434. [CrossRef]
125. van Amerongen, C.C.A.; Ofenloch, R.F.; Cazzaniga, S.; Elsner, P.; Gonçalo, M.; Naldi, L.; Svensson, Å.; Bruze, M.; Schuttelaar, M.L.A. Skin Exposure to Scented Products Used in Daily Life and Fragrance Contact Allergy in the European General Population—The EDEN Fragrance Study. *Contact Dermat.* **2021**, *84*, 385–394. [CrossRef]
126. Rios, L.M.; Moore, C.; Jones, P.R. Persistent Organic Pollutants Carried by Synthetic Polymers in the Ocean Environment. *Mar. Pollut. Bull.* **2007**, *54*, 1230–1237. [CrossRef]
127. Hirai, H.; Takada, H.; Ogata, Y.; Yamashita, R.; Mizukawa, K.; Saha, M.; Kwan, C.; Moore, C.; Gray, H.; Laursen, D.; et al. Organic Micropollutants in Marine Plastics Debris from the Open Ocean and Remote and Urban Beaches. *Mar. Pollut. Bull.* **2011**, *62*, 1683–1692. [CrossRef]
128. Martin, R.A.; Poynter, M.E. The Immunobiology of Asthma. In *Encyclopedia of Immunobiology*; Ratcliffe, M.J.H., Ed.; Academic Press: Oxford, UK, 2016; pp. 295–305. ISBN 978-0-08-092152-5.
129. Déciga-Alcaraz, A.; Tlazolteotl Gómez de León, C.; Morales Montor, J.; Poblano-Bata, J.; Martínez-Domínguez, Y.M.; Palacios-Arreola, M.I.; Amador-Muñoz, O.; Rodríguez-Ibarra, C.; Vázquez-Zapién, G.J.; Mata-Miranda, M.M.; et al. Effects of Solvent Extracted Organic Matter from Outdoor Air Pollution on Human Type II Pneumocytes: Molecular and Proteomic Analysis. *Environ. Pollut.* **2023**, *337*, 122551. [CrossRef]
130. Sun, K.; Song, Y.; He, F.; Jing, M.; Tang, J.; Liu, R. A Review of Human and Animals Exposure to Polycyclic Aromatic Hydrocarbons: Health Risk and Adverse Effects, Photo-Induced Toxicity and Regulating Effect of Microplastics. *Sci. Total Environ.* **2021**, *773*, 145403. [CrossRef]
131. Cioroiu, B.I.; Tarcau, D.; Cucu-Man, S.; Chisalita, I.; Cioroiu, M. Polycyclic Aromatic Hydrocarbons in Lung Tissue of Patients with Pulmonary Cancer from Romania. Influence According as Demographic Status and ABO Phenotypes. *Chemosphere* **2013**, *92*, 504–511. [CrossRef]
132. Moorthy, B.; Chu, C.; Carlin, D.J. Polycyclic Aromatic Hydrocarbons: From Metabolism to Lung Cancer. *Toxicol. Sci.* **2015**, *145*, 5–15. [CrossRef]
133. Kim, K.-H.; Jahan, S.A.; Kabir, E.; Brown, R.J.C. A Review of Airborne Polycyclic Aromatic Hydrocarbons (PAHs) and Their Human Health Effects. *Environ. Int.* **2013**, *60*, 71–80. [CrossRef] [PubMed]
134. Srogi, K. Monitoring of Environmental Exposure to Polycyclic Aromatic Hydrocarbons: A Review. *Environ. Chem. Lett.* **2007**, *5*, 169–195. [CrossRef] [PubMed]

toxics

Review

The Comet Assay as a Tool in Human Biomonitoring Studies of Environmental and Occupational Exposure to Chemicals—A Systematic Scoping Review

Carina Ladeira [1,2,*], Peter Møller [3], Lisa Giovannelli [4], Goran Gajski [5], Anja Haveric [6], Ezgi Eyluel Bankoglu [7], Amaya Azqueta [8], Marko Gerić [5], Helga Stopper [7], José Cabêda [9], Fernanda S. Tonin [10] and Andrew Collins [11]

1 H&TRC-Health & Technology Research Center, ESTeSL-Escola Superior de Tecnologia da Saúde, Instituto Politécnico de Lisboa, 1990-096 Lisbon, Portugal
2 NOVA National School of Public Health, Public Health Research Centre, Universidade NOVA de Lisboa, 1600-560 Lisbon, Portugal
3 Department of Public Health, Section of Environmental Health, University of Copenhagen, 1172 Copenhagen, Denmark; pemo@sund.ku.dk
4 Department NEUROFARBA, Section Pharmacology and Toxicology, University of Florence, 50121 Florence, Italy; lisa.giovannelli@unifi.it
5 Division of Toxicology, Institute for Medical Research and Occupational Health, 10000 Zagreb, Croatia; ggajski@imi.hr (G.G.); mgeric@imi.hr (M.G.)
6 Institute for Genetic Engineering and Biotechnology, University of Sarajevo, 71000 Sarajevo, Bosnia and Herzegovina; anjahaveric@ingeb.unsa.ba
7 Institute of Pharmacology and Toxicology, University of Würzburg, 97078 Würzburg, Germany; ezgi.bankoglu@uni-wuerzburg.de (E.E.B.); helga.stopper@uni-wuerzburg.de (H.S.)
8 Department of Pharmaceutical Sciences, School of Pharmacy and Nutrition, University of Navarra, 31009 Pamplona, Spain; amazqueta@unav.es
9 Guarda Nacional Republicana, Destacamento Territorial de Vila Franca de Xira, Núcleo de Proteção Ambiental, 1500-124 Lisbon, Portugal; joseluiscabeda@gmail.com
10 Pharmaceutical Care Research Group, Universidad de Granada, 18012 Granada, Spain; ffstonin@gmail.com
11 Department of Nutrition, University of Oslo, 0316 Oslo, Norway; collinsand@gmail.com
* Correspondence: carina.ladeira@estesl.ipl.pt; Tel.: +351-218980400

Citation: Ladeira, C.; Møller, P.; Giovannelli, L.; Gajski, G.; Haveric, A.; Bankoglu, E.E.; Azqueta, A.; Gerić, M.; Stopper, H.; Cabêda, J.; et al. The Comet Assay as a Tool in Human Biomonitoring Studies of Environmental and Occupational Exposure to Chemicals—A Systematic Scoping Review. *Toxics* **2024**, *12*, 270. https://doi.org/10.3390/toxics12040270

Academic Editor: Michael Petriello

Received: 27 February 2024
Revised: 31 March 2024
Accepted: 2 April 2024
Published: 5 April 2024

Abstract: Biomonitoring of human populations exposed to chemical substances that can act as potential mutagens or carcinogens, may enable the detection of damage and early disease prevention. In recent years, the comet assay has become an important tool for assessing DNA damage, both in environmental and occupational exposure contexts. To evidence the role of the comet assay in human biomonitoring, we have analysed original research studies of environmental or occupational exposure that used the comet assay in their assessments, following the PRISMA-ScR method (preferred reporting items for systematic reviews and meta-analyses extension for scoping reviews). Groups of chemicals were designated according to a broad classification, and the results obtained from over 300 original studies (n = 123 on air pollutants, n = 14 on anaesthetics, n = 18 on antineoplastic drugs, n = 57 on heavy metals, n = 59 on pesticides, and n = 49 on solvents) showed overall higher values of DNA strand breaks in the exposed subjects in comparison with the unexposed. In summary, our systematic scoping review strengthens the relevance of the use of the comet assay in assessing DNA damage in human biomonitoring studies.

Keywords: comet assay; human biomonitoring; air pollution; anaesthetics; antineoplastic drugs; heavy metals; pesticides; solvents; exposure

1. Introduction

Humans are in contact with more than 160 million chemicals, based on the World Health Organization (WHO) and United Nations (UN) compendium, while 6000 of these are responsible for 99% of the market by volume [1]. Even those chemicals that are carefully manufactured for safe use may have unwanted harmful by-products, generating potential health risks. It is important to conduct studies on environmental and occupational exposure to chemical substances and contaminants, considering the presence and severity of the adverse effects on human health [2]. Toxicological and epidemiological studies have collected biological markers (biomarkers) to evaluate the relationships between environmental or occupational chemical exposure and adverse health effects [3]. The development of molecular epidemiology introduced the concept of biomarkers of effect, strengthening the evidence of causality between chemical exposure and adverse effects, especially at an early stage before disease onset [4], and playing a pivotal role in disease prevention.

Worldwide, about 19 million people are diagnosed with some type of cancer annually, and the cancer mortality is almost 10 million [5], causing a significant financial and social burden, especially in ageing populations [6]. Since the induction of DNA damage is one of the most important steps in carcinogenesis, the biomonitoring of human populations exposed to genotoxic substances for DNA damage is potentially a useful preventive tool, as it can detect early events that can be precursors of carcinogenesis [7].

Cytogenetic methods have been extensively used for the biological monitoring of populations exposed to mutagenic and carcinogenic agents. The comet assay is widely employed in human biomonitoring for assessing DNA damage and also has applications in genotoxicity testing, environmental toxicology, and fundamental research on DNA damage and repair [7–12]. A summarised overview of the history of the assay was reviewed by Jiang et al., 2023 [13]. The alkaline comet assay identifies different types of damage resulting from recent exposure that are potentially reparable, such as single- and double-strand DNA breaks, alkali-labile lesions converted to strand breaks under alkaline conditions, and single-strand breaks associated with incomplete excision repair [14,15]; it is one of the most used methods for DNA damage biomonitoring [16]. Most human studies have focused on blood cells because they are easy to obtain, and—as they circulate in the body—the metabolic state of these cells can reflect the overall extent of body exposure [17]. However, other cell types have also been employed, such as buccal, nasal, lens epithelial, and germ cells [18,19].

The comet assay is a sensitive, rapid, versatile, and low-cost technique for quantifying and analysing DNA damage and repair at the level of individual cells [20,21], requiring small numbers of cells per sample and a relatively short time to complete a study [8]. This has made the comet assay more popular than other genotoxicity tests, such as sister chromatid exchanges, micronucleus assays, and chromosomal aberrations [13]. Thus, the comet assay is a method of choice for the measurement of DNA damage in environmental and occupational exposure studies for the assessment of the effects of chemical substances—either as single compounds or as mixtures [15]. Responding to the need for standardised protocols, a compendium of consensus protocols applying the comet assay to a variety of cells [14], as well as recommendations for describing comet assay procedures and results [22], have recently been published.

There are already some systematic reviews and meta-analyses focused on the use of the comet assay in studies of human exposure to particular classes of chemicals, such as antineoplastic drugs [23], pesticides [24], and air pollution [25], and a review published in 2009 looks at studies that employed the comet assay in the biomonitoring of environmental and occupational exposures, including radiation [18]. Despite its popularity and these systematic reviews, there is still a lack of literature and no comprehensive overview of the role of DNA damage measurement as a reliable biomarker for human monitoring programs, including different types of exposures.

This broad scoping review aims to systematically analyse evidence on the use of the comet assay in human biomonitoring studies assessing genotoxic effects from environmental or occupational exposures. Specifically, the work focuses on air pollutants, anaesthetics, antineoplastic drugs, heavy metals, pesticides, and solvents. The presentation of results, organised according to these groups of chemicals, exclusively follows alphabetical order criteria without considering the complexity of the chemical substances in each group. The reporting of "essential" information relating to the comet assay descriptors (e.g., %DNA in tail, tail length, tail moment, or visual score), the number of comets analysed per sample, and how the overall level of DNA migration is expressed (e.g., median or mean of comet scores), is necessary for scoring and data analysis of the comet assay [13,26]. It has been shown that 20–30% of published studies with comet assay results use visual scores, while 70–80% are the results from image analysis systems; tail length and tail moment used to be the most popular comet descriptors in the early 00s, but % tail DNA has become the most popular since 2010. Regarding the olive tail moment descriptor of DNA migration, it is considered to be particularly useful in describing heterogeneity within a cell population, as it can pick up variations in the DNA distribution within the tail [27], and it was very often used in the studies gathered in this scoping systematic review. More information regarding the various parameters that have appeared in scientific publications can be found in Kumaravel et al., 2009 [28].

2. Materials and Methods

The systematic scoping review was performed in accordance with the Jonna Briggs Institute and Cochrane Collaboration recommendations [29–31] and is reported following the PRISMA-ScR (preferred reporting items for systematic reviews and meta-analyses—extension for scoping reviews) checklists [32,33]. The protocol has been registered in PROSPERO—CRD42023402351. At least two authors independently conducted all steps of the study selection and data extraction. Divergences were resolved by discussion in consensus working group meetings.

2.1. Search Strategy and Eligibility Criteria

A comprehensive literature search was conducted to identify relevant studies in PubMed and Web of Science (last updated June 2023) without language limits. Searches were limited by the year of publication [from 2000, after the introduction of 'Comet Assay' as a Medical Subject Headings (MeSH) term] and to human studies. A manual search in the reference lists of the included studies was also performed, and other search engines (Google and Google Scholar) were employed.

Five distinct search strategies were developed and applied (according to the group of chemical substances under evaluation) using descriptors related to human biomonitoring and comet assay, and air pollution, anaesthetics, antineoplastic drugs, heavy metals, pesticides or solvents, combined with the Boolean operators AND and OR as follows:

- Search string for air pollution: Human Biomonitoring OR monitoring AND comet assay AND (air pollution OR diesel exhaust OR dust OR ozone OR particulate matter OR ultrafine particles OR formaldehyde OR hydrocarbon).
- Search string for anaesthetics: Human Biomonitoring OR monitoring AND comet assay AND (anaesthetic OR anaesthesia OR N_2O OR nitrous oxide OR isoflurane OR halothane).
- Search string for antineoplastic drugs: Human Biomonitoring OR monitoring AND Comet assay AND (antineoplastic drugs OR cytostatic OR cytotoxic OR cyclophosphamide OR paclitaxel OR 5-Fluororacil).
- Search string for heavy metals: Human Biomonitoring OR monitoring AND Comet assay AND (lead OR mercury OR Cadmium OR arsenic OR heavy metals).
- Search string for pesticides: Human biomonitoring OR monitoring AND comet assay AND pesticides.
- Search string for solvents: Human Biomonitoring OR monitoring AND Comet assay AND (styrene OR benzene OR toluene OR xylene OR chloroform OR tetrachloro- or trichloroethylene OR perchloroethylene OR halogenated solvents OR solvents).

Registers retrieved from the databases (PubMed and Web of Science) were transferred into Mendeley (reference manager) or Rayyan, where duplicate records were removed.

The reviewers independently performed the screening (title/abstract reading), full-text evaluation, and data extraction using Microsoft Excel sheets.

This systematic scoping review included articles meeting the following criteria (PECOS acronym):

- Population: studies evaluating human subjects with environmental or occupational exposure to chemical substances;
- Exposure: studies assessing the environmental or occupational effects of exposure to the chemical substances of interest (i.e., air pollution, anaesthetics gases, antineoplastic drugs, heavy metals, pesticides, or solvents) by means of the comet assay in biological samples;
- Comparator: non-exposed human subjects or pre-post comparative data on exposure (in case of a single-arm study);
- Outcomes: comet assay measurements such as the tail moment, tail length (µm), % tail intensity, olive tail moment, visual scoring/DNA damage index parameters, and other parameters considered;
- Study design: interventional studies (controlled trials, experimental studies) or observational comparative studies, including case-control, cohort, cross-sectional studies, and quasi-experimental studies (pre–post-test).
- Studies without data for extraction (unavailable information or an unpublished paper), conference abstracts, other study designs (reviews, case reports, letters, commentaries, and protocols), non-human studies (*in vitro* and *in vivo*), *in vitro* studies on primary human cells or cell lines, and those in non-English languages were excluded.

2.2. Data Extraction and Synthesis

A standard form (Microsoft Excel, Redmond, WA, USA) was developed by the coordinator (Carina Ladeira) and validated by all team members (co-authors) to extract data on the following: (1) authors, (2) year of publication, (3) main chemical substances in exposure, (4) country, (5) exposure assessment or biomarkers of exposure, (6) population characteristics, and (7) DNA damage measured by the comet assay. The studies were organised by the type of exposure—occupational or environmental—in each section whenever necessary. Data only available in figures were extracted, whenever possible, by a single team member.

Individual results of the studies were summarised as reported in the article, including the type of measures and units (narrative synthesis) and were sorted into one of the six categories according to the type of chemical substances (i.e., air pollution, anaesthetics, antineoplastic drugs, heavy metals, pesticides, or solvents) to properly account for their special features; flow diagrams were also presented independently.

To facilitate the comparison among studies of each group of substances, as well as ease the data interpretation and writing of the narrative text, the authors established a minimum set of methodological items that should be reported from the studies considered for analysis. In decreasing order of importance, these are (i) the existence of measurements of external exposure or markers of internal exposure; (ii) the use of additional types of biomarkers to add value to the data interpretation; and (iii) grouping subjects based on the exposure categories (e.g., work categories in occupational studies or regions in environmental studies) or studies without a control group.

3. Results

This section is divided by subheadings. It provides a concise and precise description of the experimental results, their interpretation, as well as the experimental conclusions that can be drawn.

This systematic scoping review included a total of 334 studies (128 for air pollution, 15 for anaesthetics, 19 for antineoplastic drugs, 57 for heavy metals, 65 for pesticides, and 50 for solvents) for data synthesis. The groups are arranged in alphabetical order and are described below according to the type of chemical after a brief introduction.

3.1. Air Pollution

Air pollution is currently one of the major issues in environmental and public health, recognised by leading world authorities as a risk factor associated with adverse health outcomes [34]. Both outdoor and indoor air pollution are categorised by the International Agency for Research on Cancer (IARC) as carcinogenic to humans (Group 1). Exposure to outdoor air pollutants may occur in both urban and rural areas, with the most common sources being the emissions caused by combustion processes from motor vehicles, solid fuel burning, and industry [35]. The most common air pollutants present in ambient air include particulate matter (PM) of different sizes, ozone (O_3), nitrogen dioxide (NO_2), carbon monoxide (CO), and sulphur dioxide (SO_2). Indoor air pollution can be linked to households; the release of gases or particles into the air is the primary cause of indoor air quality problems [36,37]. Regarding indoor air, one major concern is biomass smoke since it contains a number of health-damaging chemicals, including PM of different sizes, CO, oxides of nitrogen, formaldehyde, acrolein, benzene, toluene, styrene, 1,3-butadiene, and polycyclic aromatic hydrocarbons (PAHs) such as benzo(a)pyrene [38].

Workplace exposure to airborne particulates (dusts) and chemicals (including anaesthetic gases and solvents) is typically not considered to be air pollution. However, certain professions with vehicle-related exhausts have been used in studies on both gaseous and particulate components in outdoor air pollution.

Regarding specifically the search string on air pollution, it was challenging to identify studies on air pollution since the term applies to a broad spectrum of exposure situations. Thus, we have used a search string that captured a large number of papers (approximately 2500), although many of these were excluded for further review, as is shown in Figure 1. A number of papers identified in the search on air pollution were also included in the heavy metals and solvents sections due to the variety of chemicals that were studied. In addition, we have only included studies of involuntary exposure to air pollution (thus, environmental tobacco smoke was considered involuntary exposure, whereas smoking was voluntary exposure).

In our systematic scoping review, 257 articles were assessed in full-text after duplicate removal and initial screening, in which 129 were excluded, mostly because they were *in vitro* studies (n = 66), complementary papers or protocols (n = 25), without numerical comet assay data (n = 16), or not in human samples (n = 10). A total of 128 studies were included in the qualitative analysis, as summarised in Figure 1 and Table 1.

Table 1. Summary of findings from the included studies on air pollution.

Author	Year	Main Chemical Exposure	Country	Exposure Assessment or Biomarkers of Exposure	Population Characteristics	DNA Damage	Reference/DOI
				Occupational exposure			
Andersen	2018	PAH	Denmark	Urinary 1-OHP	22 professional firefighters	• **DNA strand breaks:** before (0.12 ± 0.04), after (0.13 ± 0.04); **non-sig.**	[39] 10.1002/em.22193
Andersen	2021	PAH fluorene	Denmark	Exposure levels to PAH (silicone bands, skin wipes) Exposure levels to PAHs and organophosphate esters (OPEs) Urinary excretion of PAH metabolites (OH-PAHs).	116 air force personnel (79 exposed, 37 controls)	• **DNA strand breaks** (number of lesions/10^6 bp): exposed (0.09 ± 0.04), controls (0.10 ± 0.04); **non-sig.**	[40] 10.1038/s41598-021-97382-5
Al Zabadi **	2011	PAH, VOC	France	Air concentration PAH and benzene	64 sewage workers (34 exposed, 30 unexposed)	• **% tail DNA:** exposed (8.07 ± 3.12), unexposed (2.70 ± 0.58); **sig.**	[41] 10.1186/1476-069X-10-23
Aydin	2013	Formaldehyde	Turkey	Passive air samplers (TWA8h)	92 medium-density fibreboard plants (46 exposed, 46 unexposed)	• **% tail DNA**: exposed (4.25 ± 0.29), unexposed (5.28 ± 0.22); **sig.**	[42] 10.1007/s00204-012-0961-9
Bacaksiz	2013	PAH and heterocyclic compounds	Turkey	--	60 (30 exposed asphalt workers, 30 controls)	• **% tail DNA:** exposed (24.34 ± 2.72), controls (20.04 ± 2.75); **sig.**	[43] 10.1080/09603123.2013.773586
Bagryants	2010	PAH, VOC	Czech Republic	Personal samplers, quantitative analysis of PAHs, radial diffusive samplers for VOC exposure, cotinine	120 (50 bus drivers, 20 garagemen, 50 controls)	• **% tail DNA:** bus drivers (1.60 ± 0.90), garagemen (2.42 ± 2.19), controls (1.31 ± 0.88); **sig.**	[44] 10.1016/j.toxlet.2010.08.007
Becit	2021	Marble dust	Turkey	Air samples and particle analysis	89 (48 exposed workers in marble processing plants, 41 controls)	• **% tail DNA:** exposed (1.59 ± 0.69), controls (0.95 ± 0.29); **sig.**	[45] 10.1016/j.envres.2021.111209

Table 1. *Cont.*

Author	Year	Main Chemical Exposure	Country	Exposure Assessment or Biomarkers of Exposure	Population Characteristics	DNA Damage	Reference/DOI
Barth	2016	Air pollution (outdoor)	Brazil	Urinary 1-hydroxy-pyrene (1-OHP)	82 (45 taxi drivers, 37 controls)	• **% tail DNA:** controls (8.28 ± 0.21), exposed (11.58 ± 0.35); **sig.** • **Comet tail moment:** controls (1.83 ± 0.20), exposed (2.64 ± 0.17); **sig.**	[46] 10.1007/s11356-016-7772-0
Balamur likrishnan	2014	Silica dust exposure	India	--	85 (50 exposed subjects: Group I ≤ 40 years and ≤13 years working duration (23 individuals) Group II above 40 years and above 13 years (27 individuals) working duration, 35 controls; Group I (17), Group II (18))	• **Total damaged cells:** exposed: group I (50.17 ± 14.44), group II (83.74 ± 16.20), controls: group I (22.52 ± 13.49), group II (48.55 ± 17.08); **sig.**	[47] 10.1007/s00477-013-0843-6
Bruschweiler	2016	Wood dust	Switzerland	Wood dust, PAH, and B(a)P exposure	nonsmoking wood workers (n = 31, furniture and construction workers, natural wood, 12; wooden board, 19) and controls (n = 19)	• **Comet score (visual scoring)**—median (25–75th): natural wood (11.3; 8.8–26.3), wooden board (61.5; 49.5–85), controls (11.0; 8.0–18.0); **sig.**	[48] 10.4137/EHI.S38344
Carere **	2002	Air pollution	Italy	Benzene exposure	190 (133 traffic policemen, 57 office workers as controls)	• **Comet tail moment:** exposed (0.46 ± 0.46), controls (0.36 ± 0.32); **non-sig.**	[49] 10.1016/s1383-5718(02)00108-0
Cavallo	2005	PAH	Italy	Personal air sampling, urinary OH-pyrene	41 (19 paving workers, 22 controls)	• **Comet tail moment**: control (19.5 ± 6.0), exposed (22.7 ± 7.29); **sig.**	[50] 10.1093/annhyg/mei072

Table 1. *Cont.*

Author	Year	Main Chemical Exposure	Country	Exposure Assessment or Biomarkers of Exposure	Population Characteristics	DNA Damage	Reference/DOI
Cavallo	2006	PAH	Italy	Urinary 1-hydroxy-pyrene (1-OHP)	71 (41 exposed airport personnel (group A, 24 persons, group B, 17 persons; 31 controls))	• **Comet tail moment (buccal cells):** exposed (118.87), unexposed (68.20); **sig.** • **Comet tail moment (lymphocytes):** exposed (43.01), unexposed (36.01); **sig.** only for controls and exposed subgroups (A and B)	[51] 10.1016/j.tox.2006.03.003
Cavallo	2009	PAHs, antineoplastic drugs	Italy	Exposure assessment studies cited (reported in previous papers)	163 (30 workers exposed to antineoplastic drugs, 57 workers exposed to PAHs, 76 controls)	• % **Comet** (lymphocytes): exposed (18.11), unexposed (11.24); **sig.** • **Comet tail** (lymphocytes): airport workers (21.50), controls (17.43); **sig.**; and buccal cells of airport workers (59.43), controls (34.79); **sig.** exposed (21.84), controls (16.72); **sig.** for PAH exposure	[16] 10.1002/em.20501
Cavallo	2022	Graphene	Italy	Particle number concentration (PNC, particles/cm^3) from 10 nm to 1000 nm; airborne particle matter from 250 nm to 10 mm	6 graphene workers and 11 controls	• % **tail DNA:** controls (11.20 ± 6.93), workers first biomonitoring (9.70 ± 2.88) vs. workers follow-up (14.00 ± 6.43); **sig.** • **Comet tail moment:** controls (3.80 ± 2.28), workers first biomonitoring (3.19 ± 2.03) vs. workers follow-up (3.39 ± 1.84) • **Comet tail length:** controls (20.68 ± 13.75), workers first biomonitoring (20.42 ± 5.95) vs. workers follow-up (18.90 ± 7.08)	[52] 10.1080/17435390.2022.2149359

Table 1. *Cont.*

Author	Year	Main Chemical Exposure	Country	Exposure Assessment or Biomarkers of Exposure	Population Characteristics	DNA Damage	Reference/DOI
Cebulska-Wasilewska *	2005	PAH	Czech Republic	$PM_{2.5}$ and PAH analyses	78 (40 policemen, 38 controls)	• **% tail DNA:** controls (40) winter (2.64 ± 1.37); (38) summer (2.62 ± 1.04); policemen (43) winter (2.72 ± 1.70); summer (2.91 ± 1.05); **non-sig.**	[53] 10.1016/j.mrgentox.2005.08.013
Cebulska-Wasilewska *	2007	PAH	Slovakia/Bulgaria	$PM_{2.5}$ and PAH analyses	174 policemen (99 exposed, 75 controls)	• **% tail DNA:** controls (4.06 ± 1.40), exposed (3.86 ± 1.28); **non-sig.**	[54] 10.1016/j.mrfmmm.2007.03.004
Cebulska-Wasilewska *	2007*	PAH	Slovakia/Bulgaria	Environmental PAHs	259 (144 exposed, who were municipal policemen or bus drivers; 115 controls)	• **% tail DNA:** exposed (3.7 ± 1.3), controls 3.8 ± 1.5; **non-sig.**	[55] 10.1016/j.mrfmmm.2007.03.005
Ceppi	2023	PAH and glass fibres	Slovakia	Air sampling for the PAH analysis, air fibre sampling, personal exposure monitoring for PAH, cotinine	116 (76 exposed shop floor workers, 34 controls)	• **DNA strand breaks** (mean ± SEM): exposed (77 ± 4), controls (61 ± 5); **sig.**	[56] 10.1016/j.mrgentox.2022.503572
Chen	2006	PAH (coke-oven exposure)	China	PAH analysis	363 (240 coke-oven workers and 123 controls, all males)	• **Olive tail moment:** control (0.58 ± 0.92), exposed (1.23 ± 1.12); **sig.**	[57] 10.1158/1055-9965.EPI-06-0291
Chen	2010	PCDD, metals, and silica particles,	Taiwan	Air samples analysis, metal analysis	78 (37 workers were recruited from a bottom ash recovery plant and 41 workers from fly ash treatment plants)	• **Comet tail moment:** bottom ash (2.64 ± 0.47); fly ash (7.55 ± 6.96); **sig.**	[58] 10.1016/j.jhazmat.2009.09.010
Cheng	2009	PAH (coke-oven exposure)	China	Urinary 1-hydroxypyrene (1-OHP)	158 (94 coke-oven workers and 64 controls)	• **Baseline DNA damage:** exposed (0.86; 0.77–0.97), controls (0.43; 0.35–0.52); **sig.**	[59] 10.1158/1055-9965.EPI-08-0763

Table 1. *Cont.*

Author	Year	Main Chemical Exposure	Country	Exposure Assessment or Biomarkers of Exposure	Population Characteristics	DNA Damage	Reference/DOI
Chia	2008	Zinc and copper smelting work	Taiwan	8-hydroxydeoxyguanosine (8-OH-dG) in urine (ELISA), lipid peroxidation (MDA in plasma)	67 (39 smelting workers, 28 non-exposed)	• **Comet tail moment:** exposed (0.33 ± 0.09), non-exposed (0.29 ± 0.1); **non-sig.**	[60] 10.2486/indhealth.46.174
Costa §	2008	Formaldehyde	Portugal	Air samplers (TWA8h): ranging from 1.50 and 4.43 ppm	60 (30 pathology anatomy workers, 30 controls)	• **Comet tail length:** control (41.85 ± 1.97), exposed (60.00 ± 2.31); **sig.**	[61] 10.1016/j.tox.2008.07.056
Costa §	2011	Formaldehyde	Portugal	Air sampling and FA analysis	98 (48 pathology anatomy workers, 50 non-exposed)	• **% tail DNA—mean ± SE, (range):** controls 8.01 ± 0.64 (2.83–24.40), exposed 11.76 ± 0.74 (4.72–29.67) • **Comet tail length:** controls 42 ± 1.6 (17.14–74.62), exposed 54.55 ± 2.02 (33.14–99.09); **sig.**	[62] 10.1080/15287394.2011.582293
Costa	2015	Formaldehyde	Portugal	Air sampling (TWA8h) level of exposure	171 (84 pathology anatomy workers, 87 controls)	• **% tail DNA:** control [7.5 ± 0.47 (range 0.86–24.4)] vs. exposed [11.67 ± 0.72 (range 0.23–28.07)]; **sig.**	[63] 10.1093/mutage/gev002
De Boeck	2000	Cobalt dust, hard metal dust	Belgium	Urinary 8-OH-dG	99 (24 workers exposed to cobalt dust, 27 workers exposed to hard metal dust, and 27 controls)	• **Comet tail length:** exposed cobalt 0.71 (1.38) (0.32–1.18); hard metals (0.65 (1.23) (0.36–0.90); controls 0.64 (1.25) (0.47–1.06); • **% tail DNA:** exposed cobalt 0.50 (1.44) (0.25–1.15); hard metals 0.57 (1.24) (0.38–0.77); controls 0.51 (1.35) (0.31–0.87); • **Comet tail moment:** exposed cobalt 0.37 (1.85) (0.11–1.18); hard metals 0.40 (1.45) (0.14–0.80); controls 0.34 (1.47) (0.18–0.81); **non-sig.**	[64] 10.1002/1098-2280(2000)36:2<151::aid-em10>3.3.co;2-m

Table 1. *Cont.*

Author	Year	Main Chemical Exposure	Country	Exposure Assessment or Biomarkers of Exposure	Population Characteristics	DNA Damage	Reference/DOI
Duan	2016	Diesel engine exhaust	China	Air sampling: $PM_{2.5}$, elemental carbon, NO_2, SO_2, and airborne PAHs urinary 1-OHP	207 (101 DEE-exposed workers and 106 controls)	• **% tail DNA:** controls (18.75 ± 28.29), exposed (60.02 ± 28.59); **sig.**	[65] 10.1136/oemed-2015-102919
Everatt **	2013	Perchloroethylene	Lithuania	PCE concentration in air: 31.40 ± 23.51	59 (30 dry cleaner workers, 29 control)	• **Comet tail length:** (lymphocytes): exposed (10.45 ± 6.52) vs. unexposed (5.77 ± 2.31); **sig.**	[66] 10.1080/15459624.2013.818238
Galiotte	2008	Hair dyes, waving, and straightening preparations	Brazil	--	124 hairdressers (69 exposed females, 55 unexposed)	• **Total Comet Score:** exposed (159.8 ± 71) vs. unexposed (125.4 ± 64.1); **sig.**	[67] 10.1093/annhyg/men037
Giri	2011	PAH	India	Air sampling, [B(a)P] analysis	220 (115 coal-tar workers, 105 controls)	• **Comet tail moment:** controls (0.44 ± 0.31); exposed (12.06 ± 0.56); **sig.**	[68] 10.1016/j.scitotenv.2011.07.009
Gomaa	2012	Formaldehyde	Egypt	--	45 (30 lab technicians, 15 unexposed)	• **Comet tail length** (peripheral blood): exposed (47.3 ± 8.5) vs. unexposed (12.5 ± 1.5); **sig.** • **Comet tail moment** (peripheral blood): exposed (56.1 ± 16.5) vs. unexposed (10.8 ± 1.2); **sig.**	[69]
Göethel **	2014	Air pollution, benzene, and CO	Brazil	Urinary t,t-muconic acid (t,t-MA) and 8OHdG carboxyhaemoglobin (COHb) in whole blood	99 (43 gas station staff, 34 drivers, 22 unexposed)	• **DNA damage index WBC (AU):** gas station staff (89.8 ± 21.5), drivers (94.2 ± 12.8), unexposed (48.6 ± 35.9); **sig.**	[70] 10.1016/j.mrgentox.2014.05.008

Table 1. *Cont.*

Author	Year	Main Chemical Exposure	Country	Exposure Assessment or Biomarkers of Exposure	Population Characteristics	DNA Damage	Reference/DOI
Hachesu	2019	Air pollution (traffic)	Iran	--	104 taxi drivers (11 smokers, 93 non-smokers)	• **Comet tail moment:** smokers (2.70 ± 2.48), non-smokers (3.31 ± 4.37), all (3.24 ± 4.19); • **% tail DNA:** smokers (7.12 ± 3.47), non-smokers (7.34 ± 5.67), all (7.32 ± 5.45); • **Comet tail length:** smokers (7.24 ± 3.55), non-smokers (10.37 ± 7.90), all (10.02 ± 7.59); • **Comet tail intensity:** smokers (14.79 ± 5.89), non-smokers (14.13 ± 5.06), all (14.20 ± 5.13); non-**sig.**	[71] 10.1007/s11356-019-04179-1
Huang	2012	PAH (coke-oven exposure)	China	Airborne samples analysis	298 (202 exposed coke-oven workers: bottom 67, side 57, top 78 of the coke-oven; 96 controls)	• **Olive tail moment:** controls (0.55 ± 0.93); bottom (0.98 ± 1.07); side (1.37 ± 1.07); top (1.39 ± 1.09); **sig.**	[72] 10.1016/j.toxlet.2012.04.004
Jasso-Pineda **,¥	2015	Arsenic, lead, PAH, DDT/DDE	Mexico	As and 1-OHP in urine Lead and total DDT/DDE in blood	276 children total; 191 for air pollution (65 low PAH exposure; 50 biomass combustion; 76 high PAH exposure)	• **Olive tail moment:** low exposure (2.1 ± 1.0); biomass combustion (6.6 ± 3.0); high exposure (7.5 ± 3.5); **sig.**	[73] 10.1016/j.scitotenv.2015.02.073
Jiang	2010	Formaldehyde	China	Air samplers (TWA8h): 0.83 ppm, ranging 0.08–6.30 ppm	263 (151 plywood industry workers, 112 controls)	• **Olive tail moment:** exposed (3.54 [95%CI = 3.19–3.93]), unexposed (0.93 [95%CI = 0.78–1.10]); **sig.**	[74] 10.1016/j.mrgentox.2009.09.011

Table 1. *Cont.*

Author	Year	Main Chemical Exposure	Country	Exposure Assessment or Biomarkers of Exposure	Population Characteristics	DNA Damage	Reference/DOI
Khanna	2014	Tobacco dust	India	--	61 (31 female bidi rollers, 30 controls)	• **Comet tail length:** young bidi rollers (14.67 ± 1.47) vs. older bidi rollers (22.26 ± 1.02) vs. controls (11.52 ± 2.75); **sig.**	[75] 10.4103/0971-6580.128785
Khisroon	2020	Gold jewellery fumes	Pakistan	--	94 (54 gold jewellery workers, 40 controls)	• **Total comet score (TCS):** gold jewellery workers (128.0 ± 60.6), controls (47.7 ± 21.4); **sig.**	[76] 10.1080/1354750X.2020.1791253
Kianmehr	2017	Fuel smoke	Iran	--	55 (11 exposed to natural gas, 11 exposed to diesel, 11 exposed to kerosene, 11 exposed to firewood, 11 unexposed)	• **Comet tail moment:** firewood-burning (4.40 ± 1.98), natural gas (1.35 ± 0.84), diesel (1.85 ± 1.33), kerosene (2.19 ± 2.20), unexposed (0.17 ± 0.23); **sig.** for firewood • **Comet tail length**: firewood-burning (19.35 ± 5.97), natural gas (9.91 ± 4.10), diesel (12.31 ± 4.51), kerosene (13.37 ± 5.65), unexposed (2.89 ± 1.22); **sig.** • **% tail DNA**: firewood-burning (6.21 ± 1.88), natural gas (3.89 ± 1.17), diesel (4.03 ± 1.95), kerosene (4.08 ± 1.91), unexposed (6.21 ± 1.88); **sig.**	[77] 10.1177/0748233717712408
Knudsen	2005	Diesel-powered truck exhausts	Estonia	Cited in a previous paper	92 (50 underground mine workers, 42 surface workers)	• **DNA damage** (median): Underground non-smokers 113; underground smokers 157; surface smokers 90; surface non-smokers 142; **sig.** in underground workers	[78] 10.1016/j.mrgentox.2005.03.004

Table 1. *Cont.*

Author	Year	Main Chemical Exposure	Country	Exposure Assessment or Biomarkers of Exposure	Population Characteristics	DNA Damage	Reference/DOI
Krieg	2012	JP-8 jet fuel	USA	Urinary (2-methoxy ethoxy) acetic acid (MEAA) and creatinine, benzene, and naphthalene in exhaled breath	310 (Before: low 152, moderate 42, and high exposure 116; After a 4 h work shift exposure: low 151, moderate 43, high 116)	• **% tail DNA:** before: low (75.43 ± 5.93); moderate (75.94 ± 5.95); high (75.27 ± 4.69); After: low (75.78 ± 5.89); moderate (75.60 ± 6.10); high (75.47 ± 5.03); **non-sig.** • **Olive tail moment**: before: low (5390.78 ± 1142.55); moderate (5577.56 ± 1216.76); high (5370.35 ± 950.63) After: low (5511.14 ± 1133.04); moderate (5415.14 ± 1130.05); high (5425.66 ± 984.76); **non-sig.**	[79] 10.1016/j.mrgentox.2012.05.005
Kvitko	2012	PAH, PM, pesticides, solvents	Brazil	--	For PAH and PM exposure 109 (44 coal miners, 65 controls)	• **Damage Index (DI):** exposed (18 ± 9.72), controls (5 ± 5.81); **sig.** • **Damage Frequency (FD)**: exposed (14 ± 6.90), controls (2 ± 2.08); **sig.**	[80] 10.1590/S1415-47572012000600022
Leng	2004	PAH (coke-oven exposure)	China	Urinary 1-hydroxypyrene (1-OHP)	193 (143 Coke-oven workers, 50 controls)	• **Olive tail moment**: coke-oven workers (2.6, 95% CI = /2.1/3.3), non-coke-oven workers (1.0, 95% CI = /0.8/1.2); **sig.**	[81] 10.1080/13547500400015618

Table 1. *Cont.*

Author	Year	Main Chemical Exposure	Country	Exposure Assessment or Biomarkers of Exposure	Population Characteristics	DNA Damage	Reference/DOI
León-Mejía	2011	Dust particles	Colombia	--	200 (100 exposed open-cast coal mine workers, 100 controls)	• **Comet tail length**: exposed (23.4 ± 6.5), unexposed (14.3 ± 2.5) • **% tail DNA:** exposed (13.1 ± 7.9), unexposed (2.9 ± 1.5) • **DI (damage index):** exposed (60.0 ± 39.5), unexposed (9.0 ± 6.4); **sig.**	[82] 10.1016/j.scitotenv.2010.10.049
León-Mejía	2019	Diesel exhaust (gases, PAH, PM)	Colombia	--	220 (120 exposed mechanics and 100 controls)	• **% tail DNA:** controls (23.39 ± 9.18), exposed (30.91 ± 17.52); **sig.** • **Damage index:** controls (107.05 ± 27.88), exposed (131.22 ± 48.15); **sig.**	[83] 10.1016/j.ecoenv.2018.12.067
Lin	2013	Formaldehyde	China	Air-monitoring badges	178 (96 plywood industry, 82 controls)	• **Olive tail moment:** lower exposure (0.88 ± 0.55), higher exposure (1.01 ± 0.56), controls (0.67 ± 0.55); **sig. increased with increasing levels of FA exposure**	[84] 10.1539/joh.12-0288-oa
Marczynski	2002	PAH (coke-oven exposure)	Germany	1-Hydroxypyrene (1-OHP) and sum of five hydroxyphenanthrenes (OHPHs), creatinine, and cotinine	95 19 coke-oven workers, 29 graphite-electrode-producing workers), 32 controls	• **Tail extent moment:** graphite-electrode-producing workers 7.95 ± 3.34, coke-oven workers 3.5 ± 1.72, controls 2.54 ± 0.68; **sig. increased for graphite-electrode-producing workers**	[85] 10.1093/carcin/23.2.273

Table 1. *Cont.*

Author	Year	Main Chemical Exposure	Country	Exposure Assessment or Biomarkers of Exposure	Population Characteristics	DNA Damage	Reference/DOI
Marczynski	2010	Bitumen	Germany	--	42 bitumen-exposed workers	• **DNA strand break—median (range) in** (a) **Induced sputum:** pre: 196 (158–209), and post: 202 (50–225) shift (b) **Blood:** pre: 1.7 (1.2–2.4), and post: 1.3 (1.1–1.9); **non-sig.**	[86] 10.1177/ 0960327109359635
Marczynski	2011	Vapours and aerosols of bitumen	Germany	Urinary hydroxylated metabolites of naphthalene, phenanthrene, pyrene	438 (320 exposed construction workers, 118 unexposed)	• **Olive tail moment:** exposed pre-shift (1.74 [1.26–2.57]), unexposed pre-shift (1.41 [0.98–2.30]), exposed post-shift (1.51 [1.14–2.12]), unexposed post-shift (1.19 [0.98–1.49]) • **% DNA tail:** exposed pre-shift (6.51 [4.72–9.31]), unexposed pre-shift (5.06 [3.66–8.95]), exposed post-shift (5.73 [4.04–7.97]), unexposed post-shift (4.66 [3.66–5.90]); **sig.**	[87] 10.1007/s00204-011-0682-5
Moretti	2007	PAH	Italy	Urinary 1-OHP	191 (109 graphite-electrode-producing workers, 82 controls)	• **% DNA tail:** exposed (5.28 ± 0.21), control (4.33 ± 0.22); **sig.**	[88] 10.1186/ 1471-2458-7-270
Novotna	2007	Air pollution	Czech Republic	Air samples analysis; personal air sampler. Quantitative analysis of cPAHs	65 non-smoking city policemen (54 outdoor policemen, 11 indoor policemen)	• **% DNA tail:** exposed January (7.04 ± 0.38), unexposed January (3.75 ± 0.85); exposed September (4.72 ± 0.29), unexposed September (2.65 ± 0.18); **sig.**	[89] 10.1016/j.toxlet. 2007.05.013

Table 1. *Cont.*

Author	Year	Main Chemical Exposure	Country	Exposure Assessment or Biomarkers of Exposure	Population Characteristics	DNA Damage	Reference/DOI
Oh	2006	PAH	South Korea	Urinary 1-OHP,2-naphthol, and creatinine in urine	138 (54 automobile emission inspectors, 84 controls)	• **Olive tail moment** (mononuclear cells): exposed (1.71 ± 0.23), controls (1.34 ± 0.16); **sig.** • **% tail DNA** (mononuclear cells): exposed (14.91 ± 2.37), controls (9.17 ± 2.22); **sig.** • **Olive tail moment** (polynuclear cells): exposed (3.21 ± 0.42), controls (2.76 ± 0.38); **sig.** • **% tail DNA** (polynuclear cells): exposed (15.58 ± 3.58), controls (13.35 ± 2.44); **sig.**	[90] 10.1016/j.etap.2005.08.004
Peteffi	2016	Formaldehyde	Brazil	Urinary formic acid concentrations	91 (46 exposed furniture manufacturing workers, 45 controls)	• **Damage index:** exposed (6.7), unexposed (2.0); **sig.** • **Damage frequency:** exposed (6%), unexposed (2%); **sig.**	[91] 10.1177/0748233715584250
Peteffi	2016	Formaldehyde	Brazil	Environmental FA concentrations; urinary formic acid	50 hairdresser workers	• **Damage index**: 7.00 (2.00–52.25) • **Damage frequency:** 6.50 (2.00–44.00); **sig.**	[92] 10.1007/s11356-015-5343-4
Recio-Vega	2018	PAH	Mexico	Urinary 1-OHP	70 brick factory workers (35 exposed; 35 controls)	• **Comet tail length:** controls (29.61 ± 9.0), exposed (42.07 ± 10.0); **sig.** • **Comet tail moment:** controls (4.07 ± 3.5), exposed (8.11 ± 4.8); **sig.** • **Comet tail migration:** controls (11.37 ± 8.9), exposed (23.19 ± 11.2); **sig.**	[93] 10.1007/s00420-018-1320-9

Table 1. *Cont.*

Author	Year	Main Chemical Exposure	Country	Exposure Assessment or Biomarkers of Exposure	Population Characteristics	DNA Damage	Reference/DOI
Rekhadevi	2009	wood dust	India	Wood dust levels	120 (60 carpentry workers, 60 controls)	• **Comet tail length:** *Age* < 35 controls (5.90 ± 2.62), exposed (12.42 ± 1.52); ≥35 controls (7.76 ± 1.61), exposed (15.82 ± 2.01); smoking controls (7.91 ± 1.26), exposed (16.33 ± 1.52); not smoking controls (6.52 ± 2.53), exposed (12.36 ± 1.42); Alcohol consumption yes controls (8.00 ± 1.40), exposed (6.90 ± 1.15); no alcohol consumption controls (5.80 ± 2.51), exposed (12.86± 1.69); **sig.**	[94] 10.1093/mutage/gen053
Rohr	2013	Coal dust	Brazil	--	128 (71 coal-exposed workers and 57 controls)	• **Damage index** controls 15.53 ± 8.80, exposed 33.69 ± 28.70; **sig.** • **Damage frequency** controls 12.40 ± 6.18 27.46 ± 23.75; **sig.**	[95] 10.1016/j.mrgentox.2013.08.006
Sardas	2010	Welding fumes and solvent-based paints	Turkey	--	78 (52 workers in construction, 26 controls)	• **% DNA tail:** exposed (12.34 ± 2.05) vs. unexposed (6.64 ± 1.43); **sig.**	[96] 10.1177/0748233710374463
Scheepers **	2002	Diesel exhaust (benzene, PAHs)	Estonia, Czech Republic	Analysis of air samples, urinary metabolites of PAH and benzene	92 underground miners (drivers of diesel-powered excavators) (46 underground workers, 46 surface workers)	• **DNA damage** lymphocytes (visual scoring comets): underground workers (134), surface workers (104); **non-sig.**	[97] 10.1016/s0378-4274(02)00195-9

Table 1. *Cont.*

Author	Year	Main Chemical Exposure	Country	Exposure Assessment or Biomarkers of Exposure	Population Characteristics	DNA Damage	Reference/DOI
Sellappa	2010	Cement dust exposure	India	--	164 (96 building construction workers and 68 controls)	• **Comet tail length:** **Controls:** *Age* ≤ 40 (9.90 ± 0.92); ≥41 (8.09 ± 1.18); Smoking Yes (10.40 ± 2.42), No (9.21 ± 1.32); Tobacco chewing Yes (10.12 ± 2.71), No (8.85 ± 2.33); Alcohol Consumption Yes (9.96 ± 2.44), No (9.23 ± 2.30) **Workers:** Age ≤ 40 (16.85 ± 2.08); **sig.**; ≥41 (14.12 ± 2.33); **sig.**; Smoking Yes (15.97 ± 2.61); **sig.**; No (13.71 ± 2.89); **sig.**; Tobacco chewing Yes (15.71 ± 2.34); **sig.**; No (15.71 ± 2.34); **sig.**; Alcohol Consumption Yes (14.05 ± 2.59); **sig.**; No (12.90 ± 2.98); **sig.**	[98]
Sellappa	2011	PAH	India	Urinary 1-OHP	73 (36 road pavers; 37 control)	• **Comet tail length** controls: smokers (13.3 ± 3.74); non-smokers (10.9 ± 2.85); alcohol drinkers (11.1 ± 2.92); non-drinkers (9.9 ± 2.83), **workers**: smokers (19.4 ± 4.99); **sig.** non-smokers (15.5 ± 4.94); **sig.** • alcohol drinkers (16.2 ± 2.03); **sig.** • non-drinkers (15.1 ± 3.12); **sig.**	[99]
Shen	2016	Diesel	China	Urinary OH-PAHs, urinary εdA levels	185 (86 exposed diesel engine testing workers, 99 unexposed)	• **Olive tail moment:** non-exposed (1.16 ± 2.45), exposed (5.29 ± 2.30); **sig.** • **% DNA tail:** non-exposed (2.20 ± 29.45), exposed (66.44 ± 25.93); **sig.**	[100] 10.1016/j.scitotenv.2015.10.165

Table 1. *Cont.*

Author	Year	Main Chemical Exposure	Country	Exposure Assessment or Biomarkers of Exposure	Population Characteristics	DNA Damage	Reference/DOI
Siwińska	2004	PAH	Poland	Urinary 1-hydroxypyrene (HpU)	98 coke-oven workers (49 exposed; 49 controls)	• **Comet tail length**—median with quartiles (25–75th): controls 34.6 (31.4; 40.4); exposed: 32.3 (29.0; 37.3); **sig.**	[101] 10.1136/oem. 2002.006643
Sul	2003	PAH	South Korea	Urinary 1-OH-pyrene and creatinine, 2-naphthol	95 (24 workers from automobile emission companies, 28 workers from waste incinerating company, 43 unexposed)	• **DNA damage (in T-lymphocytes):** emission inspection workers (1.41 ± 0.22), incineration workers (1.76 ± 0.27), controls (1.42 ± 0.22); **sig.** • **Comet tail moment (B-lymphocytes):** emission inspection (2.44 ± 0.32), incineration workers (2.36 ± 0.37), controls (1.40 ± 0.27); **sig.** • **Comet tail moment (granulocytes):** emission inspection (3.32 ± 0.38), incineration workers (2.85 ± 0.49), controls (2.72 ± 0.59); **sig.**	[102] 10.1016/s1383-5718(03)00095-0
Toraason	2006	1-Bromopropane	USA	Personal-breathing zone samples collected for 1–3 days up to 8 h per (TWA8h). Bromide (Br) in blood and urine.	64 workers (42 facility A (non-sprayer—low exposure 29; sprayer—high exposure 13) and 22 workers facility B (non-sprayer—low exposure 16; sprayer—high exposure 6))	• **Comet tail moment**: start of the week: low exposure A (2517 ± 641), high exposure A (2867 ± 895); low exposure B (2856 ± 359); high exposure B (3430 ± 984); end-of-week: low exposure A (3080 ± 697); **sig.** high exposure A (3178 ± 762); low exposure B (2770 ± 504); high exposure B (2974 ± 280)	[103] 10.1016/j.mrgentox. 2005.08.015

Table 1. *Cont.*

Author	Year	Main Chemical Exposure	Country	Exposure Assessment or Biomarkers of Exposure	Population Characteristics	DNA Damage	Reference/DOI
Tovalin **	2006	Air pollution (traffic), VOCs, PM2.5, ozone	Mexico	Personal occupational and non-occupational monitoring for VOCs, PM2.5, O_3	55 City traffic exposure (28 outdoor workers, 27 indoor workers)	• **Comet tail length** (WBC): outdoor workers (median 46.80 [maximum 132.41]), indoor workers (median 30.11 [maximum 51.47]); **sig.**	[104] 10.1136/oem.2005.019802
Ullah	2021	Air pollution (traffic), coal mining dust	Pakistan	--	240 (60 participants exposed to traffic pollution, 60 controls, 60 mine workers, 60 controls)	• **Comet tail length—mean (min-max):** traffic conductors 28.69 (26.83–30.55), controls 8.62 (7.98–9.26); **sig.**, coal miners 30.16 (29.06–31.26), controls 9.82 (9.42–10.22); **sig.**	[105] 10.12669/pjms.37.2.2848
van Delft	2001	PAH (coke-oven exposure)	Netherlands	Urinary 1-hydroxypyrene	72 (28 coke-oven workers, 37 controls)	• **DNA breaks:** exposed (1.3 ± 0.4), controls (1.4 ± 0.4); **non-sig.**	[106] 10.1016/S0003-4878(00)00065-X
Villarini	2008	Dust (a-quartz and other particles from blasting), gases (nitrogen dioxide, NO_2), diesel exhausts, oil mist	Italy	--	73 (39 underground workers and 34 unexposed subjects)	• **% tail DNA:** exposed (3.08 ± 0.29), control 2.85 ± 0.18; **non-sig.**	[107] 10.1080/15287390802328580
Vital	2021	Environmental tobacco smoke (occupational settings)	Portugal	Monitoring the level of indoor air contaminants, namely, particulate matter ($PM_{2.5}$), CO, and CO_2	76 (17 smoker workers (SW), 32 non-exposed non-smoker workers (NE NSW), 32 exposed non-smoker workers E NSW)	• **% tail DNA:** SW (2.94 ± 0.94); NE NSW (2.93 ± 0.70); E NSW (3.24 ± 1.34); **non-sig.** • **Comet tail length:** SW (3.30 ± 1.64); NE NSW (3.13 ± 0.80); E NSW (3.00 ± 0.90); **non-sig.**	[108] 10.3389/fpubh.2021.674142

Table 1. *Cont.*

Author	Year	Main Chemical Exposure	Country	Exposure Assessment or Biomarkers of Exposure	Population Characteristics	DNA Damage	Reference/DOI
Wang	2007	PAH (coke-oven exposure)	China	Benzo[a]pyrene-r-7, t-8, t-9, c-10-tetrahydotetrol-albumin (BPDE-Alb) adducts	309 (207 coke-oven workers exposed, 102 controls)	• **Olive tail moment:** control (0.63 ± 0.93), exposed (1.20 ± 1.10); **sig.**	[109] 10.1136/oem.2006.030445
Wang	2010	PAH (coke-oven exposure)	China	Airborne PAH monitoring and urinary 1-Hydroxypyrene	475 workers (157 low, 160 intermediates, 158 high exposure)	• **Olive tail moment** (median, 5–95 percentiles): all 0.36 (0.13–1.24), low 0.33 (0.12–1.06), intermediate 0.38 (0.17–1.74), high 0.40 (0.14–3.17); **non-sig.**	[110] 10.1158/1055-9965.EPI-09-0270
Wang	2011	PAH (cooking oil fumes)	China	Urinary 1-OHP	110 (67 kitchen workers, 43 controls)	• **Comet tail length:** exposed (8.03 [6.83–9.18]), controls (6.89 [5.89–8.16]); **sig.** • **% DNA tail:** exposed (23.9 [17.8–30.1]) vs. controls (21.3 [16.2–29.1]); **sig.**	[111] 10.1539/joh.11-0074-oa
Wultsch	2011	PAH	Austria	Cr, Mn, Ni, As, in urine, creatinine	42 waste incinerator workers (23 exposed, 19 unexposed)	• **DNA migration (tail factor)**: Group I [≥1 and ≤3 months employment] (6.7 ± 1.9), Group II [>3 and ≤8 months] (6.3 ± 1.5), Group III [>8 and ≤11 months] (6.5 ± 2.4), unexposed (7.1 ± 1.6); **non-sig.**	[112] 10.1016/j.mrgentox.2010.08.002
Yang	2007	PAH (coke-oven exposure)	China	PAH and urinary 1-OHP monitoring	101 coke-oven workers (Low (n = 33) Intermediate (n = 35) High (n = 33) exposure)	• **Olive tail Moment:** low (1.63 ± 0.46), intermediate (1.74 ± 0.69), high (2.54 ± 0.75); **sig.** between low and high	[113] 10.1289/ehp.10104

Table 1. *Cont.*

Author	Year	Main Chemical Exposure	Country	Exposure Assessment or Biomarkers of Exposure	Population Characteristics	DNA Damage	Reference/DOI
Yu	2022	PAH (coke-oven exposure)	China	Urinary monohydroxy PAHs (OH-PAHs)	332 coke-oven workers	• **Olive tail Moment**: Total participants (0.44 (0.30, 0.75)), <20 years of working (0.44 (0.28, 0.71)), (0.44 (0.32, 0.86)); **non-sig.** • **% tail DNA**: Total participants (3.20 (2.14, 5.18)), <20 years of working (3.18 (2.01, 4.88)), (3.21 (2.19, 5.68)); **non-sig.** • **Comet tail length**: Total participants (3.61 (3.24, 4.88)), <20 years of working (3.65 (3.20, 4.65)), (3.59 (3.28, 5.05)); **non-sig.** • **Comet tail moment**—median (25–75th percentile): Total participants (0.14 (0.08, 0.33)), <20 years of working (0.15 (0.08, 0.30)), (0.13 (0.09, 0.34)); **non-sig.**	[114] 10.1007/s11356-022-19828-1
Zhang	2021	PAHs (coke-oven exposure)	China	Urinary 1-hydroxypyrene (1-OHP) analysis	256 (173 male coke-oven workers, 83 male hot-rolling workers not exposed as a control group)	• **% tail DNA:** controls 4.92, exposed 40.8 • **Olive tail Moment:** controls 3.73, exposed 22.1; **sig.**	[115] 10.1016/j.envpol.2020.115956
Zendehdel ø	2017	Formaldehyde	Iran	Monitoring FA exposure	83 (49 melamine tableware workshop workers, 34 controls)	• **Olive tail moment**—median (min–max): exposed 13 (7.4–36.7), controls 8.4 (6.4–31.7); **sig.** • **Comet tail moment**—median (min–max): exposed 22.2 (12.3–65), controls 14.8 (6.4–57.7); **sig.**	[116] 10.1080/02772248.2017.1343335

Table 1. *Cont.*

Author	Year	Main Chemical Exposure	Country	Exposure Assessment or Biomarkers of Exposure	Population Characteristics	DNA Damage	Reference/DOI
Zendehdel Ø	2018	Formaldehyde	Iran	Air sampling	87 (53 melamine tableware workshop workers, 34 unexposed)	• **Comet tail moment** (whole blood): exposed (20.9 [12.3 to 65.1]), unexposed (14.8 [6.4 to 57.7]); **sig.**	[117] 10.1007/s11356-018-3077-9
Zendehdel Ø	2018	Formaldehyde	Iran	Air sampling	88 (54 melamine tableware workshop workers, 34 controls)	• **Comet tail length (median; min-max):** exposed (28.9; 13.9–81), controls 18.5 (14–71); **sig.**	[118] 10.1177 /0960327117728385
Environmental exposure							
Alvarado-Cruz	2017	Air pollution	Mexico	PM_{10} characterization, urinary levels of 1-OHP (PAHs exposure) and t,t-MA (benzene exposure)	141 children	• **Olive tail moment** (interquartile range 25–75): 33.6 (28.0–40.2); **sig.** positive association with PM10	[119] 10.1016/j.mrgentox.2016.11.007
Andersen	2019	Diesel-powered trains particles	Denmark	Levels of 1-OHP, 2-OHF, 1-NAPH, and 2-NAPH in urine	83 healthy volunteers 54 exposed to diesel, 29 exposed in electric train)	• **DNA damage** (SB lesions/10^6 bp): electric (0.12 ± 0.13), diesel (0.18 ± 0.13); **sig.**	[120] 10.1186/s12989-019-0306-4
Avogbe **	2005	PM (UFPs), benzene	Benin	Ambient UFP, urinary excretion of S-PMA	135 city traffic exposure (29 drivers, 37 roadside residents, 42 suburban, 27 rural)	• **% DNA tail** (MNBC): drivers (6.09 ± 3.46) vs. roadside residents (6.32 ± 4.00) vs. suburban (5.42 ± 2.28) vs. rural (4.26 ± 1.76); **sig.**	[121] 10.1093/carcin/bgh353
Beyoglu	2010	Indoor tobacco smoke	Turkey	--	60 children from paediatric unit (30 exposed, 30 controls)	• **% tail DNA:** exposed (10.73 ± 1.38), controls (8.16 ± 1.29); **sig.**	[122] 10.1016/j.ijheh.2009.10.001

Table 1. *Cont.*

Author	Year	Main Chemical Exposure	Country	Exposure Assessment or Biomarkers of Exposure	Population Characteristics	DNA Damage	Reference/DOI
Cetkovic	2023	Air pollution	Bosnia and Herzegov	--	33 volunteers (Summer and winter sampling)	• **Comet tail intensity:** winter (1.14 ± 0.23); summer (1.19 ± 0.19); • **Comet tail length:** winter (2.20 ± 0.14); summer (2.25 ± 0.17); • **Comet tail moment:** winter (1.03 ± 0.29); summer (1.07 ± 0.25); **non-sig.**	[123] 10.1093/mutage/geac016
Cho	2003	Hair dye fumes	Korea	--	20 volunteers (before and after hair-dyeing)	• **Comet tail moment:** before (1.47 ± 0.41); after (1.75 ± 0.29); **sig.**	[124] 10.1539/joh.45.376
Chu	2015	Air pollution	China	Personal 24 h PM2.5 exposure	301 (108 from Zhuhai, 114 from Wuhan, 79 from Tianjin)	• % tail DNA—Median (25–75th percentile): Zhuhai 1.36 (0.67, 2.66); Wuhan 2.15 (0.77, 4.63); Tianjin 2.97 (1.47, 6.32); **significance not indicated**	[125] 10.1016/j.toxlet.2015.04.007
Coronas	2009	PM	Brazil	Weekly airborne particulate matter (PM10) samples	74 healthy men recruits, 18–40 years old, living or working at the target site (37 exposed, 37 unexposed)	• **Comet tail intensity:** exposed (10.04 ± 7.13) vs. unexposed (7.09 ± 3.85); **sig.** • **Comet tail moment:** exposed (2.53 ± 2.28) vs. unexposed (0.82 ± 0.68); **sig.**	[126] 10.1016/j.envint.2009.05.001

Table 1. *Cont.*

Author	Year	Main Chemical Exposure	Country	Exposure Assessment or Biomarkers of Exposure	Population Characteristics	DNA Damage	Reference/DOI
Coronas	2016	PAHs (in PM)	Brazil	Air sampling Quantification of 16 PAHs from organic extract of PM 2.5: Acenaphthene, Acenaphthlene, Anthracene, Benzo(a)anthracene, Benzo(a)pyrene, Benzo(a)fluoranthene, Benzo(g,h,i)perylene, Indeno(1,2,3-cd)pyrene, Benzo(k)fluoranthene, Chrysene, Dibenzo(a,h) Anthracene, Phenanthrene, Fluoranthene, Fluorene, Naphthalene, and Pyrene.	62 children aged 5–12 years (42 exposed, 20 controls)	• **% DNA tail: controls** 7.2 ± 3.15 (interval 1.04–23.86), exposed 7.1 ± 2.16 (1.09–28.89); **non-sig.**	[127] 10.1016/ j.chemosphere. 2015.09.084
Danielsen	2008	Wood smoke	Sweden	Urinary 8-oxoGua, 8-oxodG	13 never-smoking subjects	• **DNA damage:** SB (per 10^6 bp): Time after exposure to filtered air: 3 h (0.071 ± 0.053), 20 h (0.085 ± 0.043); time after exposure to wood smoke: 3 h (0.042 ± 0.036), 20 h (0.035 ± 0.019); **non-sig.**	[128] 10.1016/j.mrfmmm. 2008.04.001
da Silva	2015	PAH	Brazil	--	45 children of Santo Antônio da Patrulha, Rio Grande do Sul	• **Comet tail length:** 23.1 ± 12.44 • **Comet tail intensity:** 7.3 ± 11.66 • **Comet tail moment:** 0.9 ± 2.30	[129] 10.1016/j.mrgentox. 2014.11.006

Table 1. *Cont.*

Author	Year	Main Chemical Exposure	Country	Exposure Assessment or Biomarkers of Exposure	Population Characteristics	DNA Damage	Reference/DOI
Forchhammer	2012	Wood smoke (controlled exposure)	Denmark	14, 220, or 354 μg/m^3 of particles from a well-burning modern wood stove for 3 h in a climate-controlled chamber with 2-week intervals	20 healthy non-smoking subjects (controlled exposure)	• **DNA damage (single-strand breaks)** (mean ± SEM): controls (0.16 ± 0.03 lesions/10^6 bp) (n = 18); **non-sig. effect of wood smoke**	[130] 10.1186/ 1743-8977-9-7
Gamboa	2008	PAH	Mexico	Air sampling	6–15 years old children (37) (12 from oil extraction activity; 10 from no extraction activity regions, 15 controls)	• **Comet tail length**: exposed (14.21–42.14), controls (12.25 to 0.63); **significance not indicated**	[131] 10.3390/ ijerph5050349
Gong	2014	Air pollution	China	$PM_{2.5}$ (mg/m^3): Zhuhai 68.35 (37.17–116.79); Wuhan 114.96 (86.55–153.20); Tianjin 146.60 (88.63–261.41)	307 (110 from Zhuhai, 118 from Wuhan, 79 from Tianjin)	• **% tail DNA**—median (25–75 percentile): Zhuhai 1.36 (0.65–2.59); Wuhan 1.85 (0.77–4.39); Tianjin 2.97 (1.47–6.32); **significance not indicated**	[132] 10.1016/j.toxlet. 2014.06.034
Han	2010	PAH	China	PAH metabolites (2-OHNa, 9-OHPh, 2-OHFlu, and 1-OHP) in urine	232 men from Chongqing, China.	• **% tail DNA:** 13.26%, 95% CI 7.97–18.55; • **Comet tail length** (12.25; 95% CI 0.01–24.52), • **Comet tail distribution** (7.55; 95% CI 1.28–18.83); **sig. associated with 2-OHNa**	[133] 10.1289/ehp.1002340
Hemmingsen	2015	Diesel exhaust	Sweden	3 h to diesel exhaust (276 μg/m^3) from a passenger car or filtered air, with co-exposure to traffic noise at 48 or 75 dB(A)	18 individuals with controlled exposure (3 h)	• **DNA damage** (before and after DE exposure): 0.32 ± 0.04; 0.30 ± 0.04; **non-sig.**	[134] 10.1016/j.mrfmmm. 2015.03.009

Table 1. *Cont.*

Author	Year	Main Chemical Exposure	Country	Exposure Assessment or Biomarkers of Exposure	Population Characteristics	DNA Damage	Reference/DOI
Hisamuddin	2022	PAHs (in PM)	Malaysia	Gravimetric sampling of $PM_{2.5}$ PAHs Extraction: Acenaphthene, Acenaphthlene, Anthracene, Benzo(a)anthracene, Benzo(a)pyrene, Benzo(a)fluoranthene, Benzo(g,h,i)perylene, Indeno(1,2,3-cd)pyrene, Benzo(k)fluoranthene, Chrysene, Dibenzo(a,h) Anthracene, Phenanthrene, Fluoranthene, Fluorene, Naphthalene, and Pyrene.	228 school children	• **Comet tail moment:** high traffic group (3.13 ± 0.53) vs. low traffic group (2.80 ± 0.81); **sig.**	[135] 10.3390/ ijerph19042193
Ismail	2019	Traffic-related air pollution	Malaysia	Air samples analysis	104 (52 exposed group, 52 controls)	• **Comet tail length:** exposed (35.95 ± 7.93); controls (30.32 ± 8.36); **sig.**	[136] 10.5572/ajae. 2019.13.2.106
Jasso-Pineda **	2015	Arsenic, lead, PAH, DDT/DDE	Mexico	Arsenic and 1-OHP in urine Lead and total DDT/DDE in blood	276 children (40/25 with high/low arsenic, 55/10 with high/low lead)	• **Comet tail moment:** high/low arsenic (4.5 ± 1.08/3.2 ± 0.5); **sig** high/low lead (3.7 ± 1.8/ 4.1 ± 1.5); **non-sig.**	[73] 10.1016/j.scitotenv. 2015.02.073
Jensen	2014	wood smoke exposure	Denmark	Exposure to high indoor concentrations of PM2.5 (700–3,600 μg/m^3), CO (10.7–15.3 ppm), and NO_2 (140–154 μg/m(3)) during 1 week.	11 university students	• **DNA strand breaks:** before (0.0.51 ± 0.031), after (0.061 ± 0.0.46); **non-sig.**	[137] 10.1002/em.21877

Table 1. *Cont.*

Author	Year	Main Chemical Exposure	Country	Exposure Assessment or Biomarkers of Exposure	Population Characteristics	DNA Damage	Reference/DOI
Koppen **	2007	Air pollution, PAHs, VOCs (benzene and toluene)	Belgium	Outdoor ozone concentrations, urinary concentrations of PAH, t,t′-muconic acid, o-cresol, VOCs metabolites	200 adolescents	• **% DNA tail** (WBC): 1.16 ± 0.51 • **Correlation** DNA damage/o-cresol and OH-pyrene; **sig.**	[138] 10.1002/jat.1174
Koppen **,§	2020	PAH, metals, benzene, POPs, phthalates, PM	Belgium	Ar, Cd, Cu, Ni, Pb, Tl, Cr in blood, outdoor air analysis	2283 adolescents (14–18 years old)	• **% DNA tail:** mean 2.4 [2.3–2.5]	[139] 10.1016/j.envres.2020.110002
Lemos	2020	PAHs (in PM)	Brazil	Air sampling Quantification of 16 PAHs from organic extract of PM 2.5: Acenaphthene, Acenaphthlene, Anthracene, Benzo(a)anthracene, Benzo(a)pyrene, Benzo(a)fluoranthene, Benzo(g,h,i)perylene, Indeno(1,2,3-cd)pyrene, Benzo(k)fluoranthene, Chrysene, Dibenzo(a,h) Anthracene, Phenanthrene, Fluoranthene, Fluorene, Naphthalene, and Pyrene.	54 children living in industrial areas	• **Comet tail intensity:** NW site 2.5 km from the petrochemical source of emission (10.65 ± 0.78), NWII site 35 km from the source of emission (6.73 ± 0.92), controls (7.20 ± 3.15); **sig.**	[140] 10.1016/j.envres.2020.109443
León-Mejía	2023	Coal mining	Colombia	--	270 150 individuals exposed to coal mining residues from the locality of Loma-Cesar, 120 nonexposed individuals from the City of Barranquilla	• **% DNA tail:** controls (8.11 ± 1.98), exposed (9.61 ± 1.06); **non-sig.**	[141] 10.1016/j.envres.2023.115773

Table 1. *Cont.*

Author	Year	Main Chemical Exposure	Country	Exposure Assessment or Biomarkers of Exposure	Population Characteristics	DNA Damage	Reference/DOI
Mondal	2010	Fuel smoke (biomass and liquefied petroleum)	India	$PM_{2.5}$ and PM_{10} (stationary sampling)	217 (132 biomass users, 85 liquefied petroleum gas users)	• **% DNA tail:** biomass users (21.6 ± 5.2), gas users (16.8 ± 3.3); **sig.** • **Comet tail length**: biomass users (46.6 ± 4.7) vs. gas users (44.1 ± 4.6); **sig.** • **Olive tail moment**: biomass users (4.2 ± 1.0) vs. gas users (4.2 ± 1.0); **sig.**	[142] 10.1016/j.mrgentox.2010.02.006
Mondal	2011	Fuel smoke (biomass and liquefied petroleum)	India	$PM_{2.5}$ and PM_{10} (stationary sampling)	161 premenopausal women (85 cooking with biomass; 76 control women cooking with liquid petroleum gas)	• **% DNA tail:** exposed (32.23 ± 8.31), unexposed (12.41 ± 3.87); **sig.** • **Comet tail length:** exposed (37.81 ± 11.21), unexposed (14.22 ± 3.89); **sig.** • **Olive tail moment:** exposed (7.08 ± 2.11), unexposed (3.15 ± 0.97); **sig.**	[143] 10.1016/j.ijheh.2011.04.003
Mukherjee ♀	2013	Fuel smoke (biomass and liquefied petroleum)	India	Urinary trans, trans-muconic acid	105 (56 biomass users, 49 cleaner liquefied petroleum gas users)	• **% DNA tail:** biomass users (36.2 ± 9.4), gas users (9.0 ± 4.1) • **Comet tail length**: biomass users (44.2 ± 6.0), gas users (32.3 ± 7.3) • **Olive tail moment**: biomass users (6.2 ± 2.2), gas users (1.2 ± 0.5); **sig.**	[144] 10.1002/jat.1748

Table 1. *Cont.*

Author	Year	Main Chemical Exposure	Country	Exposure Assessment or Biomarkers of Exposure	Population Characteristics	DNA Damage	Reference/DOI
Mukherjee [9]	2014	Fuel smoke (biomass and liquefied petroleum)	India	$PM_{2.5}$ and PM_{10} (stationary sampling)	150 (80 biomass users, 70 liquefied petroleum gas (LPG) users)	• **% tail DNA:** LPG users (10.1 ± 3.2), BMF users (36.2 ± 8.2); **sig.** • **Comet tail length:** LPG users (29.3 ± 4.6) vs. BMF users (45.2 ± 5.5); **sig.** • **Olive tail moment:** LPG users (1.2 ± 0.5) vs. BMF users (6.2 ± 1.9); **sig.**	[145] 10.1016/j.etap.2014.06.010
Nagiah	2015	Air pollution	South Africa	--	100 pregnant women (50 from a highly industrialised south Durban and 50 from the less industrialised north Durban)	• **Comet tail length** (25th, 75th percentile): north Durban 0.47 (0.41, 0.52); south Durban 0.55 (0.47, 0.60); **sig.**	[146] 10.1177/0960327114559992
Pacini	2003	Ozone	Italy	Air quality monitoring	119 (102 subjects from Florence, 17 controls from Sardinia)	• **% tail DNA:** Florence (45.7 ± 21.0): Sardinia (26.4 ± 6.7); **sig.**	[147] 10.1002/em.10188
Pandey	2005	Fuel smoke (biomass fuel liquefied petroleum gas)	India	--	144 volunteers (70 biomass fuel users, 74 liquefied petroleum gas (LPG) users)	• **Tail percent DNA:** LPG users (8.29 ± 0.18) vs. BMF users (11.19 ± 0.35); **sig.** • **Comet tail length:** LPG users (40.26 ± 0.88) vs. BMF users (51.15 ± 1.32); **sig.** • **Olive tail moment:** LPG users (2.77 ± 0.07) vs. BMF users (3.83 ± 0.15); **sig.**	[148] 10.1002/em.20106
Pelallo-Martínez **,¥	2014	PAH, lead, benzene, toluene	Mexico	Urinary and blood Pb, benzene, toluene, PAHs	97 children, air pollution (44 Allende, 37 Nuevo Mundo, 16 Lopez Mateos)	• **Olive tail moment** (WBC): Allende (8.3 [3.1–16.8]) vs. Nuevo Mundo (10.6 [5.6–22.9]) vs. Lopez Mateos (11.7 [7.4–15.9]); **sig.**	[149] 10.1007/s00244-014-9999-4

Table 1. *Cont.*

Author	Year	Main Chemical Exposure	Country	Exposure Assessment or Biomarkers of Exposure	Population Characteristics	DNA Damage	Reference/DOI
Pereira	2013	PAH	Brazil	PAH analysis	59 subjects from two towns of Rio Grande do Sul State (24, site 1 (exposed)—high quantity of nitro and amino derivatives of PAHs; 35 from site 2 (controls)—lesser anthropogenic influence)	• **Comet tail intensity—Mean ± SD (range):** exposed 6.7 ± 2.90 (3.25–14.40), controls 6.5 ± 2.81 (2.43–15.43) **non-sig.** • **Comet tail moment—Mean ± SD (range)**: exposed 0.8 ± 0.70 (0.31–7.53), controls 0.7 ± 0.36 (0.30–2.70); **non-sig.**	[150] 10.1016/j.ecoenv.2012.12.029
Pérez-Cadahia	2006	Air pollution	Spain	VOCs determination by dosimeters	110 (25 volunteers cleaning beaches, 20 manual workers beach, 23 high-pressure cleaners, 42 controls)	• **Comet tail length:** exposed (48.79 ± 0.10) vs. unexposed (51.47 ± 0.10); **sig.**	[151] 10.1100/tsw.2006.206
Piperakis	2000	Air pollution	Greece	--	80 healthy individuals living in urban and rural areas with different smoking habits	• **DNA damage (visual scoring):** urban non-smokers (78 ± 10.2), urban smokers (99 ± 10.9), rural non-smokers (71 ± 7.8), rural smokers (98 ± 12.5); **sig.**	[152] 10.1002/1098-2280(2000)36:3<243::aid-em8 > 3.0.co;2-
Rojas	2000	Ozone	Mexico	Ozone values	38 (27 exposed to hydrocarbons northward and 11 southward, exposed to ozone)	• **Comet tail length:** north (67.17 ± 7.93) (8) (57.77 ± 4.55) (20); south (87.56 ± 11.75) (5) (88.24 ± 13.41) (5); **sig.**	[153] 10.1016/s1383-5718(00)00035-8
Sánchez-Guerra	2012	PAH	Mexico	Urinary 1-OHP	82 children	• **Olive tail moment:** 9.52; **sig.** affected by PAH exposure	[154] 10.1016/j.mrgentox.2011.12.006
Shermatov	2012	Second hand cigarette smoking	Turkey	Urinary cotinine and creatinine	57 children (27 exposed, 27 controls)	• **DNA damage (arbitrary units):** exposed (62.14 ± 56.31), controls (6.14 ± 5.51); **sig.**	[155] 10.1007/s13312-012-0250-y

Table 1. *Cont.*

Author	Year	Main Chemical Exposure	Country	Exposure Assessment or Biomarkers of Exposure	Population Characteristics	DNA Damage	Reference/DOI
Sopian	2021	PAHs (PM)	Malaysia	60 indoor and outdoor $PM_{2.5}$ samples PAHs analysis: naphthalene (NAP), acenaphthene (ACP), acenaphthylene (ACY), anthracene (ANT), fluorene (FLU), phenanthrene (PHE), anthracene (ANT), fluoranthene (FLA), pyrene (PYR), benzo(a)anthracene (BaA), chrysene (CYR), benzo(b)fluoranthene (BbF), benzo(k)fluoranthene (BkF), benzo(a)pyrene (BaP), indeno(1,2,3-cd)pyrene (IcP), dibenzo(a,h)anthracene (DbA), and benzo(ghi)perylene (BgP)	234 children (near petrochemical industry)	• **Comet tail moment:** exposed group (27.20 ± 8.21), unexposed (21.03 ± 4.88); **sig.**	[156] 10.3390/ ijerph18052575
Torres-Dosal	2008	Wood smoke	Mexico	Urinary 1-OHP Carboxyhemoglobin determination	20 healthy volunteers (pre- and post-intervention)	• **Comet tail moment:** before (5.8 ± 1.3), after (2.8 ± 0.9); **sig.**	[157] 10.1016/j.scitotenv. 2007.10.039
Verschaeve	2007	PAH	Belgium	1-Hydroxypyrene	45 healthy subjects in different seasons	• **% tail DNA (average; mean):** June (1.67; 1.29); August (2.16; 1.25); November (1.36 1.06); February (1.26; 0.99); **sig.**	[158] 10.1002/jat.1244

Table 1. *Cont.*

Author	Year	Main Chemical Exposure	Country	Exposure Assessment or Biomarkers of Exposure	Population Characteristics	DNA Damage	Reference/DOI
Vinzents	2005	PM (UFPs)	Denmark	Personal exposure in terms of number of concentrations of UFPs in the breathing zone, using portable instruments in six 18 h periods	15 subjects bicycling in traffic or indoors on six occasions (controlled exposure)	• **DNA strand break (per 10^6 bp):** in traffic, 74 bicycling days median (range) 0.06 (0.03–0.11); indoors, 14 bicycling days, 0.06 (0.02–0.12); **non-sig.**	[159] 10.1289/ehp.7562
Wilhelm **,¥	2007	PAH, benzene, heavy metals	Germany	Monitored ambient air quality data, urinary (PAH) metabolites, benzene metabolites	935 air pollution close to industrial settings (620 exposed children, 315 unexposed)	• **Comet tail moment** (lymphocytes)—percentile 50: exposed (1.99) vs. unexposed (1.32); **sig.** • **Comet tail moment**—percentile 90: exposed (6.69) vs. unexposed (1.89); **non-sig.**	[160] 10.1016/j.ijheh. 2007.02.007
Wu	2007	Environmental tobacco smoke	Taiwan	--	291 (18 smokers, 143 environmental tobacco exposure, 130 non-smokers)	• **DNA damage score:** smokers (71.0 ± 46.6), environmental tobacco smoke-exposed (84.3 ± 44.3), non-smokers (63.5 ± 35.0); **sig.** between ETS-exposed and non-smokers	[161]
Zani **	2020	PM10, PM2.5, NO_2, CO, SO_2, benzene, and O3	Italy	Air sampling	152 pre-school children (3–6 years old)	• **% DNA tail:** 6.2 ± 4.3; • **Visual scoring:** 182.1 ± 30.9; **non-sig.**	[162] 10.3390/ ijerph17093276
Zani	2021	Air pollution	Italy	Air pollutant levels	142 children 6–8 years old (71 first winter, 71 second winter)	• **DNA damage** (visual score): first winter (173.2 ± 50.8), second winter (208.8 ± 67.1); **sig.** **Not significant association with air pollutant levels**	[163] 10.3390/ atmos12091191

Table 1. *Cont.*

Author	Year	Main Chemical Exposure	Country	Exposure Assessment or Biomarkers of Exposure	Population Characteristics	DNA Damage	Reference/DOI
Zeller	2011	Controlled exposure to formaldehyde	Germany	FA vapours (0 to 0.8 ppm) for 4 h/day over a period of five working days under strictly controlled conditions and bicycling (~80 W) four times for 15 min.	37 volunteers	• **Comet tail moment:** before exposure 0.30 ± 0.117; after exposure 0.33 ± 0.118; non-**sig.** • **Comet tail intensity:** before exposure 2.28 ± 0.492; after exposure 2.66 ± 0.646; **sig.**	[164] 10.1093/mutage/ger016

** Studies also in solvents table; ɣ Studies also in heavy metals table. * From the three papers from Cebulska-Wasilewska, the second 2007 paper (2007*) shows results compiled from the previous two papers. Thus, the second 2007 paper is not counted as an original study. § The second paper (Costa et al., 2011) is an expansion of the previous study sample with the addition of a new comet assay descriptor. Thus, one original study is counted for both papers. Ø Three papers from Zendehdel and co-workers appear to be very similar, although there are cross-references to ascertain whether these data originate from the same study. In essence, the authors appear to have reported results on different comet descriptors in separate papers, deriving, however, from the same subjects enrolled in the same biomonitoring. Thus, the papers are counted as one study. ♀ The second paper (Mukherjee, 2014) contains more subjects from six different villages as compared to the first study with studies from five villages (Mukherjee 2013). Nevertheless, the results are very similar, suggesting that the first paper describes only part of the complete dataset. Thus, we have counted the papers as one study.

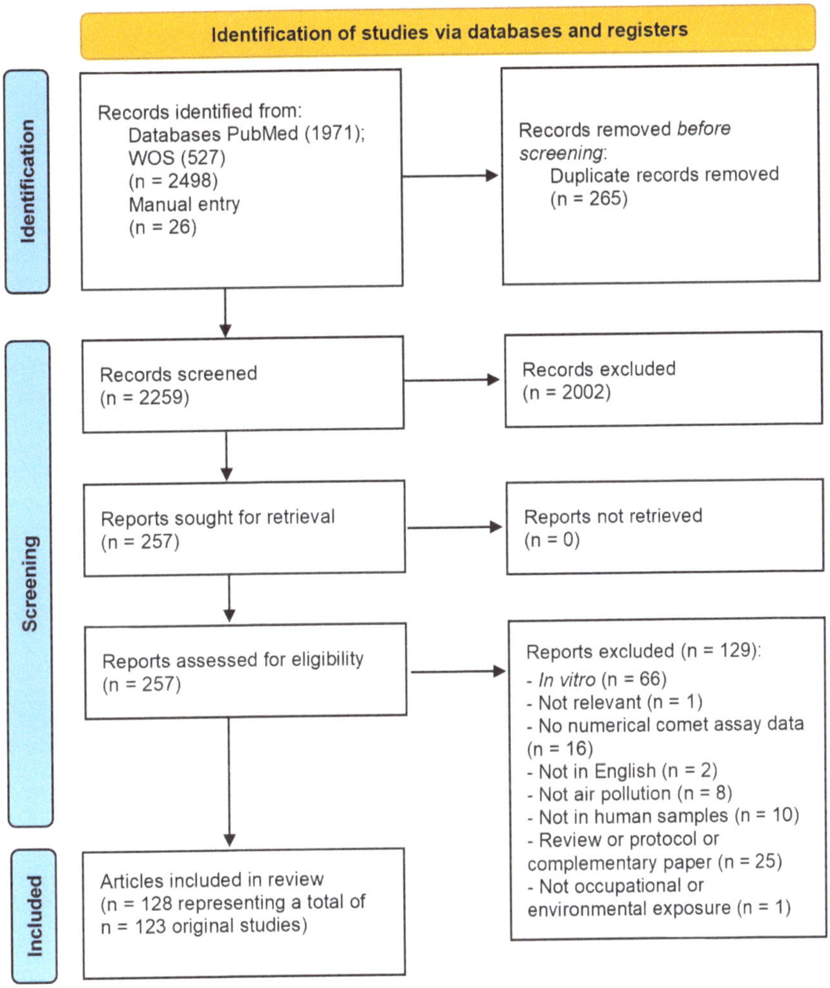

Figure 1. PRISMA flow diagram of systematic scoping review for air pollutants.

Overall, 81 studies (63.3%) evaluated occupational exposure and 47 studies (36.7%) environmental exposure. Occupational exposure to air pollutants included silica dust, welding fumes, vapours, gases, volatile organic compounds (VOCs), and metals. These studies were performed in Asia (n = 36, 44.4%), followed by Europe (n = 30, 37.0%), the Americas (n = 14, 17.3%), and Africa (n = 1, 1.2%). Fourteen (16.9%) studies assessed the effects of exposure to PAHs as the sole measured pollutants in firefighters [39], paving workers [99], airport personnel [51], policemen [53–55], coal tar workers [68], graphite-electrode-producing workers [88], automobile inspectors [90], brick factory workers [93], and automobile emission and waste incinerating companies [102,112]. Eight (9.6%) studies considered the PAH exposure combined with other chemicals, such as fluorene [40], VOCs [41,44], heterocyclic compounds [43], antineoplastic drugs [16], fibre glass [56], heavy metals, dichlorodiphenyltrichloroethane (DDT), and dichlorodiphenyldichloroethylene (DDE) [73], as well as metals, benzene, persistent organic pollutants (POPs), and others [80]. Thirteen (15.9%) studies evaluated the exposure to formaldehyde in fibreboard plants [42], pathology anatomy laboratories [61–63], the plywood industry [74,84], a furniture manufacturing facility [91], melamine tableware manufacturing workshops [116–118], and in hairdressers; one of these directly reporting formaldehyde exposure and including a control group [92] and the other assessing exposure to hair dyes and waiving and straightening products that also have formaldehyde in their composition [67]. Eleven (13.4%) studies were performed on dust, specifically marble dust [45], silica dust [47], wood dust [48], coal [95] and coal together with traffic air pollution [105], cobalt dust and other metals [64], tobacco dust [75], graphene [52,94], and two referred as dust particles [82,107]. Twelve (14.6%) studies were based on coke-oven exposure [57,59,72,81,85,106,109–111,113–115]; this type of emission usually consists of complex mixtures of dust, vapours, and gases, which can include carcinogens such as cadmium and arsenic. Eight (9.8%) studies were conducted on diesel exhaust [65,77–79,83,97,100,107], with two studies [77,79] specifically on fuel and one study on diesel exhaust and dust [107]. Seven (8.5%) studies were performed under the air pollution "umbrella", on outdoor air pollution [46], combined with benzene and CO exposure [70], traffic vehicle exhausts [71,104], traffic and coal mining [105], and in traffic policemen [49,89]. Three (3.7%) studies were made on welding fumes and solvent based paints [96], metals (zinc and copper) smelting work [60], and gold jewellery fumes [76]. Furthermore, other studies in the selected papers were found, such as polychlorinated dibenzodioxins, metals and silica [58], perchloroethylene [66], DDT, DDE together with arsenic and lead [73], bitumen [86,87], cement [98], tobacco smoke [108], and 1-bromopropane [103].

From a total of 81 studies, 65 (80.2%) performed exposure assessments by using air sampling measurements (n= 30, 46.1%) or personal air sampling devices (n = 6, 9.2%) or by using biomarkers of exposure, such as urinary 1-hydroxypyrene (1-OHP) metabolite from PAHs exposure (n= 27, 41.5%), as well as other metabolites measured in urine or blood (n = 10, 15.3%).

Significantly higher DNA damage levels, as evaluated by the comet assay, were observed in 66 of these studies (81.5%). The remaining studies (n = 15, 18.5%) did not show statically significant results, namely PAH exposure [39,54,55], coke-oven PAH exposure [106,110,111,114], smelting [60], dust [64,107], traffic air pollution [71], JP-8 jet fuel [79], diesel exhaust [97], bitumen [86], and tobacco dust [108]. The study from Cavallo [52], in six graphene workers and eleven controls, used three comet descriptors, reaching statistically significant results with % DNA in the tail but not by using the tail moment and length. The descriptors used to express the comet assay data (one or more in the same study) were as follows: % DNA in tail/tail intensity in 33 studies, tail length in 25 studies, tail moment in 21 studies, olive tail moment in 16 studies, DNA damage index in 7, and other descriptors mentioned in 13 studies.

Regarding environmental exposure to air pollutants, as with occupational exposure, there is a variety of chemical exposures, including PAHs "alone" or combined, PM, diesel exhaust, wood smoke, tobacco smoke, and others. These studies were performed in Europe (n = 17; 36.2%), followed by Asia (n = 14; 28.8%), South America (n = 14; 28.8%), and Africa (n = 2; 4.3%). Regarding exposure to PAHs, from a total of eight (17.2%) studies, five (62.5%) were performed in children [129,131,135,154,156] and the other three (37.5%) in adults [133,150,158]. From six studies conducted in children and adolescents, two studies reported a combined exposure between PAHs, metals, and VOCs [149,160], and two others besides these chemical substances were also phthalates [73,139]. The studies from Coronas [127] and Lemos [140] reported both atmospheric $PM_{2.5}$ concentrations and contents of 16 PAHs in the organic extract of $PM_{2.5}$ collected on filters. Four (8.5%) studies addressed PM exposure, PM_{10} [126], ultrafine particles in controlled exposure [159], ultrafine particles combined with benzene [121], PM_{10}, $PM_{2.5}$, gases (NO_2, CO, and SO_2), and benzene [162]. Two studies addressed diesel exhaust [120,134], while others assessed fuel smoke, specifically biomass fuel, in comparison with liquefied petroleum gas [142–145,148], while three addressed wood smoke [128,130,157] in indoor environments.

Three studies addressed involuntary exposure to tobacco smoke, namely indoor tobacco smoke [122], second-hand cigarette smoking in children [155], and environmental tobacco smoking [161]. Two studies assessed exposure to ozone [147,153], one investigated the effects of formaldehyde under experimental conditions [164], and others looked at hair dye fumes [124] and coal mining residues [141].

From a total of 44 studies, 35 (74.5%) performed exposure assessments; air sampling was measured in twelve (25.5%) studies, seven (14.9%) measured ambient PM, and four (8.5%) specifically quantified PAHs from PM extracts [127,135,140,156]. Ten (21.3%) studies measured urinary 1-OHP, an internal biomarker of PAH exposure, and 12 (25.5%) measured other metabolites in urine or blood. Three studies were on controlled exposure to diesel exhaust [134], indoor wood smoke [137], and formaldehyde [164].

Significantly higher DNA damage, as evaluated by the comet assay, was observed in 38 of these studies (79.1%). The remaining studies (n = 10, 20.8%) did not show statistically significant results, namely PAH exposure [127,140,150,156], air pollution [123], wood smoke [130,137], diesel exhaust [134], ultrafine particles [159], the mixture of PM, gases, and solvents [162], and the mixture of PAHs, metals, and phthalates [73].

The descriptors used to express the comet assay data (one or more in the same study) were as follows: % DNA in tail/tail intensity in 21 studies, tail length in 11 studies, tail moment in 14 studies, olive tail moment in 4 studies, DNA damage index in 5, and strand breaks in 6 (i.e., primary comet descriptors converted to DNA strand-break frequency by using calibration with ionising radiation).

In summary, this comprehensive analysis of various studies, both occupational and environmental, on the genotoxic effects of a variety of air pollutants indicates increased levels of DNA strand breaks in subjects exposed to these substances compared with non-exposed subjects, with a majority of statistically significant results. It is important to stress that by reducing air pollution levels to the WHO-recommended concentrations, an average person might improve their life expectancy by 2 years, and the comet assay might be useful in detecting the most vulnerable population.

3.2. *Anaesthetics*

Anaesthetics play a crucial role in medical procedures, inducing controlled sedation for surgeries and other interventions. Common gases include nitrous oxide and various halogenated agents. While patients benefit from their use, healthcare workers exposed during their professional routine are at risk of health effects [165–169]. Long-term exposure may lead to symptoms such as headaches, dizziness, and nausea and has been associated with reproductive issues, including miscarriages and fertility problems in healthcare workers. Additionally, there is a potential for liver and kidney damage, as well as an increased risk of cancer [168–172]. Available data reviewed in [166,167] suggested an association with genotoxic risks, particularly for nitrous oxide and halogenated agents, but not for propofol and its metabolites.

In our systematic scoping review on anaesthetics gases, 103 articles were identified after duplicate removal, of which 59 were excluded after screening (i.e., reading title/abstract). From the 44 that were read in full, a total of 29 were excluded (the reasons are shown in Figure 2). Finally, 15 studies were included in the qualitative analysis, as summarised in Figure 2 and Table 2.

Table 2. Summary of findings from the included studies on anaesthetics.

Author	Year	Main Chemical Exposure	Country	Exposure Assessment or Biomarkers of Exposure	Population Characteristics	DNA Damage	Reference/DOI
Occupational exposure							
Aun	2018	Isoflurane, sevoflurane, desflurane, and N_2O	Brazil	--	26 medical residents	• **Comet tail intensity:** baseline (6.1 ± 3.4) vs. half-year of exposure (7.0 ± 4.1) vs. 1 year of exposure (7.3 ± 3.3); **non-sig.**	[173] 10.1016/j.mrfmmm.2018.10.002
Baysal	2009	Halothane, isoflurane, sevoflurane, N_2O, and desflurane	Turkey	--	60 (30 anaesthesiologist, certified registered nurse anaesthetist, surgeons, 30 controls)	• **DNA damage (arbitrary unit):** exposed (19.7 ± 16.6) vs. controls (8.8 ± 4.1); **sig.**	[174] 10.1016/j.clinbiochem.2008.09.103
Chandrasekhar	2006	Halothane, isoflurane, sevoflurane, sodium pentothal, N_2O, Desflurane, and enflurane	India	--	99 (45 exposed operating room staff, 54 controls)	• **Comet tail length:** exposed (16.08) vs. controls (7.04); **sig.**	[175] 10.1093/mutage/gel029
El-Ebiary	2013	Halothane, Isoflurane, (sevoflurane), and N_2O (as pure, liquefied compressed, medical grade nitrous oxide gas)	Egypt	--	60 [40 operating room staff (anaesthetists, nurses, technicians), 20 controls]	• **% DNA tail:** controls (1.78 ± 0.71) vs. staff (3.69 ± 1.05) [anaesthetists (3.7 ± 1.02) vs. surgeons (3.63 ± 1.16) vs. technicians (4.2 ± 0.96) vs. nurses (3.51 ± 0.95)]; **sig.** for total exposed group, and for subgroups, **non-sig.** between subgroups	[176] 10.1177/0960327111426584

Table 2. *Cont.*

Author	Year	Main Chemical Exposure	Country	Exposure Assessment or Biomarkers of Exposure	Population Characteristics	DNA Damage	Reference/DOI
Figueiredo	2022	Inhalational of aesthetic isoflurane	Brazil	Workplace exposure assessment: waste anaesthetic gases (WAG), isoflurane, monitoring	76 (39 professionals working in a veterinary hospital, 37 matched controls)	• **% DNA tail (according to age):** <31, control (6.0 ± 4.7 [3.8–7.8]) vs. exposed (9.8 ± 7.3 * [6.4–12.8]), $p = 0.03$; **sig** ≥31, control (7.2 ± 3.8 [5.0–10.1]) vs. exposed (8.4 ± 6.4 [4.7–11.0]), $p = 0.55$ **not-sig.** • **% DNA tail (according to age and exposure time):** <31, exposure < 5 years (8.9 ± 5.4 [7.1–11.1]) vs. (9.9 ± 4.5 [8.2–11.5]), $p = 0.69$; **not sig** ≥31, exposure ≥ 5 years (4.1 ± 2.2 [2.8–3.4]) vs. (9.7 ± 6.6 * [7.7–12.5]), $p = 0.01$ **sig.**	[177] 10.1007/ s11356-022-20444-2
Izdes *	2009	N_2O, isoflurane, sevoflurane, and desfluran	Turkey	--	74 [19 office workers, 17 anaesthesia nurses, 19 nurses—antineoplastic drugs; 19 controls (unexposed office workers)]	• **Total comet scores (TCS):** anaesthesia nurses (18.58 ± 5.03), control (6.84 ± 3.16); **sig.**	[178] 10.1539/joh.m8012
Izdes	2010	Waste anaesthetic gases (N_2O, isoflurane, sevoflurane, and desflurane)	Turkey	--	80 [40 nurses, 40 controls (unexposed health care workers)]	• **Tail intensity:** anaesthesia nurses (8.36 ± 2.16) vs. unexposed controls (3.77 ± 0.97); **sig.**	[179] 10.1080/ 19338244.2010.486421
Khisroon	2020	Mixture not specified	Pakistan	--	99 (50 exposed, 49 unexposed)	• **Total Comet Score (TCS)**: exposed (128.4 ± 44.3) vs. unexposed (50.5 ± 20.8); **sig.**	[180] 10.1136/ oemed-2020-106561

Table 2. *Cont.*

Author	Year	Main Chemical Exposure	Country	Exposure Assessment or Biomarkers of Exposure	Population Characteristics	DNA Damage	Reference/DOI
Rozgaj	2009	Sevoflurane, isoflurane, and N_2O	Croatia	--	100 (50 room staff [anaesthetists, nurses, technicians], 50 controls)	• **Comet tail length:** exposed (21.04 ± 7.30) vs. unexposed (17.57 ± 3.39); **sig.** • **Comet tail moment:** exposed (0.58 ± 0.40) vs. unexposed (0.51 ± 0.32); **non-sig.**	[181] 10.1016/ j.ijheh.2007.09.001
Sardas *	2006	N_2O, isoflurane, sevoflurane, and desflurane	Turkey	--	34 [17 exposed anaesthesiology staff, 17 controls (unexposed office workers)]	• **TCS (total comet score):** exposed (21.5 ± 5.0) vs. unexposed (8.6 ± 4.7); **sig.**	[182] 10.1007/ s00420-006-0115-6
Souza	2016	Waste anaesthetic gases (isoflurane, sevoflurane, desflurane, and N_2O)	Brazil	Concentrations of halogenated anaesthetics (isoflurane, sevoflurane, and desflurane) and N_2O using a sample flow rate of 10 L/min	60 (30 anaesthesiologists, 27 internal medicine physicians)	• **Tail moment:** Comet assay (control 0.31 ± 0.27; exposed 0.34 ± 0.30); **non-sig.**	[183] 10.1016/ j.mrfmmm.2016.09.002
Szyfter §	2004	Sevoflurane, halothane, and isoflurane	Poland	Analysis of N_2O, volatile anaesthetics and organic solvents in the ambient air of operating rooms	49 [29 operating room staff (anaesthetists, nurses, technicians), 20 controls]	• **Average migration (μM) of PBL DNA:** exposed (41.57 ± 9.00) vs. controls (43.21 ± 8.00); **non-sig.**	[184]

Table 2. *Cont.*

Author	Year	Main Chemical Exposure	Country	Exposure Assessment or Biomarkers of Exposure	Population Characteristics	DNA Damage	Reference/DOI
Szyfter §	2016	N_2O, halothane, isoflurane, and sevoflurane	Poland	Concentration of waste anaesthetic gases (N_2O, halothane, isoflurane, and sevoflurane)	200 (100 anaesthetists, 100 controls)	• **Comet length:** exposed (43.21 ± 8.00) vs. unexposed (41.57 ± 9.02); **non-sig.**	[185] 10.1007/s13353-015-0329-y
Wrońska-Nofer	2009	N_2O, sevoflurane or isoflurane and halogenated hydrocarbons	Poland	Air N_2O (breathing zone sampling) and volatile anaesthetics (individual dosimeters)	167 medical staff members (84 exposed male anaesthetists and 55 nurses, and 83 unexposed controls without a history of working in operating rooms)	• **DNA damage score**: low exposure (29.5± 1.94) vs. high exposure (34.3 ± 2.73) vs. unexposed (24.0 ± 1.54); **sig.**	[186] 10.1016/j.mrfmmm.2009.03.012
Wrońska-Nofer	2012	N_2O	Poland	Air N_2O (stationary monitoring sampling) halogenated anaesthetics and toxic solvents, 8 individual dosimeters)	72 (36 exposed nurses in operating rooms, 36 unexposed nurses)	• **DNA damage score:** exposed (31.1 ± 1.5) vs. unexposed (23.3 ± 1.5); **sig.**	[187] 10.1016/j.mrfmmm.2011.10.010

* The studies have partially overlapping populations of unexposed controls (i.e., healthy office workers). Comet assay results of 16 of the 19 subjects in the second study were obtained in the first study. There is no information regarding the reuse of comet data in the group of exposed nurses. § The papers report the same result, 41.57 ± 9.00 (median = 40.22), although in different groups in the 2016 paper as compared to the 2004 paper. Furthermore, the dataset with a mean of 43.21 ± 8.00 is reported in both papers but for different groups and with a different median (43.28 versus 42.28). In both cases, the results are surprisingly similar, considering that one study uses 29/20 subjects in each group, whereas the other study uses 100/100 subjects (exposed/unexposed). The authors have not clarified whether or not the same data have been reported twice.

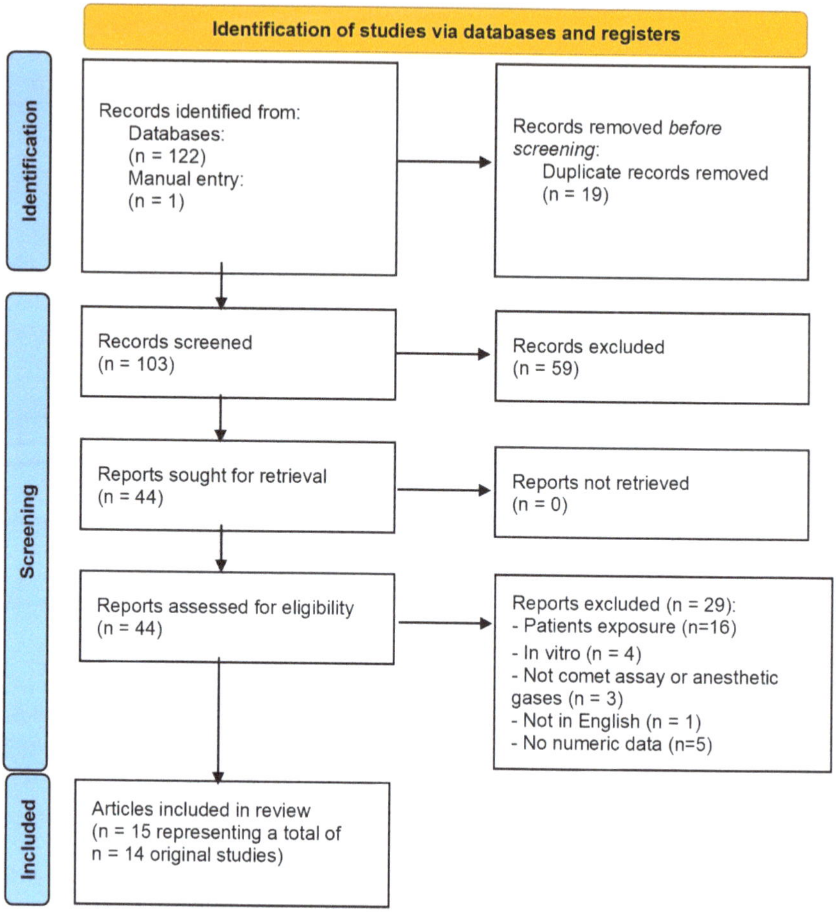

Figure 2. PRISMA flow diagram of systematic scoping review for anaesthetics.

Most of the studies were conducted in Asia (mainly Turkey, n = 6; 40.0%), followed by Europe (mainly Poland, n = 5; 33.3%) and South America (Brazil, n = 3; 20.0%), with only one study conducted in Africa (Egypt; 6.7%). A total of 15 studies of occupational exposure were conducted on medical room staff during their working shifts (anaesthesiologists, nurses, and technicians). Regarding exposure assessment, six studies [177] conducted workplace exposure assessments and two studies [174] measured the oxidative status of the subjects, not a specific biomarker of exposure to anaesthetic gases. It was verified that occupational exposure can lead to DNA-damaging effects (n = 11, 73.3%) and that younger exposed professionals with higher workloads tend to display higher levels of DNA damage [174–182,186,187]. Only four of the reviewed papers showed no significant effects of occupational exposure [173,183–185]. In general, the studies that found positive results also mention the need for further research in this area and for the protection of workers dealing with anaesthetics. The descriptors used to express the comet assay data were as follows: % DNA in tail/tail intensity in five studies, tail length and DNA damage index in three studies each, tail moment in two studies, and other descriptors in three studies.

In summary, the overall results from the application of the comet assay in the study of anaesthetics indicate that exposure may have genotoxic effects, contributing to a better understanding of the potential risks to healthcare workers and thus strongly supporting the need for a mitigation of the risks.

3.3. Antineoplastic Drugs

Antineoplastic drugs, also known as cytotoxic or cytostatic drugs, are a heterogeneous group of chemicals that share an ability to inhibit tumour growth by disrupting cell division and killing actively growing cells [188]. Although patients may benefit from these treatments, there is still a major health concern regarding the use of some drugs classified as carcinogenic, mutagenic, or teratogenic agents [188,189]. Moreover, hospital workers can be exposed to antineoplastic drugs during drug preparation, administration, and contact with contaminated workplace, surfaces, medical equipment, clothing, and patient excreta [190–193].

Evidence has shown that occupational exposure to antineoplastic drugs is associated with an increased risk of acute health effects, including hair loss, headaches, and hypersensitivity; adverse reproductive outcomes, such as infertility, spontaneous abortions, and congenital malformations; and certain cancers [194–199].

In our systematic scoping review on occupational exposure to antineoplastic drugs, 68 articles were identified after the removal of duplicates, of which 47 were excluded after screening (reading title/abstract). From the 21 articles read in full, 2 were excluded because they did not present comet assay data. Nineteen studies of occupational exposure [12,16,191,196,200–214] remained for qualitative analysis, as summarised in Figure 3 and Table 3.

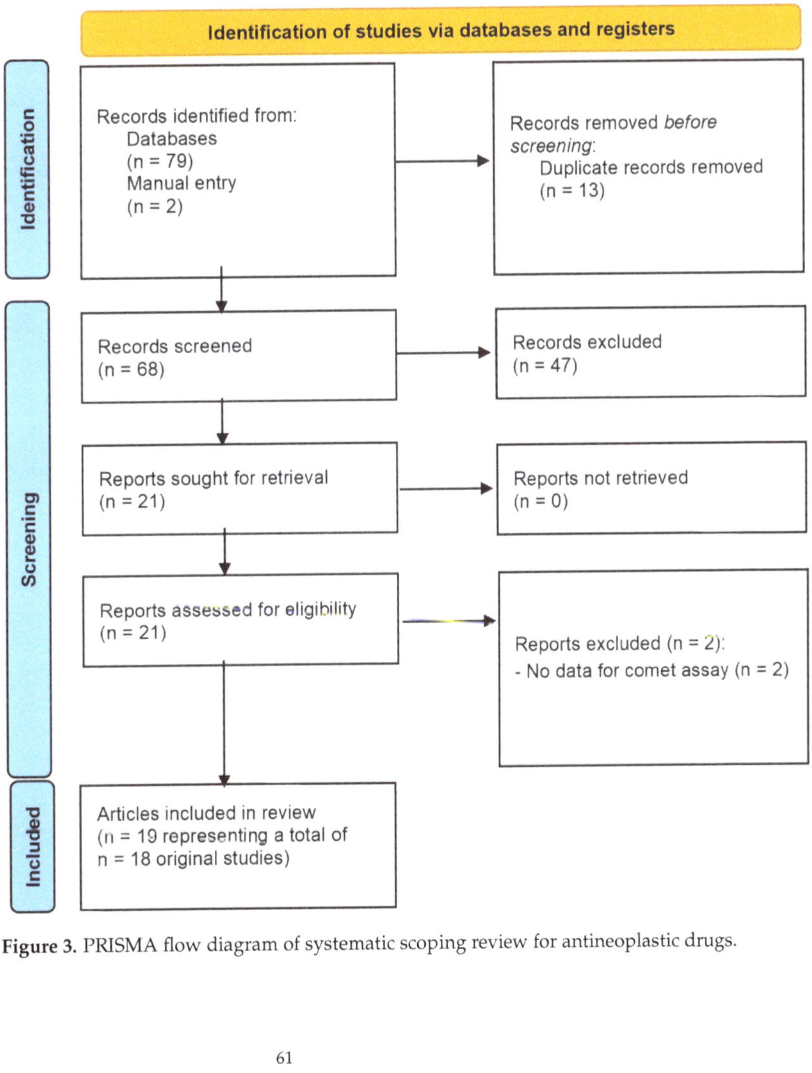

Figure 3. PRISMA flow diagram of systematic scoping review for antineoplastic drugs.

Table 3. Summary of findings from the included studies on antineoplastic drugs (occupational exposure).

Author	Year	Main Chemical Exposure	Country	Exposure Assessment or Biomarkers of Exposure	Population Characteristics	DNA Damage	Reference/DOI
Aristizabal-Pachon	2002	Antineoplastic drugs	Colombia	--	80 (40 exposed, 40 unexposed) hospital workers	• **Comet tail length—Mean**: exposed (4.62 ± 1.477 μm) vs. unexposed (2.41 ± 0.577); **sig.**	[212] 10.1007/ s43188-019-00003-7
Buschini	2013	Antineoplastic drugs	Italy	--	137 (63 exposed, 74 unexposed) nurses	• **% DNA tail—Mean:** exposed (0.95 ± 0.03) vs. unexposed (0.99 ± 0.03); **non-sig.**	[209] 10.1136/ oemed-2013-101475
Cavallo	2009	Antineoplastic drugs	Italy	--	106 (30 exposed, 76 unexposed) hospital workers	• **% DNA tail in lymphocytes:** exposed (10.72 ± 7.04) vs. unexposed (11.24 ± 8.6); **non-sig.** • **Comet tail moment in lymphocytes—Mean:** exposed (16.86 ± 9.13) vs. unexposed (16.72 ± 7.17), $p > 0.05$. • **% DNA tail in buccal cells:** exposed (10.02 ± 6.1) vs. unexposed (13.78 ± 9.80); **non-sig.** • **Comet tail moment in buccal cells—Mean:** exposed (34.58 ± 25.98) vs. unexposed (32.31 ± 12.79); **non-sig.**	[16] 10.1002/em.20501
Connor	2010	Antineoplastic drugs	USA	Fixed-location and personal breathing zone air samples Cyclophosphamide, ifosfamide, paclitaxel, 5-fluorouracil, and cytarabine surface contamination Urinary cyclophosphamide and paclitaxel.	121 (68 exposed, 53 unexposed) hospital workers	• **% DNA in tail:** exposed (53.06 ± 7.32) vs. unexposed (53.12 ± 7.5); **non-sig.** • **Olive Tail Moment—Mean**: exposed (2.540 ± 652) vs. unexposed (2.518 ± 715); **non-sig.**	[207] 10.1097/ JOM.0b013e3181f72b63

Table 3. *Cont.*

Author	Year	Main Chemical Exposure	Country	Exposure Assessment or Biomarkers of Exposure	Population Characteristics	DNA Damage	Reference/DOI
Cornetta	2008	Antineoplastic drugs	Italy	-	90 (83 exposed and 73 unexposed) hospital workers	• **Comet %DNA tail:** • exposed (1.16 ± 0.82) vs. unexposed (0.77 ± 0.47); **Sig.**	[204] 10.1016/ j.mrfmmm.2007.08.017
Hongping	2006	Vincristine	China	--	30 (15 exposed, 15 unexposed) workers from a plant production	• **Comet tail length—Mean**: exposed (1.72 ± 0.15 µm) vs. unexposed (0.71 ± 0.01 µm); **Sig.** • **Comet tail moment—Mean:** exposed (0.29 ± 0.03 µm) vs. unexposed (0.17 ± 0.05 µm); **Sig.**	[214] 10.1016/ j.mrfmmm.2006.02.003
Huang	2022	Antineoplastic drugs	China	--	455 (305 exposed, 150 unexposed) nurses	• **Comet Tail moment—Mean**: exposed (0.62) vs. unexposed (0.46); **Sig.** • **Comet Olive Tail moment**—Mean: exposed (1.10) vs. unexposed (0.92); **Sig.** • **Comet Tail length—Mean**: exposed (6.17) vs. unexposed (5.16); **Sig.** • **% DNA in tail**: exposed (4.06) vs. unexposed (3.52); **Sig.**	[213] 10.1136/ oemed-2021-107913
Kopjar *	2009	Antineoplastic drugs	Croatia	--	100 (50 exposed, 50 unexposed) healthcare workers	• **Comet tail length—Mean**: exposed (17.46 ± 0.08 µm) vs. unexposed (14.00 ± 0.02); **Sig.**	[191] 10.1016/ j.ijheh.2008.10.001

Table 3. *Cont.*

Author	Year	Main Chemical Exposure	Country	Exposure Assessment or Biomarkers of Exposure	Population Characteristics	DNA Damage	Reference/DOI
Kopjar *	2001	Antineoplastic drugs	Croatia	--	70 (50 exposed, 20 unexposed) hospital workers	• **Comet tail length—Mean**: exposed (17.46 ± 1.99 μm) vs. unexposed (12.55 ± 0.82 μm); **Sig.** • %**DNA tail—Mean:** exposed (81.49 ± 4.31%) vs. unexposed (76.01 ± 3.70%); **Sig.** • **Comet tail moment:** exposed (14.31 ± 2.16 μm) vs. unexposed (9.78 ± 0.91 μm); **Sig.**	[196] 10.1093/mutage/16.1.71
Ladeira	2015	Antineoplastic drugs	Portugal	Cyclophosphamide, 5-Fluorouracil, and Paclitaxel surface contamination	92 (46 exposed, 46 unexposed) hospital workers	• **% DNA tail:** exposed (15 ± 1.40) vs. unexposed (12.41 ± 1.24); **Non-sig.**	[210] 10.3934/genet.2015.3.204
Laffon	2005	Antineoplastic drugs (cyclophosphamide, cisplatin, doxorubicin, mitomycin C, 5-fluorouracil, methotrexate)	Portugal	--	52 (30 exposed, 22 unexposed) nurses	• **Comet tail length—Mean:** exposed (46.46 ± 0.09 μm) vs. unexposed (42.68 ± 0.10 μm); **Sig.**	[12] 10.1002/ajim.20189
Maluf	2000	Antineoplastic drugs	Brazil	--	24 (12 exposed, 12 unexposed, plus a historic control of 34 non-exposed workers) hospital workers	• **DNA damage index** (visual scoring): exposed (20.83 ± 10.19) vs. unexposed (8.08 ± 5.16); **sig.**	[200] 10.1016/S1383-5718(00)00107-8

Table 3. *Cont.*

Author	Year	Main Chemical Exposure	Country	Exposure Assessment or Biomarkers of Exposure	Population Characteristics	DNA Damage	Reference/DOI
Oltulu	2019	Antineoplastic drugs	Turkey	--	59 (29 exposed, 30 unexposed) hospital workers	• **DNA damage index** (visual scoring 0–200): exposed (2.00 IQR 0.00–3.00) vs. unexposed (0.00 (0.00–2.25); **non-sig.**	[211] 10.33808/clinexphealthsci.563988
Rekhadevi	2007	Antineoplastic drugs	India	Urinary cyclophosphamide	120 (60 exposed nurses and 60 unexposed subjects)	• **Comet tail length lymphocytes mean:** • Exposed (13.66 ± 2.37) vs. unexposed (6.21 ± 0.0.92); **sig.**	[203] 10.1093/mutage/gem032
Rombaldi	2008	Antineoplastic drugs	Brazil	-	40 (20 exposed and 20 unexposed) hospital workers	• **Comet Damage Index:** • exposed (18.89 ± 8.62) vs. unexposed (6.21 ± 2.78); **sig.**	[205] 10.1093/mutage/gen060
Sasaki	2008	Antineoplastic drugs	Japan	--	224 (121 exposed, 57 highly exposed [antineoplastic preparation], 46 unexposed) female nurses	• **Comet tail length in log units:** exposed (0.764 ± 0.121) vs. unexposed (0.711 ± 0.089); **Sig.** • **Comet tail moment in log units:** exposed (0.312 ± 0.253) vs. unexposed (0.253 ± 0.237); **Non-sig.**	[206] 10.1539/joh.50.7
Ursini	2006	Antineoplastic drugs	Italy	5-Fluorouracil, cytarabine, gemcitabine, cyclophosphamide, and ifosfamide surface contamination Biological monitoring of α-Xuoro-β-alanine in urine (metabolite of 5-Xuorouracile)	65 (30 exposed, 35 unexposed) hospital workers	• **Comet tail moment buccal cells—Mean:** pharmacy technicians (32.6 ± 18.2 μm) vs. hospital nurses (43.2 ± 36.0 μm) vs. ward nurses (27.4 ± 13.9 μm) vs. unexposed (28.6 ± 12.4 μm); **Non-sig.** • **Comet tail moment lymphocytes—Mean:** pharmacy technicians (20.8 ±10.1 μm) vs. hospital nurses (15.5 ± 9.0 μm) vs. ward nurses (14.7 ± 7.9 μm) vs. unexposed (16.1 ± 8.1μm); **Non-sig.**	[201] 10.1007/s00420-006-0111-x

Table 3. *Cont.*

Author	Year	Main Chemical Exposure	Country	Exposure Assessment or Biomarkers of Exposure	Population Characteristics	DNA Damage	Reference/DOI
Villarini	2011	Antineoplastic drugs	Italy	5-Fluorouracil and cytarabine surface contamination Urinary cyclophosphamide	104 (52 exposed, 52 unexposed) healthcare workers	• **Comet tail length—Mean:** exposed (2.73 ± 0.28) vs. unexposed (1.67 ± 0.14); **Sig.**	[208] 10.1093/mutage/geq102
Yoshida	2006	Antineoplastic drugs (cyclophosphamide, dacarbazine, isophosphamide, aclarubicin, amrubicin, bleomycin, daunorubicin, doxorubicin, pirarubicin, carboplatin, cisplatin, docetaxel, etoposide, irinotecan, paclitaxel, vinblastine, vincristine, vinorelbine, rituximab)	Japan	*umu* assay from surface contamination	37 (19 exposed, 18 unexposed) female nurses	• **Comet tail length lymphocytes—Median**: exposed (8.5, ranging 4.5–13.6 μm) vs. unexposed (5.1, ranging 3.5–10.3 μm); **Sig.**	[202] 10.1539/joh.48.517

* Updated studies from the same author/group of authors. In the first paper, the authors report the mean and SD as 17.46 ± 1.99 and 12.55 ± 0.82 for the exposed and controls, respectively. However, these data are at odds with the calculated SEM in the 2009 paper (i.e., 0.08 and 0.02 in exposed and controls, respectively). Based on the reported group size, the SEMs should be 0.28 (exposed, n = 50) and 0.18 (controls, n = 20), respectively.

From a total of 19 studies, around half were conducted in Europe (n = 9, 47.4%), 6 (31.6%) in Asia, and 4 (21.1%) in the Americas. All the studies were from occupational settings; one was from a production plant [214], and all the others (n = 18, 94.7%) involved hospital workers. Only four studies (21.1%) presented exposure assessment data from surface contamination [201,207,208,210], and one study (5.2%) tested the genotoxicity of 19 antineoplastic drugs used in the hospital ward and 8 wipe-samples from the workbench after handling antineoplastic drugs, using the *umu* assay [202]. The study by Connor (2010) measured fixed-location air samples and personal breathing zone air samples [207]. For biological monitoring of exposure, four studies (21.1%) performed urinary measurements, and three of these studies (15.8%) also made exposure assessments [201,207,208]. The study from Rombaldi (2008) measured the serum endpoints of oxidative stress, such as superoxide dismutase (SOD), catalase (CAT) and thiobarbituric acid-reactive substances (TBARS) [205]; however, these are not considered specific biomarkers of exposure.

The results of studies on the genotoxic effects of antineoplastic drugs using the comet assay in occupationally exposed workers are inconsistent, but a slightly positive association exists. Overall, 13 studies (68.4%) showed a statistically significant increase in DNA damage in the exposed group compared with the controls [12,191,200–206,208,211–214]. Ursini (2006) showed positive results for both biological matrices under study—lymphocytes and buccal cells [201].

In five studies (26.3%), the levels of DNA damage did not differ statistically between the exposed and non-exposed groups [16,196,207,209,210], although in two of them (10.5%), a trend towards an increase in DNA damage was observed in the exposed group [196,210], while in one study, the % DNA in the tail in both lymphocytes and buccal cells was marginally higher in the control subjects [16]. It is important to mention that antineoplastic drugs are well-known cross-linking agents, which can reduce the ability of DNA with strand breaks to migrate in an electric field. The presence of a cross-linking agent could have hidden an increase in the DNA migration associated with the induction of DNA strand breaks [208]. The study of Hongping (2006) reported mixed results because there was a significant increase in the comet tail length and a non-significant increase in the comet tail moment in the exposed group [214]. The parameters used to express the comet assay results were as follows: tail length and tail moment in nine studies each, % DNA in the tail in seven studies, and the DNA damage index in three studies (some studies cited more than one parameter). When assessing the potential hazards of antineoplastic drugs in an occupational setting, it is also important to consider the use of personal protective equipment. Well-educated staff, adequate protection, and the use of automated systems for drug handling significantly decrease the possibility of contamination and exposure, thus affecting the comet assay results.

In summary, this comprehensive analysis of various studies on the genotoxic effects of antineoplastic drugs indicates increased levels of DNA strand breaks in subjects exposed to these drugs compared with non-exposed subjects, showing a majority of statistically significant results.

3.4. Heavy Metals

Several heavy metals pose significant health risks, particularly to industrial workers (as these substances are frequently used in this context) and residents in nearby areas. While the harmful effects of acute exposure to heavy metals are well-documented, there is a growing concern about their long-term effects and effects of combined exposures, especially considering their persistent nature, meaning that even minimal exposure to heavy metals may be detrimental to health, with particular risks of neurological disorders and cancer. Moreover, studies have demonstrated that metal ions interact with cellular components, including DNA, and that this can result in an altered structure and mutations, as well as cell death and carcinogenesis [215–217].

Our systematic scoping review gathered 979 reports (971 from the databases and 8 by manual entry, excluding duplicates). After the preliminary screening by title and abstract,

889 documents were excluded as they did not refer to human biomonitoring. From the 90 potentially eligible studies, 33 were excluded (mostly for not presenting comet assay data, study design flaws, DNA repair rather than DNA damage, etc.). The remaining 57 studies were assessed for qualitative analysis, as summarised in Figure 4 and Table 4.

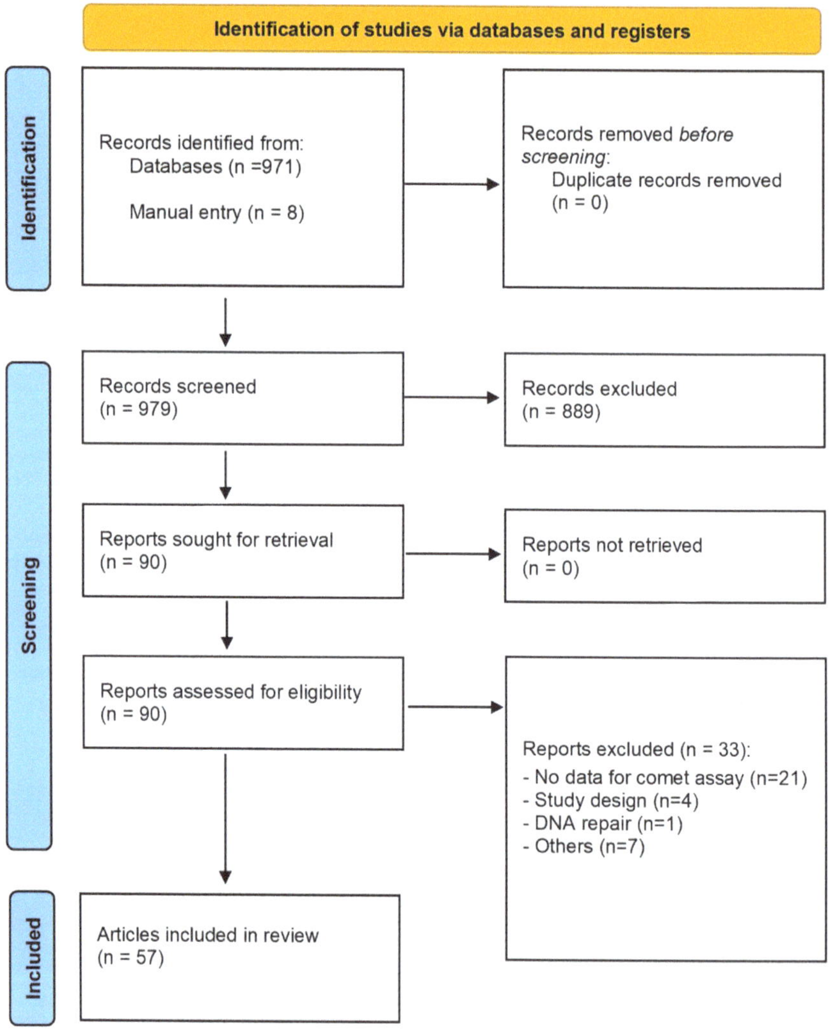

Figure 4. PRISMA flow diagram of systematic scoping review for heavy metals.

From a total of 57 studies, 24 studies (42.1%) were conducted in Asia, 19 studies (31.6%) in Europe and 14 studies (24.6%) in the Americas. There were 37 studies (64.9%) of occupational exposure, mainly from industry settings, and 18 studies (31.6%) of environmental exposure, of which 3 (16.7%) were in children, 2 (11.1%) in adolescents and 11 (61.1%) in the general population. Two studies were classified as both occupational and environmental.

Table 4. Summary of findings from the included studies on heavy metals.

Author	Year	Main Chemical Exposure	Country	Exposure Assessment or Biomarkers of Exposure	Population Characteristics	DNA Damage	Reference/DOI
Occupational exposure							
Aksu	2019	Cr, Cu, Cd, Ni, Pb	Turkey	Cr, Mn, Ni, Cu, As, Cd, Pb in blood	96 (48 welders, 48 controls)	• **Comet tail intensity** (lymphocytes): exposed (6.52 ± 3.13) vs. unexposed (2.31 ± 1.09); **sig.**	[218] 10.1016/ j.mrgentox.2018.11.006
Balachandar	2010	Chromium	India	Cr in air and urine Cr in air	108 (36 leather tanning industry workers, 36 environmental exposure subjects, 36 controls)	• **Comet tail length:** occupational exposure (4.21 [3.21–10.98]) vs. environmental exposure (3.98 [2.98–11.27]) vs. controls (3.01 [2.68–9.40]); reported to be **sig.** for exposed workers	[219] 10.1007/ s00420-010-0562-y
Batra	2010	Lead	India	Pb in blood	220 (110 workers occupationally exposed to lead, 110 controls)	• % **DNA tail:** exposed (14.80 ± 1.31) vs. unexposed (6.12 ± 1.80); **sig.**	[220] 10.7860/JCDR/ 2020/43682.13572
Cavallo	2002	Antimony	Italy	Airborne Sb_2O_2; personal air samplers	46 (23 workers assigned to different fire-retardant treatment tasks in the car upholstery industry, 23 controls)	• **Comet tail moment:** control: 16 ± 7 (SD), exposed group A: 14 ± 8, exposed group B: 19 ± 9, **non-sig.**	[221] 10.1002/em.10102
Chinde	2014	Lead	India	Pb in blood	400 (200 lead–acid storage battery recycling and manufacturing industry workers, 200 controls)	• % **DNA tail:** exposed (12.97 ± 2.33) vs. unexposed (4.80 ± 2.57); **sig.**	[222] 10.1007/ ts11356-014-3128-9

Table 4. *Cont.*

Author	Year	Main Chemical Exposure	Country	Exposure Assessment or Biomarkers of Exposure	Population Characteristics	DNA Damage	Reference/DOI
Coelho	2013	Lead, Cd, As	Portugal	Metalloids levels in blood	122 (41 miners, 41 subjects living near a mine, 40 controls)	• **% DNA tail:** occupational exposure (18.73 ± 7.60) vs. environmental exposure (25.58 ± 2.75) vs. unexposed (12.40 ± 3.04); **sig.**	[223] 10.1016/ j.envint.2013.08.014
Danadevi	2003	Lead	India	Pb, Cd in blood	81 (45 workers employed in a secondary Pb recovery unit, 36 controls)	• **Damage index (DI, visual scale—AU)**: exposed (44.6 ± 8.5) vs. unexposed (21.1 ± 11.7); **sig.**	[224] 10.1016/s0300-483x(03)00054-4
Danadevi	2004	Cr, Ni	India	Cr, Ni in blood	204 (102 welders, 102 controls)	• **Comet tail length:** controls: 8.9 ± 3.2, welders: 23.1 ± 3.9, **sig.**	[225] 10.1093/mutage/geh001
De Boeck	2000	Cobalt	Belgium, Norway, Finland, Sweden, England	Co in urine	99 (35 cobalt dust, 29 carbide-cobalt, 35 unexposed)	• **% DNA tail:** Co (0.50 ± 1.44) vs. hard metals (0.57 ± 1.24) vs. unexposed (0.51 ± 1.35); **non-sig.** • **Comet tail length:** Co (0.71 ± 1.38) vs. hard metals (0.65 ± 1.23) vs. unexposed (0.64 ± 1.25); **non-sig.** • **Comet tail moment:** Co (0.37 ± 1.85) vs. hard metals (0.40 ± 1.45) vs. unexposed (0.34 ± 1.47); **non-sig.**	[64] 10.1002/1098-2280(2000)36:2<151::aid-em10>3.3.co;2-m
De Olivera	2012	Copper (and other metals)	Brazil	Cu in blood	22 (11 copper-smelter, 11 controls)	• **Damage index (DI, visual scale—AU)** exposed (17.6 ± 10.2) vs. unexposed (4.29 ± 2.53); **sig.**	[226] 10.1177/ 0748233711422735

Table 4. *Cont.*

Author	Year	Main Chemical Exposure	Country	Exposure Assessment or Biomarkers of Exposure	Population Characteristics	DNA Damage	Reference/DOI
De Restrepo	2000	Lead	Colombia	Lead in air Pb in blood	56 (43 workers of electric battery factories exposed to lead compounds, 13 controls)	• **Comet tail length:** Group I >40 μg/dL (55.60 [42.52–68.70]) vs. Group II 41–80 μg/dL (65.60 [52.50–78.63]) vs. Group III 81–120 μg/dL (60.53 [50.50–70.60]) vs. Group IV >120 μg/dL (85.90 [69.21–102.53]); **sig.** between the lowest and highest concentration groups.	[227] 10.1002/1097-0274(200009)38:3<330::aid-ajim13>3.0.co;2-z
Fracasso	2002	Lead	Italy	Pb, Cd in blood	66 (37 battery plant workers, 29 controls)	• **% DNA tail:** exposed (58.4 ± 15.8) vs. unexposed (40.9 ± 15.6); **sig.** • **Comet tail length:** exposed (117.1 ± 32.8) vs. unexposed (106.6 ± 25.3); **non-sig.** • **Comet tail moment:** exposed (69.0 ± 25.5) vs. unexposed (45.5 ± 19.4); **sig.**	[228] 10.1016/s1383-5718(02)00012-8
Gambelunghe	2003	Chromium	Italy	Cr urine	39 (19 chrome-plating workers, 20 controls)	• **Comet tail moment:** exposed (0.42 ± 0.21) vs. unexposed (0.42 ± 0.21); **sig.**	[229] 10.1016/s0300-483x(03)00088-x
García-Lestón	2011	Lead	Portugal	Lead in blood Zn protoporphyrin, δ-aminolaevulinic acid dehydratase activity	108 (70 workers in plants using inorganic lead, 38 controls)	• **% DNA tail:** exposed (4.3) vs. unexposed (5.3) **non-sig.**	[230] 10.1016/ j.mrgentox.2011.01.001
Grover	2010	Lead	India	4.5 $\mu g/m^3$ Pb in air Pb in blood and urine	180 (90 workers of secondary Pb recovery unit, 90 controls)	• **Comet tail length:** exposed (17.86 ± 0.88) vs. unexposed (8.15 ± 0.63); **sig.**	[231] 10.1016/ j.ijheh.2010.01.005

Table 4. *Cont.*

Author	Year	Main Chemical Exposure	Country	Exposure Assessment or Biomarkers of Exposure	Population Characteristics	DNA Damage	Reference/DOI
Hernandez-Franco	2022	Lead	Mexico	Pb in blood	53 (37 battery recycling workers, 16 controls)	• **Comet tail length:** control: 36, exposed: 40 μm; **non-sig.**	[232] 10.3390/ijerph19137961
Iarmarcovai	2005	Lead, cadmium	France	Al, Cd, Cr, Co, Pb, Mn, Ni, Zn in blood and urine	57 (27 welders, 30 controls)	• **Olive tail moment:** exposed (4.5 ± 1.7) vs. unexposed (2.8 ± 0.8); **sig.**	[233] 10.1093/mutage/gei058
Kašuba	2012	Lead, cadmium	Croatia	Pb, Cd in blood	60 (30 pottery-glaze workers, 30 controls)	• **Comet tail intensity:** exposed (3.21 ±0.73) vs. unexposed (1.54 ± 0.73); **sig.** • **Comet tail moment:** exposed (0.55 ±0.16) vs. unexposed (0.21 ± 0.02); **sig.** • **Comet tail length:** exposed (16.66 ± 1.20) vs. unexposed (14.10 ± 0.2); **sig.**	[234] 10.1007/s00420-011-0726-4
Kašuba	2020	Lead	Croatia	Pb in blood ALAD activity and EP level	98 (50 manufacture lead workers, 48 unexposed)	• **Comet tail length:** exposed (16.15 ± 5.33) vs. unexposed (14.27 ± 1.23); **non-sig.** • **Comet tail Intensity:** exposed (2.64 ± 3.22) vs. unexposed (1.61 ± 0.74); **non-sig.**	[235] 10.2478/aiht-2020-71-3427

Table 4. *Cont.*

Author	Year	Main Chemical Exposure	Country	Exposure Assessment or Biomarkers of Exposure	Population Characteristics	DNA Damage	Reference/DOI
Kayaalti	2015	Lead	Turkey	Pb in blood	61 occupationally exposed to lead workers (36 low exposure, 25 high exposure)	• **Tail intensity:** Low: 46,908.41 ± 11,596.55, exposed: 62,219.17 ± 21,180.57; **sig.** • **Comet tail moment.** Low: 4.00 ± 0.62, exposed: 4.90 ± 1.26; **sig.** • **"DNA tail" (presumably tail length)** Low: 85.58 ± 24.24, exposed: 103.94 ± 34.22; sig. (all data are mean and SD)	[236] 10.1080/ 19338244.2013.787964
Khisroon	2021	Cd, Cr, Fe, Mn, Ni, Pb	Pakistan	Cd, Cr, Fe, Mn, Ni, Pb in scalp hair	118 (59 welders, 59 controls)	• **DNA damage index:** exposed (121.8 ± 10.7) vs. controls (56.5 ± 17.6); **sig.**	[237] 10.1007/ s12011-020-02281-x
Liu	2017	Indium	China	In in urine In in ambient	120 (57 indium exposed workers, 63 controls)	• **Comet tail length**: exposed (16.36 ± 7.56) vs. unexposed (10.80 ± 5.63); **sig.** • **% DNA tail**: exposed (5.01 ± 3.08) vs. unexposed (2.69 ± 1.61); **sig.**	[238] 10.1093/toxsci/kfx017
Meibian-Zhang	2008	Chromium	China	Cr in air Cr in blood and urine	90 Exposure group I: 30 tannery workers exposed to trivalent chromium from tanning department; exposure group II: 30 tannery workers from finishing department; 30 controls.	• **Olive tail moment:** moderate exposure (3.43 [2.31–8.29]) vs. high exposure (5.33 [2.90–8.50]) vs. unexposed (2.04 [0.09–3.83]); **sig.**	[239] 10.1016/ j.mrgentox.2008.04.011

Table 4. *Cont.*

Author	Year	Main Chemical Exposure	Country	Exposure Assessment or Biomarkers of Exposure	Population Characteristics	DNA Damage	Reference/DOI
Minozzo	2010	Lead	Brazil	Lead in blood	106 (53 workers in recycling of automotive batteries, 53 controls)	• **Damage index (DI, visual scale—AU)**: exposed (21.70 ± 27.85) vs. unexposed (2.57 ± 2.79); **sig.**	[240] 10.1016/ j.mrgentox.2010.01.009
Muller	2022	Chromium	Brazil	Cr, Pb, As, Ni, V in blood	100 (50 male chrome-plating workers, 50 unexposed)	• **% DNA tail** (alkaline CA): exposed (10.10 ± 2.16) vs. unexposed (8.31 ± 1.32); **sig.**	[241] 10.1080/ 01480545.2020.1731527
Olewińska	2010	Lead	Poland	Lead (PbB) and zinc protoporphyrin (ZPP) in blood	88 (62 metalworkers exposed to lead, 26 controls)	• **% DNA tail:** exposed (60.3 ± 14) vs. unexposed (37.1 ± 17.6); **sig.**	[242]
Palus	2003	Lead, cadmium	Poland	Pb, Cd in blood	106 (44 Pb exposed, 22 Cd exposed, 40 unexposed)	• **Damage index (DI, visual scale—AU)**: Pb-exposed (15.6 ± 4.1) vs. Cd-exposed (19.6 ± 5.2) vs. unexposed (11.3 ± 5.0); **sig.**	[243] 10.1016/s1383-5718(03)00167-0
Palus	2005	Arsenic	Poland	As concentration in dust and fumes As in urine	155 (71 copper-smelter workers, 80 controls)	• **Comet tail moment:** control: 2.1 (0.0–30.0) and workers: 13.2 (0.0–140.0); **sig.**	[244] 10.1002/em.20132
Pandeh	2017	Fe	Iran	Iron status (including serum iron)	56 (30 steel company workers, 26 controls)	• **Tail length**: 15.88 (8.94–20.44) vs. 6.17 (5.57–8.07); **sig.** • **% DNA tail**: 8.98 (5.81–11.37) vs. 3.97 (30.7–4.84); **sig.** • **Tail moment**: 3.42 (1.60–6.01) vs. 0.68 (0.53–0.93); **sig.** • **Tail intensity:** 24.59 (11.74–29.53) vs. 20.19 (17.50–22.26); **sig.**	[245] 10.1007/ s11356-017-8657-6

Table 4. *Cont.*

Author	Year	Main Chemical Exposure	Country	Exposure Assessment or Biomarkers of Exposure	Population Characteristics	DNA Damage	Reference/DOI
Pawlas	2017	Lead	Poland	Cd, Zn in blood	116 (78 lead and zinc-smelter and battery recycling plan workers, 38 controls)	• **% DNA tail:** exposed (14.1 ± 8.8) vs. unexposed (16.2 ± 12.8); **non-sig.** • **Comet tail moment:** exposed (6.5 ± 8.4) vs. unexposed (10.2 ± 15.7); **non-sig.** • **Comet tail length:** exposed (28.4 ± 13.5) vs. unexposed (31.9 ± 24.4); **non-sig.**	[246] 10.17219/acem/64682
Pérez-Cadahía	2008	Lead	Spain	Al, Ni, Cd, Pb, Zn in blood	240 (61 oil collectors, 59 hired workers, 60 high-pressure machine workers, 60 unexposed)	• **% DNA tail:** exposed—all groups (0.18 ± 0) vs. unexposed (0.09 ± 0); **sig.**	[247] 10.4137/ehi.s954
Rashid	2018	Cd, Zn	Pakistan	Cd, Zn in blood	60 (35 traffic police wardens, 25 controls)	• **Comet tail length:** exposed (4.65 ± 1.70) vs. unexposed (2.07 ± 1.26); **sig.**	[248] 10.1016/ j.scitotenv.2018.02.254
Singh	2016	Lead	India	Pb in blood	70 (35 welders, 35 unexposed)	• **Comet tail length:** exposed (29.21 ± 8.8) vs. unexposed (1.47 ± 0.5); **sig.**	[249] 10.1177/ 0748233715590518
Wang	2018	Pb	China	Pb in blood	267 146 electronic waste processing workers, 121 controls)	• **% DNA tail:** exposed (6.5 ± 0.9) vs. unexposed (1.8 ± 0.3); **sig.**	[250] 10.1016/ j.envint.2018.04.027

Table 4. *Cont.*

Author	Year	Main Chemical Exposure	Country	Exposure Assessment or Biomarkers of Exposure	Population Characteristics	DNA Damage	Reference/DOI
Wani	2017	Lead, Zn	India	Pb in blood Zn in blood	130 (92 occupationally exposed to lead or lead and zinc, 38 unexposed controls were selected from neighbouring with similar age)	• **Comet tail length:** Exposed in lowest employment time group: 8.36 ± 2.16; unexposed in lowest employment time group: 6.91 ± 1.67; exposed in highest employment time group: 20.15 ± 3.53; unexposed in highest exposure time group: 12.99 ± 3.75; **sig.** (All)	[251] 10.1007/ s11356-017-8569-5
Vuyyuri	2006	Arsenic	India	As in blood	365 (200 glass workers, 165 controls)	• **Comet tail length:** exposed (14.95 ± 0.21) vs. unexposed (8.29 ± 0.71): **sig.**	[252] 10.1002/em.20229
Wultsch	2011	As, Mn, Ni, Cr	Austria	Cr, Mn, Ni, As in urine	42 (23 waste incinerator workers, 19 controls)	• **DNA migration (tail factor)**: Group I [≥1 and ≤3 months employment] (6.7 ± 1.9) vs. Group II [>3 and ≤8 months] (6.3 ± 1.5) vs. Group III [>8 and ≤11 months] (6.5 ± 2.4) vs. unexposed (7.1 ± 1.6); **non-sig.**	[112] 10.1016/ j.mrgentox.2010.08.002
Zhang	2011	Chromium	China	Cr in air Cr in blood	250 (157 electroplating workers, 93 unexposed)	• **% DNA tail:** exposed (3.69 [0.65–16.2]) vs. unexposed (0.69 [0.04–2.74]); **sig.** • **Comet tail moment:** exposed (1.13 [0.14,6.77]) vs. unexposed (0.14 [0.01–0.39]); **sig.** • **Comet tail length:** exposed (11.77 [3.46, 52–19]) vs. unexposed (3.26 [3.00, 4.00]); **sig.**	[253] 10.1186/ 1471-2458-11-224

Table 4. *Cont.*

Author	Year	Main Chemical Exposure	Country	Exposure Assessment or Biomarkers of Exposure	Population Characteristics	DNA Damage	Reference/DOI
Zhijian Chen	2006	Lead	China	Pb in air Pb in blood	50 storage battery workers (25 exposed, 25 unexposed)	• **Comet tail moment:** exposed (1.48 ± 3.43) vs. unexposed (0.49 ± 1.35); **sig.** • **Comet tail length:** exposed (2.42 ± 0.45) vs. unexposed (1.02 ± 0.55); **sig.**	[254] 10.1016/j.tox.2006.03.016
Environmental exposure							
Andrew	2006	Arsenic	USA, Mexico	As in drinking water	24 subjects (12 low exposure, 12 high exposure)	• **Comet tail moment:** low (1.4 ± 0.5) vs. high (2.6 ± 0.6); **sig.**	[255] 10.1289/ehp.9008
Banerjee	2008	Arsenic	India	As in water As in urine, nail, hair	90 (30 exposed subjects with skin lesions, 30 without skin lesions, 30 controls)	• **Olive tail moment:** exposed no skin lesions (2.76 ± 1.39) vs. exposed with skin lesions (2.51 ± 1.40) vs. unexposed (0.55 ± 0.83); **sig.** • **Comet tail length:** exposed no skin lesions (11.85 ± 5.51) vs. exposed with skin lesions (13.54 ± 4.38) vs. unexposed (2.20 ± 0.72); **sig.**	[256] 10.1002/ijc.23478
Basu	2005	Arsenic	India	As in water As in urine, nails, hair	60 volunteers (30 high-level exposure, 30 controls)	• **Comet tail length:** exposed (86.501 ± 5.135) vs. unexposed (21.25 ± 1.004); **sig.** • **DNA damage index** • **Exposed** (**1.212** ± 0.049) vs. controls (0.579 ± 0.043); **sig.**	[257] 10.1016/j.toxlet.2005.05.001
Cruz-Esquivel	2019	As, Hg	Colombia	As, Hg in blood	100 volunteers (50 exposed, 50 unexposed)	• **% DNA tail:** exposed (36.03 ± 5.9) vs. unexposed (13.1 ± 2.1); **sig.**	[258] 10.1007/s11356-019-04527-1

Table 4. *Cont.*

Author	Year	Main Chemical Exposure	Country	Exposure Assessment or Biomarkers of Exposure	Population Characteristics	DNA Damage	Reference/DOI
David	2021	Cd, Cr, Zn	Pakistan	Ni, Cd, Zn, Cr in blood	232 children (134 living at brick kiln industries, 98 controls)	• **% DNA tail:** exposed (15.02 ± 0.56) vs. unexposed (10.33 ± 0.55); **sig.**	[259] 10.1080/ 19338244.2020.1854645
Franken	2017	PAHs, metals	Belgium	Cr, Cd, Ni in urine As in blood MeHg in hair	598 adolescents (14–15 years old)	• **% DNA tail** (geometric mean): 4.1 (3.9–4.3)	[260] 10.1016/ j.envres.2016.10.012
Jasso-Pineda	2012	Lead, arsenic	Mexico	Pb in blood As in urine	85 exposed subjects (48 high area, 12 middle area, 25 low area)	• **Comet tail moment:** low (2.5 ± 0.4) vs. middle (3.5 ± 0.4) vs. high (5.2 ± 0.6); **sig.**	[261] 10.1007/ s12011-011-9237-0
Jasso-Pineda *	2015	Arsenic, lead, PAH, DDT/DDE	Mexico	As and 1-OHP in urine Lead and total DDT/DDE in blood	276 children (40/25 with high/low arsenic, 55/10 with high/low lead)	• **Comet tail moment:** high/low arsenic (4.5 ± 1.08/3.2 ± 0.5) sig.; high/low lead (3.7 ± 1.8/ 4.1 ± 1.5) **non-sig.**	[73] 10.1016/ j.scitotenv.2015.02.073
Jasso-Pineda	2007	Lead, As	Mexico	As, Pb, Cd, Cu, and Zn in soil Pb in blood, As in urine	60 children (12 low area, 28 medium area, 20 high area exposure)	• **Comet tail moment:** low exposure (3.9 ± 0.2) vs. medium exposure (5.4 ± 0.2) vs. high exposure (4.8 ± 0.3); **sig. (high versus low)**	[262] 10.1002/ ieam.5630030305
Khan	2012	Chromium	India	Cr in blood	200 volunteers (100 exposed, 100 unexposed)	• **Comet tail length:** exposed (27.39 ± 9.50) vs. unexposed (8.89 ± 2.49); **sig.**	[263] 10.1016/ j.scitotenv.2012.04.063
Koppen *,§	2020	PAHs, metals, benzene, POPs, phthalates	Belgium	Ar, Cd, Cu, Ni, Pb, Tl, Cr in blood Outdoor air	2283 adolescents (14–18 years old)	• **% DNA tail (alkaline CA)**: mean 2.4 [2.3–2.5] (**positively associated with blood metals**)	[139] 10.1016/ j.envres.2020.110002

Table 4. *Cont.*

Author	Year	Main Chemical Exposure	Country	Exposure Assessment or Biomarkers of Exposure	Population Characteristics	DNA Damage	Reference/DOI
Lourenço	2013	Uranium	Portugal	U, Zn, Mn in blood	84 volunteers (54 exposed, 30 unexposed)	• **DNA damage index:** Stratification in three age groups: • <40 years: control sites 42.84 ± 28.6 and Cunha Baixa 82.11 ± 42.84; **non-sig.** • 40–60 years: control sites 28.6 ± 21.42 and Cunha Baixa 135.7 ± 74.9; **sig.** • >60 years: control site 35.7 ± 14.3 Cunha Baixa 71.4 ± 64.3; **sig.**	[264] 10.1016/j.tox.2013.01.011
Mendez-Gomez	2008	As, Pb	Mexico	As, Cd, Pb in air (playground) and drinking water, As in urine, Pb in blood	65 subjects (living near a smelter facility, 22 near, 22 intermediate, 21 distant)	• **Tail length:** 28.6 (19.2–48.0), 25.3 (11.8–43.4), 29.2 (12.3–48.0); **non-sig.**	[265] 10.1196/annals.1454.027
Pelallo-Martinez *,§	2014	Lead	Mexico	Pb in blood	97 volunteers 44 Allede, 37 Mundo Nuevo, 16 Lopez Mateo)	• **Olive tail moment:** Allende (8.3 [3.1–16.8]) vs. Mundo Nuevo (10.6 [5.6–22.9]) vs. Lopez Mateo (11.7 [7.4–15.9]); sig.	[149] 10.1007/s00244-014-9999-4
Sampayo-Reyes	2010	Arsenic	Mexico	As in water As in urine	286 subjects (five villages)	• **% DNA tail:** low exposure (22.90 ± 1.17) vs. medium exposure (32.76 ± 2.55) vs. high exposure (35.80 ± 3.05); **sig.**	[266] 10.1093/toxsci/kfq173
Staessen	2001	Lead, cadmium	Belgium	Pb, Hg in blood Hg in urine	200 exposed volunteers (100 in Peer, 42 in Wilrijk, 58 in Hoboken)	• **% DNA tail:** Peer (1.02 ± 0.44) vs. Wilrijk (1.70 ± 0.49) vs. Hoboken: (1.01 ± 0.42); **sig.**	[267] 10.1016/s0140-6736(00)04822-4

Table 4. *Cont.*

Author	Year	Main Chemical Exposure	Country	Exposure Assessment or Biomarkers of Exposure	Population Characteristics	DNA Damage	Reference/DOI
Wu	2009	Lead	Taiwan	Lead in blood	154 volunteers (71 immigrant women from China, 83 native women from Taiwan)	• **% DNA tail:** native (33.5 ± 11.7) vs. immigrant (31.3 ± 9.8); **non-sig.**	[268] 10.1016/ j.scitotenv.2009.07.025
Yanez	2003	Lead, arsenic	Mexico	As, Pb in soil and house dust Pb in blood, As in urine	55 children (20 exposed, 35 unexposed)	• **Comet tail moment** (geometric mean): exposed (6.8 [5.2–8.9]) vs. unexposed (3.2 [2.6–3.9]); sig. • **Comet tail length** (geometric mean): exposed (67.6 [58.3–79.3]) vs. unexposed (41.7 [35.8–48.6]); **sig.**	[269] 10.1016/ j.envres.2003.07.005

* Studies also in air pollution table; § Studies also in solvents table.

In the present review, the term "heavy metals" was used as a descriptor of the exposure, but it should be noted that most (if not all) of the studies refer to complex mixtures of metals (i.e., co-exposures). Moreover, it is likely that the study populations are exposed to multiple metals and maybe other hazardous substances, even though only one or a few metals have been assessed. Thus, confounding is a possibility in studies where genotoxicity is thought to be attributed to a specific type of heavy metal. Certain studies appear to have an exploratory approach (e.g., the Flemish biomonitoring studies on environmental exposures) [139,260,267], whereas other studies target specific agents (e.g., chromium in studies on welders).

Moreover, the definition of heavy metals is inconsistent in the literature. For instance, one rationale states that these are elements with a higher molecular weight than elementary iron, which is suitable as it includes arsenic and excludes substances such as sodium and aluminium. However, it also includes copper and nickel, which is problematic because these metals could also be regarded as transition metals [270]. Certain ions of elements in the fourth period of the periodic table catalyse the conversion of hydrogen peroxide to hydroxyl radicals, which is an important mechanism of their genotoxic effect [271]. This contrasts with "classic" heavy metals, such as lead, mercury, and cadmium, which are not chemical catalysts, while their mechanism of action is related to the inhibition of enzymes. It should also be emphasised that the oxidation state, chemical form (e.g., organic versus inorganic), solubility and particle size (in case of inhalation exposure) are key factors to be considered when assessing the genotoxic hazard of metals [217].

Lead is the heavy metal that has been assessed in most studies in this review (n = 36; 63.2%) [73,139,149,218,220,222–224,227,228,230–237,240–243,246,247,249–251,254,260–262,264,265,267–269], followed by arsenic (n = 18; 31.6%) [73,112,139,218,223,241,244,252,255–258,260–262,265,266,269], chromium (n = 14; 24.6%) [112,139,218,219,225,229,233,239,241,253,259,260,263,265], cadmium (n = 12; 21.1%) [139,218,233,234,237,243,247,248,259,260,264,265], and nickel (n = 11; 19.3%) [112,139,218,225,233,237,241,247,259,260,264]. A few studies have assessed the genotoxicity of other metals, such as iron [226,237,245,250], cobalt [64], iridium [238], antimony [221], and uranium [264]. In studies on lead exposure, this metal has either been the only element assessed (n = 16; 44.0%), or it has been measured in combination with other metals (n = 20; 56.0%). In the latter group, arsenic (n = 10), cadmium (n = 10), and nickel (n = 8) are the most prevalent co-exposures. The group of studies with metals other than lead is dominated by studies on arsenic (n = 8) and chromium (n = 7).

Overall, 20 studies (34.5%) have assessed lead exposure. Sixteen studies have only assessed lead exposure. Four studies have measured exposure to lead and other metals. In these four studies, exposure groups have had different levels of lead exposure, whereas there has been the same level of exposure to other metals. Thus, there is only an exposure contrast of lead in these four studies [234,250,251,268]. In 16 studies (80.0%), a significant increase in DNA strand breaks in lead-exposed subjects was observed [220,222,224,227,228,231,234,236,240,242,243,249–251,254,268], whereas 4 studies have shown unaltered levels of strand breaks [230,232,235,246]; 1 study additionally showed increased DNA damage in exposed subjects although they were not exposed to lead alone [149]. Assessment of the studies with a measurement of multiple types of heavy metals indicates that five of them (22.7%) have found consistency between increased exposure and DNA strand breaks [233,261,262,269], whereas nine (40.9%) demonstrated no effect on this outcome [73,139,218,223,241,247,260,265,267]. One study (4.5%) had unaltered levels of lead exposure yet increased levels of DNA strand breaks in subjects from a uranium mining district who were exposed to manganese and uranium [264]. Interestingly, there seems to be an over-representation of positive test results in lead-exposed subjects in studies that have assessed mainly lead (80.0%, 16 out of 20 studies; [220,222,224,227,228,231,234,236,240,242,243,249–251,254,268] versus [230,232,235,246]) compared to studies with a more elaborate exposure assessment (35.7%, 5 out of 13 studies; [233,261,262,269] versus [73,139,218,223,241,247,260,265,267]).

All studies with only an arsenic exposure assessment found increased levels of DNA strand breaks (n = 5) [244,252,255–257,266]. In studies with multiple metal exposures, there are many that have found statistically significant effects of arsenic exposure on levels of DNA strand breaks (n = 7) [73,241,258,261,262,269]. However, some studies with the assessment of multiple metals have not found elevated arsenic exposure (and therefore no association between exposure and DNA damage) or no association between arsenic exposure and levels of DNA strand breaks [112,139,218,260,265].

Five out of the thirteen (38.5%) studies were restricted to the effects of chromium exposure [219,229,239,253,263], as well as four studies, including chromium and other metals [225,233,237,259], found increased levels of DNA strand breaks in the exposed population. Conversely, two studies found no genotoxic effect [139,260], one study showed an increased level of DNA strand breaks in subjects who were not exposed to chromium [218], and two studies found unaltered levels of DNA strand breaks in subjects without chromium exposure contrast [112,241]. The parameters used to express the comet assay data (one or more in the same study) were as follows: % DNA in tail/tail intensity in 24 studies, tail length in 22 studies, tail moment in 16 studies, DNA damage index in 5 studies, and olive tail moment in 4 studies.

In summary, this comprehensive analysis of various studies on the genotoxic effects of heavy metals indicates increased levels of DNA strand breaks in subjects exposed to lead, arsenic, and chromium compared with the non-exposed subjects. Interestingly, studies that primarily examined lead exposure exhibited a higher proportion of positive results in comparison with the studies with broader exposure assessments, suggesting a potential bias in favour of detecting lead-related effects. Moreover, some contradictory results among the chromium studies might suggest that the impact of this metal on DNA strand breaks may be insignificant at low exposure levels and that other factors may contribute to this outcome. Further research is necessary to fully understand the potential effects of some metals (alone or combined with other metals and substances) regarding DNA damage.

3.5. Pesticides

Pesticides represent a large group of substances which are used in pest control, broadly classified based on target organisms (e.g., insecticides, herbicides, and fungicides), chemical structure (e.g., organochlorines, organophosphates, carbamates, and pyrethroids), or the mechanism of action and toxicity [272]. Although over 80% of pesticide use is attributed to agriculture, a significant percentage (around 20%) of these substances is employed in public health protection programs (e.g., to protect plants from pests, weeds, or diseases, and humans from vector-borne diseases), maintenance of non-agricultural areas as urban green spaces and sports fields, production of pet shampoos, building and food cover materials, as well as paints for boat protection [273–275].

Recent data from the Food and Agriculture Organization (FAO) suggest that in the past 30 years, negligible changes in the land area used for agriculture occurred, but that the use of active substances in pesticides significantly increased—from 1.8 million—to 3.5 million tons annually, which corresponds to an increase from 1.22 kg/ha to 2.26 kg/ha of land [276]. Since pesticides are designed to improve crop yields, they are intentionally and diffusely applied to large areas, making their control difficult. Considering that the adverse nature of these compounds includes persistency (some can persist for even years in the environment) and lipophilicity (enabling biomagnification through the food web), their residues can be found in soil, freshwater, groundwater, air, and food [272,277,278]. Additionally, over 95% of pesticides have a harmful effect on non-target organisms, which include humans, as their mechanisms of action include inhibition of neural signals by disrupting the sodium/potassium balance, cholinesterase inhibition, opening sodium channels, blockage of receptors, or competition for hormonal receptors [278,279].

Humans can ingest, inhale, and absorb pesticides through the skin. Most individuals are exposed to low concentrations of pesticides in food, water, and the general environment; however, specific populations can experience a high concentration of exposure due to their occupation (e.g., open-field and greenhouse farmers, pesticide industry workers, public health agents, and pest exterminators) [273,278,280]. Moreover, due to their high body surface area to weight ratio, specific physiology, and behaviour, children represent a population vulnerable to developing health effects from pesticide exposure [281].

Apart from the environmental effects [275,279], pesticide exposure is associated with several human health effects, such as asthma, diabetes, Parkinson's disease, cognitive impairment, reproductive health effects, immunotoxicity, cardiotoxicity, leukaemia, and different types of cancer [273,278,280,282–286]. However, it is difficult to establish a firm link between pesticide exposure and DNA damage due to complex exposure assessment, control for other effect-changing variables, as well as a lack of adequate studies and inconsistent epidemiological data [287].

Our systematic scoping review gathered 90 reports assessed for eligibility, of which 25 were eliminated, mostly because they lacked comet assay data. Finally, 65 reports (representing 59 studies, some being published in more than one article) were included in the qualitative analysis—see Figure 5 and Table 5.

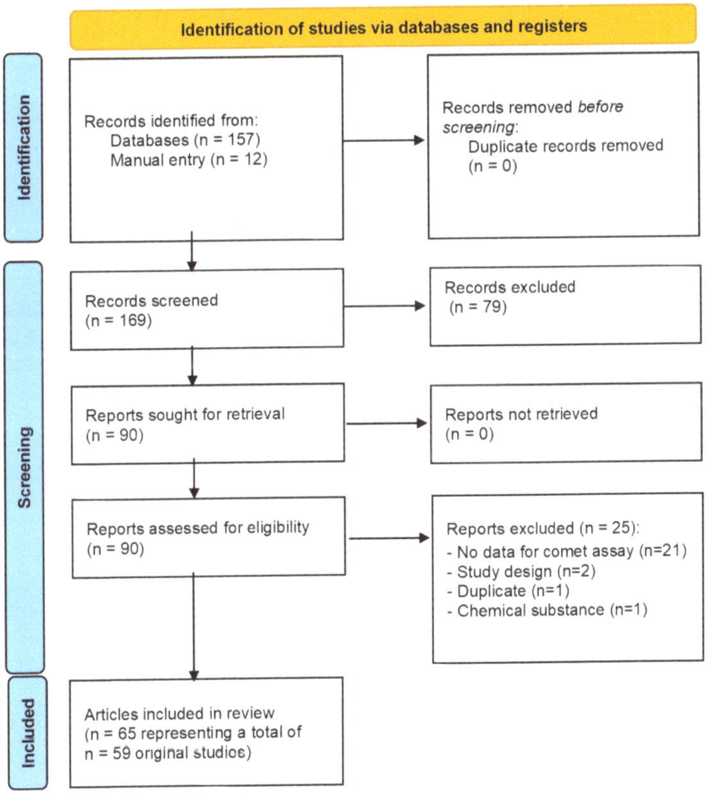

Figure 5. PRISMA flow diagram of systematic scoping review for pesticides.

Considering that around 2 million tons of pesticides from a total global production of 3.5 million tonnes (57.1%) is used in the Americas and Asia [276], it was expected that most included studies would have been performed in these regions (n = 55; 84.6%). Effectively, from a total of 65 studies, 30 studies (46.2%) were conducted in Asia (mainly India), 25 studies (38.5%) in the Americas (mainly Brazil), 8 studies (12.3%) in Europe, and 2 studies (3.1%) in Africa. The majority of the studies compared levels of DNA damage between non-exposed subjects and agriculture workers (n = 45; 69.2%) and pesticide industry workers (n = 11; 16.9%). In addition, a few of the studies assessed health agents who are occupationally exposed to these compounds (n = 3; 4.6%) or focused on the environmental exposure of children (n = 6; 9.2%).

Table 5. Summary of findings from the included studies on pesticides.

Author	Year	Main Chemical Exposure	Country	Exposure Assessment [Mean Concentration Pesticides (ppm)] or Biomarkers of Exposure	Population Characteristics	DNA Damage	Reference
Occupational exposure							
Abhishek	2010	--	India	--	67 (40 exposed, 27 unexposed agricultural workers	• **%DNA tail:** exposed (10.56 ± 3.63) vs. unexposed (5.18 ± 2.60); **sig.** • **Damage Index:** exposed (150.25 ± 60.84) vs. unexposed (31.37 ± 27.85); **sig.**	[288] 10.1089/rej.2009.0931
Aiassa	2019	Glyphosate, cypermethrin, chlorpyrifos	Argentina	--	52 (30 exposed, 22 unexposed) agricultural workers	• **Comet tail moment—Mean:** exposed (3206 ± 785.4 µm) vs. unexposed (269.7 ± 67.91 µm); **sig.**	[289] 10.1007/s11356-019-05344-2
Ali	2018	Cyhalothrin, endosulfan, deltamethrin	Pakistan	Serum concentrations: Deltamethrin: exposed (0.54 ± 0.22) vs. unexposed (0.28 ± 0.13); $p < 0.01$ Endosalfan: exposed (1.07 ± 0.52) vs. unexposed (0.36 ± 0.12); $p < 0.001$ Cyhalothrin: exposed (1.04 ± 0.38) vs. unexposed (0.33 ± 0.15); $p < 0.01$	138 (69 exposed, 69 unexposed) cotton-picking workers	• **Comet tail length—Before**: exposed (14.64 ± 2.68 µm) vs. unexposed (9.6 ± 2.31 µm); sig.—After: exposed (18.29 ± 2.75 µm) vs. unexposed (9.8 ± 2.40 µm); **sig.** • **Comet tail length—Mean**: exposed (16.47 ± 2.65 µm) vs. unexposed (9.7 ± 2.34 µm); **sig.**	[290] 10.1080/01480545.2017.1343342
Alves	2016	Dithiocarbamate, carbamate, dicarboximide, organophosphate, neonicotinoid, pyrethoid. isoxazolidinone, dinitroaniline	Brazil	List of compounds commonly used in the area	137 (77 exposed, 60 unexposed) tobacco farmers	• **Damage index:** exposed (28.01 ± 21.43) vs. unexposed (9.72 ± 7.50); **sig.** • **Damage frequency:** exposed (19.54 ± 13.03) vs. unexposed (6.75 ± 4.73); **sig.**	[291] 10.1590/0001-3765201520150181

Table 5. *Cont.*

Author	Year	Main Chemical Exposure	Country	Exposure Assessment [Mean Concentration Pesticides (ppm)] or Biomarkers of Exposure	Population Characteristics	DNA Damage	Reference
Arshad	2016	Carbamates, organophosphates, pyrethroids	Pakistan	Blood malathion levels: detected in 72% of the exposed blood samples with na average value of 0.14 mg/L (range 0.01–0.31 mg/L)	58 (38 exposed, 20 unexposed) pesticide-manufacturing workers	• **Comet tail length—Mean:** exposed (7.04 ± 0.21 µm) vs. unexposed (0.94 ± 0.2 µm); **sig.** • Malathion correlated with TL	[292] 10.1016/j.shaw.2015.11.001
Benedetti	2013	Organophos phorouscarbamates, pyrethroids, organochlorines	Brazil	BuChE—U/L: exposed (8231 ± 1368) vs. unexposed (8068 ± 920); $p > 0.05$ List of compounds used by volunteers	127 (81 exposed, 46 unexposed) agricultural workers	• **Damage index (0–400):** exposed (38.5 ± 19.9) vs. unexposed (19.6 ± 10.3); **sig.** • **% damage frequency:** exposed (23.1 ± 9.4) vs. unexposed (13.3 ± 6.4); **sig.**	[293] 10.1016/j.mrgentox.2013.01.001
Bhalli	2006	Organophosphates, carbamates, pyrethroids	Pakistan	--	64 (29 exposed, 35 unexposed) pesticide-manufacturing workers	• **Comet tail length—Mean**: exposed (20.0 ± 2.87 µm) vs. unexposed (7.4 ± 1.48 µm); **sig.**	[294] 10.1002/em.20232
Bhalli	2009	Carbamate, organophosphate, organochlorine, pyrethroids	Pakistan	Cypermethrin, cyhalothrin, deltamethrin, and endosulfan serum levels before and after spraying	97 (47 exposed, 50 unexposed) agricultural workers	• **Comet tail length—Mean**: exposed (before: 14.90 ± 2.99 µm and after: 19.00 ± 3.63 µm) vs. unexposed (6.54 ± 1.73 µm); both comparisons; **sig.**	[295] 10.1002/em.20435

Table 5. *Cont.*

Author	Year	Main Chemical Exposure	Country	Exposure Assessment [Mean Concentration Pesticides (ppm)] or Biomarkers of Exposure	Population Characteristics	DNA Damage	Reference
Bian	2004	Pyrethroids (fenvalerate), organophosphorus compounds (phoxim), carbamates (carbaryl)	China	Fenvalerate concentration 21.55 × 10^{-4} mg/m^3 (operation site) vs. 1.19 × 10^{-4} mg/m^3 (control site), and dermal contamination 1.59 mg/m^2 higher than control	63 (21 exposed, 23 internal controls, 19 external controls) pesticide-manufacturing workers	• **Olive tail moment—Mean of comet sperm:** exposed (3.80 [1.10–5.90]) vs. internal controls (1.50 [0.65–3.05]) (p = 0.016) vs. external controls (2.00 [0.60–2.80]); **sig.** • **%DNA tail:** exposed (11.30 [2.85–18.45]) vs. Internal controls (5.60 [1.98–10.5]) (p = 0.044) vs. External controls (5.10 [1.50–7.10]); **sig.**	[296] 10.1136/oem.2004.014597
Carbajal-López	2016	Organochlorines, organophosphorus, carbamates, pyrethroids	Mexico	List of compounds commonly used in the area	171 (111 exposed, 60 unexposed) agricultural workers	• **Comet tail length—Mean**: exposed (190.77 ± 10.4 µm) vs. unexposed (106.08 ± 2.6 µm); **sig.**	[297] 10.1007/s11356-015-5474-7
Cayir	2019	Propineb, captan, boscalid, pyraclostrobin. cycloxydim, cypermethrin. alphacypermethri, deltamethrin, chlorpyrifos. permethrin	Turkey	Pesticides exposure assessment List of compounds used by the volunteers	86 (41 exposed, 45 unexposed) greenhouse workers	• **Damage index—Median AU (0–400):** exposed (8.72 [min–max: 1.62–25.09]) vs. unexposed (3.47 [min–max: 0.00–14.57]); **sig.**	[298] 10.1080/1354750X.2019.1610498
Chen	2014	Fungicides. herbicides. inseticides	China	Pesticides exposure assessment	337 (83 low exposure, 113 high exposure, 141 unexposed) fruit growers	• **Comet tail moment—Mean:** low exposed (2.18 ± 0.05 µm) vs. high exposed (2.14 ± 0.04 µm) vs. unexposed (1.28 ± 0.01 µm); **sig.**	[299] 10.1155/2014/965729.

Table 5. *Cont.*

Author	Year	Main Chemical Exposure	Country	Exposure Assessment [Mean Concentration Pesticides (ppm)] or Biomarkers of Exposure	Population Characteristics	DNA Damage	Reference
Costa	2014	Fungicides, herbicides, inseticides	Portugal	Urinary metabolites: organic farmers PYR 0.06 ± 0.05, OP/CRB 1.86 ± 0.30, THIO 62.56 ± 5.60; pesticide workers PYR 0.08 ± 0.03, OP/CRB 2.23 ± 0.19, THIO 54.33 ± 3.16, unexposed PYR 0.13 ± 0.04, OP/CRB 1.54 ± 0.23, THIO 51.83 ± 3.28 BuChE—U/L: exposed farmers (6245.62 ± 191.41) vs. exposed pesticide workers (7063.66 ± 202.31) vs. unexposed (6425.44 ± 224.15); $p = 0.943$ List of compounds used by volunteers	182 (36 organic farmers, 85 pesticide workers, 61 unexposed) agricultural workers	• **%DNA tail:** exposed pesticide workers (15.05 ± 0.85) vs. unexposed (8.03 ± 0.73); **sig.**	[300] 10.1016/ j.toxlet.2014.02.011
da Silva	2008	Carbamates and organophosphates	Brazil	--	173 (108 exposed, 65 unexposed) agricultural workers	• **Comet Damage Index—Mean:** unexposed (4.42 ± 5.85) vs. exposed < 3 days ago (20.44 ± 11.19) vs. exposed > 3 days ago (20.14 ±12.23); **sig.**	[301] 10.1093/mutage/ gen031

Table 5. *Cont.*

Author	Year	Main Chemical Exposure	Country	Exposure Assessment [Mean Concentration Pesticides (ppm)] or Biomarkers of Exposure	Population Characteristics	DNA Damage	Reference
da Silva	2012	--	Brazil	--	167 (111 exposed, 56 unexposed) tobacco farmers	• **Damage index (0–400):** exposed pesticide applicators (17.35 ± 14.40) vs. exposed harvest (23.85 ± 17.70) vs. unexposed (5.91 ± 6.86); **sig.** • **% damage frequency:** exposed pesticide applicators (11.64 ± 9.02) vs. exposed harvest (16.15 ± 11.59) vs. unexposed (4.02 ± 4.65); **sig.**	[302] 10.1016/ j.jhazmat.2012.04.074
da Silva	2014	Organophosphorate, carbamate, dithiocarbamate, pyrethroid	Brazil	BuChE activity—did not differ between exposed and unexposed	60 (30 exposed, 30 unexposed) tobacco farmers	• **Damage frequency:** exposed (10.57 ± 7.83) vs. unexposed (4.97 ± 4.76); **sig.**	[303] 10.1016/ j.scitotenv.2014.05.018
Dalberto	2022	Neonicotinoid. pyrethroid, carbamate, organophosphate	Brazil	List of compounds used by the volunteers	241 (84 exposed harvest, 72 exposed grading, 85 unexposed) tobacco farmers	• **Visual score (0–400)—Mean**: unexposed (15.3 ± 13.6) vs. harvest (37.4 ± 23.0) vs. grading (26.4 ± 19.6); **sig.**	[304] 10.1016/ j.mrgentox.2022.503485

Table 5. *Cont.*

Author	Year	Main Chemical Exposure	Country	Exposure Assessment [Mean Concentration Pesticides (ppm)] or Biomarkers of Exposure	Population Characteristics	DNA Damage	Reference
Dhananjayan	2019	Organophosphorus, organochlorine, synthetic pyrethroid, benzoylurea, limonoid, benzoylphenylurea, organosulfite, quinazoline, stereoisomers, triazole, copper compounds, diphenyl ether, phosphanoglycine, chlorophenoxy-acetic, ammonium salt, bipyridilium	India	AchE activity—U/mL: exposed (2.86 ± 0.75) vs. unexposed (3.93 ± 0.87); $p < 0.001$ BuChE activity—U/mL: exposed (2.02 ± 0.74) vs. unexposed (2.60 ± 0.74); $p < 0.001$	143 (77 exposed, 66 unexposed) tea garden workers	• **Comet tail length—Mean:** exposed (9.45 ± 5.28 μm) vs. unexposed (2.09 ± 0.95 μm); **sig.** • **Olive tail moment—Mean:** exposed (4.15 ± 2.18 μm) vs. unexposed (0.59 ± 0.44 μm); **sig.** • **%DNA tail:** exposed (13.1 ± 8.17) vs. unexposed (2.26 ± 1.63); **sig.**	[305] 10.1016/j.mrgentox.2019.03.002
Dutta and Bahadur	2019	Organophosphates, carbamates, pyrethroids	India	AchE activity—μmol/min/mL: exposed (6.43 ± 1.85) vs. unexposed (11.81 ± 3.40); $p \leq 0.001$ BuChE activity—μmol/min/mL: exposed (3.50 ± 1.89) vs. unexposed (4.73 ± 1.84); $p \leq 0.001$	155 (95 exposed, 60 unexposed) tea garden workers	• **Comet tail length—Mean:** exposed (45.98 ± 4.25 μm) vs. unexposed (15.14 ± 2.99 μm); **sig.** • **Olive tail moment—Mean:** exposed (6.41 ± 0.78 μm) vs. unexposed (2.32 ± 0.36 μm); **sig.** • **%DNA tail:** exposed (17.23 ± 1.05) vs. unexposed (5.99 ± 0.82); **sig.**	[306] 10.1016/j.mrgentox.2019.06.005
Franco	2016	Pyrethroids, carbamates, organophosphates, organochlorines, benzoylureas	Brazil	--	249 (161 exposed, 88 unexposed) community health agents	• **Olive tail moment—Mean:** exposed (7.8 ± 10.4) vs. unexposed (4.7 ± 3.8); **sig.**	[307] 10.1007/s11356-016-7179-y

Table 5. *Cont.*

Author	Year	Main Chemical Exposure	Country	Exposure Assessment [Mean Concentration Pesticides (ppm)] or Biomarkers of Exposure	Population Characteristics	DNA Damage	Reference
Garaj-Vrhovac and Želježić *	2000	Atrazine, alachlor, cyanazine, dichlorophenoxyacetic acid, malathion	Croatia	--	20 (10 exposed, 10 unexposed) pesticide-manufacturing workers	• **Comet tail length—Mean:** exposed after high exposure period (50.1 ± 9.4 µm) vs. exposed after no exposure period (17.2 ± 0.4 µm) vs. unexposed (13.3 ± 1.5 µm); **sig.** • **Comet tail moment—Mean**: exposed after high exposure period (60.8 ± 18.2 µm) vs. exposed after no exposure period (13.8 ± 0.4 µm) vs. unexposed (10.5 ± 1.1 µm); **sig.**	[308] 10.1016/s1383-5718(00)00092-9
Garaj-Vrhovac and Želježić *	2001	Atrazine, alachlor, cyanazine, 2,4-dichlorophenoxyacetic acid, malathion	Croatia	--	40 (20 exposed, 20 unexposed) pesticide-manufacturing workers	• **Comet tail length—Range**: exposed after high exposure period (16.3–95.2 µm) vs. exposed after no exposure period (11.0–30.5 µm) vs. unexposed (6.3–20.4 µm); **sig.** • **Comet tail moment—Range**: exposed after high exposure period (11.7–85.1) vs. exposed after no exposure period (6.35–25.4) vs. unexposed (5.0–15.1); **sig.**	[309] 10.1016/s0300-483x(01)00419-x
Garaj-Vrhovac and Želježić *	2002	Atrazine, alachlor, cyanazine, 2,4-dichlorophenoxyacetic acid, malathion	Croatia	--	30 (10 exposed, 20 unexposed) pesticide-manufacturing workers	• **Comet tail length—Mean**: exposed (50.13 ± 9.44 µm) vs. unexposed (13.06 ± 1.36 µm); **sig.** • **Comet tail moment—Mean**: exposed (60.85 ± 18.17 µm) vs. unexposed (10.33 ± 1.21 µm); **sig.**	[310] 10.1002/jat.855

Table 5. *Cont.*

Author	Year	Main Chemical Exposure	Country	Exposure Assessment [Mean Concentration Pesticides (ppm)] or Biomarkers of Exposure	Population Characteristics	DNA Damage	Reference
Godoy et al.	2019	Organochlorines, carbamates, pyrethroids	Brazil	List of compounds used by the volunteers	163 (74 exposed, 89 unexposed) agricultural workers	• **Comet tail length—Median**: exposed (14.75 ± 18.97 µm) vs. unexposed (9.68 ± 5.49 µm); **sig.** • **Olive tail moment—Mean:** exposed (6.08 ± 8.79 µm) vs. unexposed (3.87 ± 3.16 µm); **sig.** • **%DNA tail:** exposed (21.63 ± 20.23) vs. unexposed (14.73 ± 8.93); **sig.**	[311] 10.1007/s11356-019-05882-9
Grover	2003	Organophosphates, carbamates, pyrethroids	India	–	108 (54 exposed, 54 unexposed) pesticide-manufacturing workers	• **Comet tail length—Mean:** exposed non-smokers (18.26 ± 2.13 µm) vs. unexposed non-smokers (7.03 ± 2.39 µm); **sig.** Mean: exposed smokers (19.75 ± 2.22 µm) vs. Unexposed smokers (10.34 ± 2.38 µm); **sig.**	[312] 10.1093/mutage/18.2.201
Kahl	2018	Glyphosate, flumetralin, clomazone, imidacloprid, sulfentrazone, dithiocarbamate, magnesium aluminium phosphide, fertilizers	Brazil	–	242 (121 exposed, 121 unexposed) tobacco farmers	• **Damage index (0–400):** exposed (22.1 ± 1.6) vs. unexposed (4.6 ± 0.4); **sig.**	[313] 10.1016/j.ecoenv.2018.04.052

Table 5. *Cont.*

Author	Year	Main Chemical Exposure	Country	Exposure Assessment [Mean Concentration Pesticides (ppm)] or Biomarkers of Exposure	Population Characteristics	DNA Damage	Reference
Kasiotis	2012	Chlorpyrifos, captan, myclobutanil, propargite, acetamiprid, cypermethrin, deltamethrin	Greece	Serum levels: Myclobutanil: 1.12–5.54 ppb Cypermethrin: 22.92–30.32 ppb Deltamethrin: <LOD–30.96 ppb Propargite, chlorpyrifos, captan, acetamiprid <LOD	19 (all exposed) fruit growers	• **%DNA tail:** before exposure (12.10) vs. after exposure (24.17); **sig.** • **%DNA tail:** workers with detectible residues vs. non-detectible; **sig.**	[314] 10.1016/j.toxlet. 2011.10.020
Kaur	2011	Carbamates, organophosphates, pyrethroids	India	List and frequency of compounds used by the volunteers	260 (210 exposed [60 of them selected for follow-up], 50 unexposed) agricultural workers	• **Comet tail length—Mean:** fresh exposed (72.22 + 20.76 µm) vs. unexposed (46.92 + 8.17 µm) vs. followed-up (66.67 + 24.07 µm); **sig.**	[315] 10.4103/0971-6866.92100
Kaur and Kaur §	2020	Organophosphates, carbamates. pyrethroids	India	--	450 (225 exposed, 225 unexposed) agricultural workers	• **Comet tail length—Mean:** exposed (111.03 ± 24.7 µm) vs. unexposed (45.89 ± 11.00 µm); **sig.** • **Total comet DNA migration:** exposed (86.05 ± 16.9 µm) vs. unexposed (44.55 ± 8.07 µm); **sig.** • **Frequency of cells showing DNA migration:** exposed (53.27 ± 14.9) vs. unexposed (15.89 ± 7.89); **sig.**	[316] 10.1007/ s11033-020-05600-6

Table 5. *Cont.*

Author	Year	Main Chemical Exposure	Country	Exposure Assessment [Mean Concentration Pesticides (ppm)] or Biomarkers of Exposure	Population Characteristics	DNA Damage	Reference
Kaur and Kaur §	2020	Organophosphates, carbamates, pyrethroids	India	--	450 (225 exposed, 225 unexposed) agricultural workers	• **Comet tail length—Mean**: exposed (111.03 ± 24.7 µm) vs. unexposed (45.89 ± 11.00 µm); **sig.** • **Total comet DNA migration** (µm): exposed (86.05 ± 16.9) vs. unexposed (44.55 ± 8.07); **sig.** • **Frequency of cells showing DNA migration:** exposed (53.27 ± 14.9) vs. unexposed (15.89 ± 7.89); **sig.**	[317] 10.1080/1354750X.2020.1794040
Kaur and Kaur §	2021	Organophosphates, carbamates, pyrethroids	India	List of compounds used by the volunteers	450 (225 exposed, 225 unexposed) agricultural workers	• **Comet tail length—Mean**: exposed (86.05 ± 16.9 µm) vs. unexposed (44.55 ± 8.07 µm); **sig.**	[318] 10.1016/j.mrgentox.2020.503302
Khayat	2013	Glyphosate, fenpropathrin, carbofuran	Brazil	List of pesticide mixtures	73 (41 exposed, 32 unexposed) agricultural workers	• **Comet tail length—Median:** exposed (4.9 ± 1.81 µm) vs. unexposed (3.82 ± 2.34 µm); **sig.** • **Comet tail moment—Median:** exposed (0.18 ± 0.13 µm) vs. unexposed (0.02 ± 0.04 µm); **sig.** • **Olive tail moment—Median:** exposed (0.54 ± 0.21 µm) vs. unexposed (0.09 ± 0.13 µm); **sig.** • **%DNA tail:** exposed (5.71 ± 1.63) vs. unexposed (1.13 ± 1.25); **sig.**	[319] 10.1007/s11356-013-1747-1

Table 5. *Cont.*

Author	Year	Main Chemical Exposure	Country	Exposure Assessment [Mean Concentration Pesticides (ppm)] or Biomarkers of Exposure	Population Characteristics	DNA Damage	Reference
Lebailly	2003	Fungicide captan	France	UK Predictive Operator Exposure Model suggested 14.4 mg (0.9–66.1 mg) of captan absorbed. List of other compounds used a day before	19 (all exposed) fruit growers	• **Comet tail moment—Mean:** exposed in the morning (4.35 ± 1.11) vs. exposed the morning day after (4.80 ± 2.57); **sig.** • **%DNA damage:** exposed in the morning (10%, ranging 2–21%) vs. exposed the morning day after (13%, ranging 5–49%); **sig.**	[320] 10.1136/oem.60.12.910
Liu Y	2006	Organophosphates, carbamates, pyrethroid insecticides, fungicides, growth regulator	China (Taiwan)	List of pesticides used, area of use, and frequency of use	197 (43 low exposure, 48 high exposure, 106 unexposed) agricultural workers	• **Comet tail moment—Mean:** low exposed (1.92 ± 0.04 μm) vs. high exposed (2.35 ± 0.06 μm) vs. unexposed (1.33 ± 0.03 μm); **sig.**	[321] 10.1158/1055-9965.EPI-05-0617
Muniz	2008	Organophosphonate	USA	Adjusted urinary dialkylphosphate (DAP) metabolite levels: sum methyl DAP (μmol/L): Farmworker 1.03 ± 37%, Applicator 0.774 ± 36%, Control 0.126 ± 42%	31 (10 farmworkers, 12 applicators, 9 unexposed) agricultural workers	• **Comet tail length—Mean:** exposed applicator (7.674 ± 0.295 μm) vs. exposed farmer (7.478 ± 0.312 μm) vs. unexposed (4.509 ± 0.312 μm); **sig.** • **Comet tail moment—Mean:** exposed applicator (3.643 ± 0.111 μm) vs. exposed farmer (3.200 ± 0.11 μm) vs. unexposed (2.354 ± 0.118 μm); **sig.**	[322] 10.1016/j.taap.2007.10.027
Naravaneni, Jamil	2007	Carbamates, organophosphates, pyrethroids	India	AchE activity- U/mL: exposed (253.5 ± 21.7) vs. unexposed (311.1 ± 7.99); $p < 0.001$	370 (210 exposed, 160 unexposed) agricultural workers	• **Comet tail length—Mean:** exposed (26.13 ± 4.21 μm) vs. unexposed (7.61 ± 1.85 μm); **sig.**	[323] 10.1177/0960327107083450

Table 5. *Cont.*

Author	Year	Main Chemical Exposure	Country	Exposure Assessment [Mean Concentration Pesticides (ppm)] or Biomarkers of Exposure	Population Characteristics	DNA Damage	Reference
Paiva	2011	Organochlorates, organophosphates, pyrethroids, carbamates	Brazil	List of compounds used by the volunteers	63 (16 exposed region A, 16 exposed region B, 31 unexposed) agricultural workers	• **Damage index (0–400):** exposed region A (14.15 ± 0.95) vs. exposed region B (18.83 ± 0.68) vs. unexposed (5.63 ± 2.77); **sig.** • **% damage frequency:** exposed region A (10.16 ± 0.92) vs. exposed region B (9.56 ± 0.82) vs. unexposed (4.22 ± 0.81); **sig.**	[324] 10.1002/em.20647
Paz-y-Miño	2004	Fungicides, herbicides, inseticides	Ecuador	List of compounds used by the volunteers	66 (45 exposed, 21 unexposed) agricultural workers	• **Comet tail length—Mean**: exposed (31.58 ± 3.22 µm) vs. unexposed (25.94 ± 7.77 µm); **sig.**	[325] 10.1016/j.mrgentox. 2004.05.005
Prabha, Chadha	2017	--	India	--	100 (50 exposed, 50 unexposed) pesticide-manufacturing workers	• **Comet tail length—Mean:** exposed (26.27 ± 0.83 µm) vs. unexposed (15.89 ± 0.39 µm); **sig.**	[326] 10.1080/09723757. 2015.11886263
Ramos	2021	Glyphosate, dichlorophenoxy-acetic acid, atrazine, cypermethrin, deltamethrin,	Brazil	--	360 (180 exposed, 180 unexposed) agricultural workers	• **%DNA tail:** exposed (18.4 ± 8.1%) vs. unexposed (15.8 ± 7.7%); **sig.**	[327] 10.1016/j.scitotenv. 2020.141893
Remor	2009	Fungicides, herbicides, inseticides	Brazil	ALA-D and BuChE activity—lower in exposed group	57 (37 exposed, 20 unexposed) agricultural workers	• **Damage index (0–400):** exposed (21.38 ± 14.80) vs. unexposed (3.10 ± 1.59); **sig.** • **% damage frequency:** exposed (16.38 ± 11.68) vs. unexposed (2.35 ± 1.31); **sig.**	[328] 10.1016/j.envint. 2008.06.011

Table 5. *Cont.*

Author	Year	Main Chemical Exposure	Country	Exposure Assessment [Mean Concentration Pesticides (ppm)] or Biomarkers of Exposure	Population Characteristics	DNA Damage	Reference
Rohr	2011	Bipyridyl, organophosphates, copper sulfate, carbamates	Brazil	Pesticide exposure assessment List of compounds used by the volunteers	173 (108 exposed, 65 unexposed) agricultural workers	• **Damage Index:** exposed (150.25 ± 60.84) vs. unexposed (31.37 ± 27.85); **sig.** • **Damage index (0–400):** exposed (20.26 ± 11.76) vs. unexposed (4.42 ± 5.85); **sig.** • **% damage frequency:** exposed (10.97 ± 3.76) vs. unexposed (1.91 ± 2.09); **sig.**	[329] 10.1002/em.20562
Saad-Hussein	2017	Malathion, chloropyrifos, dimethoate, carbofuran	Egypt	List of compounds commonly used in the area	101 (51 exposed, 50 unexposed) agricultural workers	• **Comet tail length—Median:** exposed (14.59, ranging from 2 to 37 µm) vs. unexposed (8.50, ranging from 1 to 19 µm); **sig.** • **Comet tail moment—Median:** exposed (0.73, ranging from 0.12 to 1.48 µm) vs. unexposed (0.08, ranging from 0.05 to 1.48 µm); **sig.** • **%DNA tail:** exposed (4.21%, ranging from 0.83 to 17.84) vs. unexposed (0.18%, ranging from 0.00 to 5.61); **sig.**	[330] 10.1016/j.mrgentox.2017.05.005

Table 5. *Cont.*

Author	Year	Main Chemical Exposure	Country	Exposure Assessment [Mean Concentration Pesticides (ppm)] or Biomarkers of Exposure	Population Characteristics	DNA Damage	Reference
Saad-Hussein	2019	Malathionchloropyrifos, dimethoate, carbofuran	Egypt	BuChE activity—U/L: rural exposed (2836 ± 189) vs. rural unexposed (3444.9 ± 148.4) vs. urban exposed (2653.2 ± 112.6) vs. urban unexposed (3040.8 ± 83.4)	200 (50 rural exposed, 50 urban exposed, 50 rural unexposed, 50 urban unexposed) agricultural workers	• **Comet tail length—Mean:** rural exposed (17.84 ± 1.07 µm) vs. rural unexposed (8.4 ± 0.72 µm) vs. urban exposed (16.95 ± 2.15 µm) vs. urban unexposed (7.55 ± 0.70 µm); **sig.** • **Comet tail moment—Mean:** rural exposed (0.73 ± 0.05 µm) vs. rural unexposed (0.08 ± 0.001 µm) vs. urban exposed (0.30 ± 0.05 µm) vs. urban unexposed (0.08 ± 0.002 µm); **sig.** • **%DNA tail:** rural exposed (4.57 ± 0.40%) vs. rural unexposed (0.84 ± 0.19%) vs. urban exposed (3.11 ± 0.54%) vs. urban unexposed (0.89 ± 0.21%); **sig.**	[331] 10.1016/j.mrgentox.2018.12.004
Sapbamrer	2019	Organophosphates, glyphosate, paraquat	Thailand	--	56 (all exposed) agricultural workers	• **Comet tail length—Median:** pre-application (5.66, ranging from 4.55 and 6.58 µm); post-application (5.67, ranging from 4.63 and 6.55 µm); **non-sig.** • **Comet tail moment—Median:** pre-application (2.84, ranging from 2.63 and 3.20 µm); pos-application (2.83, ranging from 2.66 and 3.27 µm); **non-sig.**	[332] 10.1007/s11356-019-04650-z

Table 5. *Cont.*

Author	Year	Main Chemical Exposure	Country	Exposure Assessment [Mean Concentration Pesticides (ppm)] or Biomarkers of Exposure	Population Characteristics	DNA Damage	Reference
Simoniello	2008	Thiophthalimide, inorganic-copper, dithiocarbamate-inorganic zinc, organophosphorus, carbamate, pyrethroid, organophosphorus, organochlorine, chloronicotinyl, phosphonoglycine	Argentina	List of compounds used by volunteers	84 (27 farmers, 27 pesticide workers, 30 unexposed) agricultural workers	• **Damage Index:** exposed farmers (221.66 ± 19.95) vs. exposed pesticide workers (215.29 ± 15.06) vs. unexposed (113.20 ± 13.68); **sig.**	[333] 10.1002/jat.1361
Simoniello	2010	Thiophthalimide, inorganic-copper, dithiocarbamate-inorganic zinc, organophosphorus, carbamate, pyrethroid, organophosphorus, organochlorine, chloronicotinyl, phosphonoglycine	Argentina	AchE activity—U/L: exposed farmers (7651.52 ± 2062.07) vs. exposed pesticide workers (6740.33 ± 1454.48) vs. unexposed (9045.54 ± 2191.56); $p < 0.05$ BuChE activity—U/L: exposed farmers (6313.86 ± 1268.26) vs. exposed pesticide workers (6777.77 ± 1281.84) vs. unexposed (6993.31 ± 1131.92); $p > 0.05$	. 123 (23 farmers, 18 pesticide workers, 82 unexposed) agricultural workers	• **Damage Index—Mean**: exposed farmers (224.73 ± 20.56) vs. exposed pesticide workers (212.94 ± 14.79) vs. unexposed (113.56 ± 16.01); **sig.**	[334] 10.3109/13547500903276378

Table 5. *Cont.*

Author	Year	Main Chemical Exposure	Country	Exposure Assessment [Mean Concentration Pesticides (ppm)] or Biomarkers of Exposure	Population Characteristics	DNA Damage	Reference
Singh	2011	Pirimiphos methyl, chlorpyrifos, temephos, malathion	India	AchE activity—KAU/L: exposed (3.45 ± 0.95) vs. unexposed (9.55 ± 0.35); $p < 0.001$ Pesticides exposure index	140 (70 exposed, 70 unexposed) pesticide-manufacturing workers	• **Comet tail moment—Median:** exposed (14.48 ± 2.40 µm) vs. unexposed (6.42 ± 1.42 µm); **sig.** • **%DNA tail:** exposed (60.43 ± 5.16) vs. unexposed (31.86 ± 6.35); **sig.**	[335] 10.1016/j.etap.2010.11.005
Singh	2011	Organophosphate	India	Pesticides exposure index	230 (115 exposed, 115 unexposed) pesticide-manufacturing workers	• **Comet tail moment—Median**: exposed (14.41 ± 2.25 µm) vs. unexposed (6.36 ± 1.41 µm); **sig.**	[336] 10.1016/j.mrgentox.2011.06.006
Singh	2012	Organophosphate	India	AchE activity—KAU/L: exposed (3.76 ± 1.06) vs. unexposed (9.33 ± 0.52); $p < 0.001$ PONase activity nmol/min/mL: exposed (180.97 ± 37.59) vs. unexposed (246.70 ± 43.23) Pesticides exposure index	268 (134 exposed, 134 unexposed), Community health agents	• **Comet tail moment—Median:** exposed (14.32 ± 2.17) vs. unexposed (6.24 ± 1.37); **sig.**	[337] 10.1016/j.mrgentox.2011.11.001
Singh	2011	Organophosphate	India	AchE activity—KAU/L: exposed (3.71 ± 1.04) vs. unexposed (9.33 ± 0.52); $p < 0.001$ PONase activity nmol/min/mL: exposed (181.76 ± 37.10) vs. unexposed (246.70 ± 43.24) Pesticides exposure index	284 (150 exposed, 134 unexposed) community health agents	• **Comet tail moment—Median:** exposed (14.37 ± 2.15) vs. unexposed (6.24 ± 1.37); **sig.**	[338] 10.1016/j.taap.2011.08.021

Table 5. *Cont.*

Author	Year	Main Chemical Exposure	Country	Exposure Assessment [Mean Concentration Pesticides (ppm)] or Biomarkers of Exposure	Population Characteristics	DNA Damage	Reference
Valencia-Quintana	2021	Organophosphate, carbamate, organochlorine. piretroides	Mexico	AchE activity—U/L: exposed (52.35 ± 10.04) vs. unexposed (35.32 ± 11.07); $p \leq 0.006$ BuChE activity—U/L: exposed (297.73 ± 60.78) vs. unexposed (231.76 ± 81.60); $p \leq 0.047$ List of compounds used by the volunteers	80 (54 exposed, 26 unexposed) agricultural workers	• **Comet tail length—Mean**: exposed (78.80 ± 25.00 µm) vs. unexposed (55.62 ± 13.88 µm); **sig.** • **Comet tail moment—Mean:** exposed (6.34 ± 5.02 µm) vs. unexposed (1.89 ± 1.24 µm); **sig.** • **Olive tail moment—Mean**: exposed (6.31 ± 9.73 µm) vs. unexposed (0.24 ± 1.18 µm); **sig.**	[339] 10.3390/ijerph18126269
Varona-Uribe	2016	Organochlorines, organophosphorus, carbamates, ethylenethiourea	Colombia	Blood/serum/urine concentrations: Organophosphorus (8 substances) range 0.56–21.05; Carbamates (2 substances) range 0.03–0.04; Dithiocarbamates (1 substance) 0.90; Organochlorines (14 substances) range 0.42–46.36	223 (all exposed) agricultural workers	• **Comet tail length—Median:** exposed (17.79, ranging from 3.24 and 232.83 µm). • **%DNA tail:** exposed (6.53%, ranging from 0.15% to 97.96%)	[340] 10.1080/19338244.2014.910489
Venkata	2017	Carbamates, organochlorine, organophosphorus, pyrethroid	India	AchE activity—U/L: exposed (1090.76 ± 71.28) vs. unexposed (1290.80 ± 78.68); $p = 0.02$ List of compounds used by the volunteers	212 (106 exposed, 106 unexposed) tea garden workers	• **Comet tail length—Mean:** exposed (15.61 ± 2.54 µm) vs. unexposed (7.40 ± 1.86 µm); **sig.**	[341] 10.1080/1354750X.2016.1252954

Table 5. *Cont.*

Author	Year	Main Chemical Exposure	Country	Exposure Assessment [Mean Concentration Pesticides (ppm)] or Biomarkers of Exposure	Population Characteristics	DNA Damage	Reference
Wilhelm	2015	Fungicides, herbicides, inseticides	Brazil	List of compounds commonly used in the area	74 (37 exposed, 37 unexposed) floriculturists	• **% DNA tail:** exposed (4.22 ± 3.89) vs. unexposed (1.51 ± 2.55); **sig.** • **Damage index:** exposed (4.73 ± 4.27) vs. unexposed (1.95 ± 3.88); **sig.**	[342] 10.1007/s11356-014-3959-4
Wong [¥]	2008	Organophosphates, carbamates, pyrethroid insecticides, fungicides, growth regulator	China (Taiwan)	List of pesticides used, area of use, and frequency of use	241 (62 low exposure, 73 high exposure, 106 unexposed) fruit growers	• **Comet tail moment—Mean:** low exposed (2.03 ± 0.05 μm) vs. high exposed (2.31 ± 0.06 μm) vs. unexposed (1.33 ± 0.03 μm); **sig.**	[343] 10.1016/j.mrgentox.2008.06.005
Yadav	2011	Organophosphates	India	List of compounds used by the volunteers	62 (33 exposed, 29 unexposed) agricultural workers	• **Comet tail length—Mean:** exposed (52.18 ± 3.74 μm) vs. unexposed (7.01 ± 1.47 μm); **sig.** • **Comet tail moment—Mean:** exposed (16.91 ± 2.14 μm) vs. unexposed (1.04 ± 0.32 μm); **sig.** • **Olive tail moment—Mean:** exposed (15.58 ± 1.57 μm) vs. unexposed (1.82 ± 0.32 μm); **sig.** • **% DNA in tail:** exposed (27.45 ± 1.64) vs. unexposed (9.04 ± 0.67); **sig.**	[344] 10.1080/09723757.2011.11886131

Table 5. *Cont.*

Author	Year	Main Chemical Exposure	Country	Exposure Assessment [Mean Concentration Pesticides (ppm)] or Biomarkers of Exposure	Population Characteristics	DNA Damage	Reference
Zepeda-Arce	2017	Organochlorines, carbamates, pyrethroids	Mexico	AchE—U/g Hb: moderate exposed (19.4) vs. high exposed (20.5) vs. unexposed (18.8); $p > 0.05$ BuChE—U/L: moderate exposed (5943.97) vs. high exposed (4333.2) vs. unexposed (6673.27); $p > 0.05$ MDA concentration (nmol/mL): moderate exposed (0.98) vs. high exposed (1.0) vs. unexposed (0.97); $p = 0.79$. Pesticides exposure assessment List of compounds used by the volunteers	208 (186 moderate exposure, 60 high exposure, 22 unexposed) agricultural workers	• **Comet tail moment—Median**: moderate exposed (7.8) vs. high exposed (9.8) vs. unexposed (7.5); **non-sig.** • **Olive tail moment—Median:** moderate exposed (2.9) vs. high exposed (3.4) vs. unexposed (2.8); **non-sig.**	[345] 10.1002/tox.22398
Želježić, Garaj-Vrhovac *	2001	Atrazine, alachlor, cyanazine, 2,4-dichlorophenoxyacetic acid, malathion	Croatia	--	40 (20 exposed, 20 unexposed) pesticide-manufacturing workers	• **Comet tail length—Mean**: exposed after high exposure period (50.1 ± 9.44 µm) vs. exposed after no exposure period (17.2 ± 0.44 µm) vs. unexposed (13.3 ± 1.47 µm); **sig.** • **Comet tail moment—Mean**: exposed after high exposure period (60.8 ± 18.17) vs. exposed after no exposure period (13.8 ± 0.39 µm) vs. unexposed (10.5 ± 1.13); **sig.**	[346] 10.1093/mutage/16.4.359

Table 5. *Cont.*

Author	Year	Main Chemical Exposure	Country	Exposure Assessment [Mean Concentration Pesticides (ppm)] or Biomarkers of Exposure	Population Characteristics	DNA Damage	Reference
				Environmental exposure			
Alvarado-Hernandez	2013	Organochlorine	Mexico	17 analysed pesticides (detection range 58–100% in maternal blood, and 66–100% in umbilical cord blood) Most abundant in maternal blood: Heptachlor epoxide: 3764 ng/g lipids; Oxychlordane: 1672 ng/g lipides; Beta-HCH: 1320 ng/g lipides. Most abundant in umbilical cord blood: Heptachlor epoxide: 8707 ng/g lipides; Oxychlordane: 1411 ng/g lipides; Beta-HCH: 2815 ng/g lipides.	50 mother–infant pairs, pregnant women and their infants from rural areas	• **Olive tail moment—Mean** maternal blood (7.36 ± 6.45 μm) vs. cord blood (8.87 ± 5.04); **sig.**	[347] 10.1002/em.21753

Table 5. *Cont.*

Author	Year	Main Chemical Exposure	Country	Exposure Assessment [Mean Concentration Pesticides (ppm)] or Biomarkers of Exposure	Population Characteristics	DNA Damage	Reference
Dwivedi	2022	Organochlorines	India	10 analysed pesticides: maximum concentration found for aldrin (3.26 mg/L) in maternal blood and dieldrin (2.69 mg/L) in cord blood	221 (104 preterm delivery, 117 full-term delivery) pregnant women and their infants from rural areas	• **Comet tail length—Mean** (maternal blood): larger preterm (18.29 ± 2.75 µm) vs. small preterm (16.42 ± 1.58 µm) vs. full-term appropriate for gestational age (8.10 ± 1.60 µm) vs. full-term small for gestational age (9.8 ± 2.31 µm); Mean (cord blood): larger preterm (14.64 ± 1.88 µm) vs. small preterm (12.12 ± 1.27 µm) vs. full-term appropriate for gestational age (7.40 ± 1.82 µm) vs. full-term small for gestational age (8.3 ± 1.52 µm); **sig.** • **Olive tail moment—Mean** (maternal blood): larger preterm (3.93 ± 0.52 µm) vs. small preterm (2.16 ± 0.81 µm) vs. full-term appropriate for gestational age (0.68 ± 0.31 µm) vs. full-term small for gestational age (0.99 ± 0.45 µm); mean (cord blood): larger preterm (2.81 ± 0.51 µm) vs. small preterm (1.05 ± 0.55 µm) vs. full-term appropriate for gestational age (0.55 ± 0.37 µm) vs. full-term small for gestational age (0.62 ± 0.35 µm); **sig.**	[348] 10.1016/j.envres.2021.112010

Table 5. *Cont.*

Author	Year	Main Chemical Exposure	Country	Exposure Assessment [Mean Concentration Pesticides (ppm)] or Biomarkers of Exposure	Population Characteristics	DNA Damage	Reference
How	2014	Organophosphates	Malaysia	Blood cholinesterase levels—unexposed (79.55 ± 13.48) vs. exposed (56.32 ± 12.35)	180 (95 exposed, 85 unexposed) children exposed lived < 2 km from paddy farmland	• **Comet tail length—Mean**: exposed (8.45 ± 3.89 µm) vs. unexposed (4.38 ± 1.66 µm); **sig.**	[349] 10.1080/1059924X.2013.866917
Kapka-Skrzypczak	2019	Carbetamide, carbofuran, chloridazon, dodemorph, cyclopropanecarboxamide, permethrin	Poland	Sweat pesticides (19 positive samples) for carbetamide, carbofuran, chloridazon, dodemorph, cyclopropanecarboxamide, permethrin AchE activity and BuChE activity significantly lower in exposed group	200 children (108 exposed, 92 unexposed), lived <1 km from the nearest orchards, cultivated fields, greenhouses	• **Comet tail length—Mean**: exposed (23.39 ± 8.26% in blood samples and 24.10 ± 8.43% in sweat-positive samples) vs. unexposed (19.84 ± 7.70%); **sig.** • **Mean FPG-sensitive sites**: exposed (7.30 ± 5.65% in blood samples and 4.79 ± 4.05% in sweat-positive samples) vs. unexposed (3.05 ± 4.05%); **sig.**	[350] 10.1016/j.mrgentox.2018.12.012
Leite	2019	--	Paraguay	Plasma cholinesterase activity did not differ among groups	84 children (43 exposed, 41 unexposed). Children exposed were born < 1 km from fumigated fields and have been living in that location for >5 years	• **Comet tail length—Mean**: exposed (59.1 µm) vs. unexposed (37.2 µm); **sig.** • **Comet tail moment—Mean**: exposed (32.8 µm) vs. unexposed (14.4 µm); **sig.** • **%DNA tail:** exposed (45.2%) vs. unexposed (27.6%) • **%DNA head:** exposed (54.8%) vs. unexposed (72.4%)	[351] 10.4103/ijmr.IJMR_1497_17
Sutris	2016	Dimethyphosphate, diethylphosphate, dimethylthiophosp, diethylthiophosph, dimethylthiophosph diethyldithiphosph	Malaysia	Urine organophosphate metabolites: 46.7% positive results: dimethyphosphate (46.7%), diethylphosphate (16.7%), dimethylthiophosphate (3.3%)	180 children (all exposed) living on agricultural island	• **Comet tail length—Median:** 37.1 (IQR 17.5 to 54.5) µm; pesticide-positive volunteers: 43.5 (30.9–68.1) µm vs. negative volunteers: 24.7 (9.5–48.1) µm; **sig.**	[352] 10.15171/ijoem.2016.705

*, §, ¥—updated studies from the same author/group of authors.

Regarding exposure assessment, it is important to highlight that the exposure assessment in the reviewed papers was highly heterogeneous. Only 12 studies (18.5%) had a good exposure assessment (including blood, urine, or skin analyses for pesticide residues), while around one-third (n = 21; 32.3%) presented a medium exposure assessment by evaluation of the enzymatic activities related to possible pesticide exposure (usually AchE or BuChE), or by using a model to predict the exposure. Almost half of the studies (n = 32; 49.2%) had no exposure assessment or simply provided a list of pesticides that volunteers might have been in contact with.

The effects measured by the comet assay were nearly consistent among studies (n = 63/65 reports; 96.9%), showing significantly higher DNA damage outcomes for the exposed populations. Only two papers did not find significant changes in these measures, both assessing agricultural workers either using a moderate- vs. high-exposure groups approach [345] or a before–after pesticide application design [320]. The descriptors used to express the comet assay data (one or more in the same study) were as follows: tail length in 33 studies, tail moment in 22 studies, % DNA in tail/tail intensity in 17 studies, DNA damage index in 13, olive tail moment in 11 studies, and other descriptors in 10 studies.

In summary, despite the high variability in the number of pesticides and classes of compounds (with different effects and mechanisms of action), the findings indicate that human populations exposed to pesticides have higher levels of DNA damage. However, the evaluation of exposure as well as the impact of the factors affecting the comet assay results (e.g., smoking, family history of cancer, other potential carcinogens exposure, UV exposure, and body mass index) [353] were scarcely considered.

3.6. Solvents

Organic solvents, such as benzene, toluene, and xylene (BTX), are a group of chemicals widely used in several occupational settings and are common components of air pollution (volatile organic compounds, VOCs) as a result of traffic and industry emissions. Although these substances are (highly volatile) ground-water contaminants, exposure occurs mainly via inhalation, either in occupational settings or through outdoor/indoor environments. Exposure to organic solvents, often in mixtures, is linked to different types of organ toxicities, such as neurological, hepatic, and respiratory [354–357]. Genotoxic effects of these substances have been repeatedly reported as attributable to the generation of oxidative stress and reactive metabolites able to form DNA adducts [358]. These mechanisms are also associated with carcinogenesis, and some organic solvents are well-known carcinogens: benzene is classified by the IARC as Group 1 (carcinogenic to humans), and styrene and perchloroethylene as Group 2A (probably carcinogenic to humans). Epidemiological studies reported an increased cancer risk for workers exposed to organic solvents, such as painters (sufficient evidence for mesothelioma and cancers of the urinary bladder and lung) [359] and shoemaking (leukaemia, nasal, and bladder cancer) [360] and petrochemical industry workers (mesothelioma, skin melanoma, multiple myeloma, and cancers of the prostate and urinary bladder) [361].

Our systematic scoping review identified 183 articles—180 from databases and 3 by manual entry, of which 75 were eliminated as duplicates. After the preliminary screening by title and abstract, 51 documents were excluded. From the articles eligible for full-text assessment, seven were excluded (mostly because they did not present comet assay data). A total of 50 studies were finally included in the qualitative analysis, as summarised in Figure 6 and Table 6.

Table 6. Summary of findings from the included studies on solvents.

Author	Year	Main Chemical Exposure	Country	Exposure Assessment or Biomarkers of Exposure	Population Characteristics	DNA Damage	Reference/DOI
Occupational exposure							
Al Zabadi **	2011	PAHs, VOCs	France	Air concentration PAH and benzene	64 sewage workers (34 exposed, 30 unexposed)	• **% DNA tail** (urine genotoxicity): exposed (8.07 ± 3.12) vs. unexposed (2.70 ± 0.58); **sig.**	[41] 10.1186/1476-069X-10-23
Azimi	2017	Perchloroethylene	Iran	--	59 dry cleaners (33 exposed, 26 unexposed)	• **% DNA tail** (lymphocytes): exposed (23.03; ranging 5.73 to 48.85) vs. unexposed (8.77; ranging 3.05 to 21.03); **sig.** • **Comet tail length:** exposed (25.85; ranging 6.63 to 67.2) vs. unexposed (5.61; ranging 2.65 to 18.53); **sig.** • **Comet tail moment:** exposed (7.07; ranging 0.42 to 44.29) vs. unexposed (1.03; ranging 0.14 to 5.12); **sig.**	[362] 10.15171/ ijoem.2017.1089
Buschini	2003	Styrene	Italy	Passive air samplers (TWA8h) Urinary excretion of MA and PGA	62 workers in polyester resins and fibreglass-reinforced plastics factories (48 exposed, 14 unexposed)	• **Comet tail moment** (peripheral WBC): unexposed (TM 7.4 ± 0.5, TM99 12.4 ± 4.9) vs. exposed (TM7.8 ± 0.8, TM99 34.1 ± 14.0); **sig.**	[363] 10.1002/em.10150
Careree **	2002	Benzene and other aromatic hydrocarbons	Italy	Passive air samplers (TWA7h)	190 traffic policemen (133 exposed, 57 unexposed)	• **Comet tail moment** (PBMNC) in subgroups by sex and smoking status: exposed (0.46 ± 0.46) vs. controls (0.36 ± 0.32); **non-sig.**	[49] 10.1016/s1383-5718(02)00108-0

Table 6. *Cont.*

Author	Year	Main Chemical Exposure	Country	Exposure Assessment or Biomarkers of Exposure	Population Characteristics	DNA Damage	Reference/DOI
Cassini	2011	Paint complex mixtures	Brazil	--	62 painters (33 exposed, 29 unexposed)	• **DNA damage** (Arbitrary Units, WBC): unexposed (30.11 ± 2.08) vs. exposed (71.42 ± 2.77); **sig.**	[364] 10.2478/s13382-011-0030-2
Cavallo	2018	Styrene	Italy	Passive air samplers (4–7 h) Urinary excretion of MA and PGA	39 workers in fibreglass-reinforced plastics factories (11 workers on open moulding plastic process, 16 workers on closed moulding plastic process, 12 controls)	• **Comet Tail moment** (lymphocyte SBs): all workers (6.11 ± 3.16) vs. controls (8.53 ± 2.49); **non-sig.**	[365] 10.1016/j.toxlet.2018.06.006
Cavallo	2021	VOC	Italy	Personal VOCs exposure Urinary VOCs metabolites	35 (17 shipyard painters, 18 unexposed)	• **% DNA tail** (lymphocytes): exposed (17.68 ± 4.35) vs. unexposed (11.56 ± 2.62); **sig.**	[366] 10.3390/ijerph18094645
Cok	2004	Toluene, other VOCs	Turkey	Urinary hippuric acid and o-cresol	40 (20 male glue sniffers, 20 smoking habit matched controls)	• **Total Comet score (visual)** (lymphocytes): exposed (142.45 ± 9.61) vs. controls (103.30 ± 2.81); **sig.**	[367] 10.1016/j.mrgentox.2003.10.009
Costa	2012	Styrene	Portugal	Styrene in workplace air Urinary mandelic and phenylglyoxylic acids	152 (75 workers from a fibreglass factory, 77 unexposed)	• **Comet tail length (PBMNC):** exposed (49.39 ± 0.84) vs. unexposed (47.43 ± 0.52); **sig.**	[368] 10.1080/15287394.2012.688488
Costa-Amaral	2019	Benzene	Brazil	Benzene and toluene in air Urinary excretion of MA and S-PMA	86 (51 employees of filling stations, 35 controls)	• **% DNA tail (leukocytes):** exposed (21.34 ± 20.32) vs. controls (28.73 ± 17.72); non-**sig.**	[369] 10.3390/ijerph16122240

Table 6. *Cont.*

Author	Year	Main Chemical Exposure	Country	Exposure Assessment or Biomarkers of Exposure	Population Characteristics	DNA Damage	Reference/DOI
de Aquino	2016	Xylene, other organic solvents	Brazil	--	29 technicians in pathology laboratory (18 exposed, 11 unexposed)	• **DNA damage** (Arbitrary Units, WBC): exposed (19.61 ± 7.95) vs. unexposed (8.36 ± 6.47); **sig.**	[370] 10.1590/0001-3765201620150194
Everatt **	2013	Perchloroethylene	Lithuania	PCE concentration in air: 31.40 ± 23.51	59 dry cleaning workers (30 exposed, 29 unexposed)	• **Comet tail length** (lymphocytes): exposed (10.45 ± 6.52) vs. unexposed (5.77 ± 2.31); **sig.**	[66] 10.1080/15459624.2013.818238
Fracasso	2010	Benzene	Italy	Personal passive air samplers Urinary excretion of MA and S-PMA	133 (33 petrochemical industry operators, 28 service station staff, 21 gasoline pump staff, 51 unexposed)	• **Comet tail intensity** (lymphocytes): exposed (2.78 ± 0.92) vs. unexposed (2.26 ± 0.56); **sig.**	[371] 10.1016/j.toxlet.2009.04.028
Fracasso	2009	Styrene	Italy	Personal passive air samplers Urinary excretion of MA and S-PMA	63 workers in fibreglass-reinforced plastics factories (34 exposed, 29 unexposed)	• **Comet tail length** (lymphocytes): exposed (3.47 ± 1.14) vs. unexposed (2.44 ± 0.48); **sig.**	[372] 10.1016/j.toxlet.2008.11.010
Godderis	2004	Styrene	Belgium	Urinary mandelic acid: 201.57 mg/g creatinine ± 148.32 in exposed workers	88 workers in fibreglass-reinforced plastics factories (44 exposed, 44 unexposed)	• **% DNA tail** (PBMNC): exposed (0.80 ± 0.31) vs. unexposed (0.80 ± 0.34); **non-sig.**	[373] 10.1002/em.20069
Göethel **	2014	Benzene and CO	Brazil	Urinary t,t-muconic acid (t,t-MA) and 8OhdG Carboxyhaemoglobin (COHb) in whole blood	99 (43 gas station staff, 34 drivers, 22 unexposed)	• **DNA damage index** (Arbitrary Units): gas station staff (89.8 ± 21.5) vs. drivers (94.2 ± 12.8) vs. unexposed (48.6 ± 35.9); **sig.**	[70] 10.1016/j.mrgentox.2014.05.008

Table 6. *Cont.*

Author	Year	Main Chemical Exposure	Country	Exposure Assessment or Biomarkers of Exposure	Population Characteristics	DNA Damage	Reference/DOI
Hanova	2010	Styrene	Czechia	Styrene concentration at workplace and in blood	122 hand lamination workers in a plastics factory (71 exposed, 51 unexposed)	• **Comet assay** (lymphocytes): 1.20 ± 0.70 SSB/109 Da, subjects exposed to low: 0.77 ± 0.39 SSB/109 Da, and high: 0.51 ± 0.41 SSB/109; **sig. but negative effect**	[374] 10.1016/ j.taap.2010.07.027
Heuser	2005	Toluene, n-hexane, acetone, MEK	Brazil	Urinary hippuric acid	70 (29 solvent-based adhesive workers, 16 water-based adhesive workers, 25 controls)	• **DNA damage** (Arbitrary Units, lymphocytes): exposed (8.46 ± 7.79) vs. controls (2.82 ± 2.87); **sig.**	[375] 10.1016/j.mrgentox. 2005.03.002
Heuser	2007	Toluene, n-hexane, acetone, MEK	Brazil	Urinary hippuric acid	94 footwear workers (39 exposed, 55 unexposed)	• **DNA damage** (Arbitrary Units, lymphocytes): exposed (2.13 ± 2.45 and 8.35 ± 7.85) vs. controls (3.44 ± 3.24); **sig.**	[376] 10.1016/ j.tox.2007.01.011
Keretetse	2008	BTX	South Africa	Air samplers (TWA)	40 (20 petrol station staff, 20 controls)	• **Comet tail intensity** (lymphocytes): exposed (15.06 ± 9.10) vs. unexposed (6.30 ± 3.37); **sig.**	[377] 10.1093/annhyg/ men047
Ladeira	2020	Styrene, xylene	Portugal	Styrene and xylene air-monitoring campaigns (NIOSH 1501)	34 workers in polymer producing factory (17 exposed, 17 unexposed)	• **% DNA tail** (PBMNC): exposed (23.83 ± 20.84) vs. unexposed (5.99 ± 5.01); **sig.**	[378] 10.1016/ j.yrtph.2020.104726
Laffon	2002	Styrene	Spain	Urinary mandelic acid: average exposures of 16.76 ± 5.9, 17.51 ± 4.64, 19.33 ± 9.95 ppm)	44 workers in fiberglass-reinforced plastics factory (14 exposed, 30 unexposed)	• **Comet tail length** (PBMNC): exposed (48.68 ± 0.33) vs. unexposed (43.34 ± 0.18); **sig.**	[379] 10.1016/s0300-483x(01)00572-8

Table 6. *Cont.*

Author	Year	Main Chemical Exposure	Country	Exposure Assessment or Biomarkers of Exposure	Population Characteristics	DNA Damage	Reference/DOI
Lam	2002	Benzene	China	--	718 workers in elevator manufacturing factory (359 workers manufacturing, 205 department staff, 154 controls)	• **Tail moment** (lymphocytes): non-exposed 0.53 (0.49–0.56), exposed: 0.74 (0.68–0.80); **sig.**	[380] 10.1016/s1383-5718(02)00010-4
Li	2017	Benzene, toluene	China	Air levels of benzene and toluene Urinary S-phenylmercapturic acid (SPMA) and S-benzylmercapturic acid (SBMA)	196 (96 petrochemical staff, 100 controls)	• **% DNA tail** (WBC): exposed (6.51 ± 2.03) vs. controls (5.84 ± 2.24); **sig.**	[381] 10.1080/1354750X.2016.1274335
Londoño-Velasco	2016	Organic solvents	Spain	--	104 (52 painters, 52 unexposed)	• **% DNA tail** (lymphocytes): exposed (11.09 ± 0.65) vs. unexposed (7.29 ± 0.31); **sig.**	[382] 10.3109/15376516.2016.1158892
Martino-Roth	2003	Organic solvents, lead	Brazil	--	40 (10 car painters, 10 storage staff, 20 controls)	• **Comet tail length** (buccal cells): car painters (33.85± 0.507) vs. matched controls (30.73 ± 0.162) vs. storage staff (34.18 ± 0.484) vs. matched controls (30.54 ± 0.136); **sig.**	[383]
Migliore ¥	2006	Styrene	Italy	Urinary excretion styrene metabolites, mandelic, and phenylglyoxylic acids (MAPGA)	67 workers in fibreglass-reinforced plastics factory (42 exposed, 25 unexposed)	• **% DNA tail** (sperm): exposed (11.02 ± 2.99) vs. unexposed (7.42 ± 2.30); **sig.**	[384] 10.1093/mutage/gel012

Table 6. *Cont.*

Author	Year	Main Chemical Exposure	Country	Exposure Assessment or Biomarkers of Exposure	Population Characteristics	DNA Damage	Reference/DOI
Migliore [¥]	2002	Styrene	Italy	Urinary concentration of mandelic acid (MA)	73 workers in fibreglass-reinforced plastics factory (46 exposed, 27 unexposed)	• **% DNA tail** (sperm): exposed (10.09 ± 3.0) vs. unexposed (7.4 ± 2.30); **sig.** • **Olive tail moment:** exposed (1.5 ± 0.6) vs. unexposed (0.8 ± 0.4); **sig.**	[385] 10.1093/humrep/17.11.2912
Moro	2012	Toluene	Brazil	Urinary levels of hippuric acid (HA)	61 painters (34 exposed, 27 unexposed)	• **Damage index** (visual score, WBC): exposed (60.4 ± 3.6) vs. unexposed (39.4 ± 2.5); **sig.**	[386] 10.1016/j.mrgentox.2012.02.007
Navasumrit	2005	Benzene	Thailand	Personal benzene exposure by diffusive badges Urinary metabolites, blood benzene	148 (28 children in Chonburi, 41 children in Bangkok, 29 gasoline service staff in Bangkok, 23 factory staff, 27 controls)	• **Olive tail moment** (WBC): children Chonburi (0.13 ± 0.01) vs. children Bangkok (0.22 ± 0.01); **sig.** • **Olive tail moment** (WBC): gasoline service (0.24 ± 0.01) vs. factory staff (0.44 ± 0.06) vs. controls (0.24 ± 0.01); **sig.**	[387] 10.1016/j.cbi.2005.03.010
Pandey	2008	BTX	India	Benzene monitoring in air Benzene, toluene, and xylene in blood samples	200 petrol pump workers (100 exposed, 100 unexposed)	• **% DNA tail** (lymphocytes): exposed (11.92 ± 2.74) vs. unexposed (7.79 ± 1.17); **sig.** • **Comet tail length:** exposed (54.61 ± 7.81) vs. unexposed (50.33 ± 9.83); **sig.**	[388] 10.1002/em.20419

Table 6. *Cont.*

Author	Year	Main Chemical Exposure	Country	Exposure Assessment or Biomarkers of Exposure	Population Characteristics	DNA Damage	Reference/DOI
Poça	2021	Benzene in gasoline	Brazil	Urinary t,t-muconic acid	349 (154 exposed filling station workers, 95 convenience store workers, 100 unexposed office workers)	• **DNA damage** (Whole blood): the filling and convenience store workers had significantly higher DNA damage (Class 1) than the comparison group ($p \leq 0.001$); **sig.** • **Comet assay** (Whole blood): office workers [class 0 (96.00), class 1 (4.00), class 2 (0.00), class 3 (0.00)]; convenience store workers [class 0 (94.00), class 1 (5.33), class 2 (0.00), class 3 (0.00)]; filling station [class 0 (94.67), class 1 (5.33), class 2 (0.00), class 3 (0.00)]	[389] 10.1016/j.mrgentox.2021.503322
Rekhadevi	2010	BTX	India	Monitoring of ambient and breathing zone air BTX in blood	400 (200 fuel station staff, 200 controls)	• **Tail length WBC:** exposed (25.10 ± 2.28) vs. controls (10.27 ± 1.52); **sig.**	[390] 10.1093/annhyg/meq065
Roma-Torres	2006	BTX	Portugal	Urinary t,t-Muconic acid (t,t-MA), hippuric acid (HA), and methylhippuric acid (MHA)	78 (48 petroleum unit workers, 30 controls)	• **Comet tail length** (WBC): exposed (52.90 ± 0.85) vs. controls (48.09 ± 0.74); **sig.**	[391] 10.1016/j.mrgentox.2005.12.005
Sakhvidi	2022	Benzene found in petroleum compounds	Iran	Air sampling for benzene	32 petroleum products workers exposed to benzene, 32 non-exposed administrative	• **Tail length (TL), tail density (TD), tail momentum (TM), percentage of tail in the DNA (%DNA), and %TAC** (WBC): in control group were 78.59, 8.35, 1.20, 10.05, and 25.58, and in the exposure group were 59.21, 75.74, 57.74, 3.5, and 16.58, respectively; **sig.**	[392] 10.1007/s11356-022-19015-2

Table 6. *Cont.*

Author	Year	Main Chemical Exposure	Country	Exposure Assessment or Biomarkers of Exposure	Population Characteristics	DNA Damage	Reference/DOI
Sardas	2010	Welding fume, solvent base paint	Turkey	--	78 (26 welders, 26 painters, 26 controls)	• **% DNA tail** (lymphocytes): all exposed (12.34 ± 2.05) vs. controls (6.64 ± 1.43); $p < 0.05$ • **% DNA tail:** welders (13.59 ± 1.89) vs. painters (11.10 ± 1.35); **sig.**	[96] 10.1177/0748233710374463
Scheepers **	2002	Diesel exhaust (benzene, PAHs)	Estonia, Czech Republic	Analysis of air samples Urinary metabolites of PAH and benzene	92 underground miners (drivers of diesel-powered excavators) (46 underground workers, 46 surface workers)	• **DNA damage** (lymphocytes, visual scoring): underground workers (134) vs. surface workers (104); **non-sig.**	[97] 10.1016/s0378-4274(02)00195-9
Sul *	2002	Benzene	South Korea	Urinary t,t-muconic acid (t,t-MA), and creatinine	81 printing factory (41 exposed, 41 unexposed)	• **Olive tail moment** (lymphocytes and granulocytes): exposed (1.75 ± 0.29) vs. unexposed (1.47 ± 0.41); **sig.** • **Comet tail moment** (lymphocytes): exposed (3.86 ± 0.71) vs. unexposed (1.51 ± 0.39); **sig.** • **Comet tail moment** (granulocytes): exposed (3.61 ± 0.75) vs. unexposed (2.60 ± 0.59); **sig.**	[393] 10.1016/s0378-4274(02)00167-4
Sul *	2005	Benzene	South Korea	Personal sampler benzene Urinary trans, trans-muconic acid (t,t-MA), phenol, creatinine	61 subjects (working in printing, shoemaking, production of methylene di-aniline (MDA), nitrobenzene, carbomer, and benzene)	• **Olive tail moment**: 1.73 ± 0.81 • **Correlation** levels of benzene/DNA damage in lymphocyte of workers; **sig.**	[394] 10.1016/j.mrgentox.2004.12.011

Table 6. *Cont.*

Author	Year	Main Chemical Exposure	Country	Exposure Assessment or Biomarkers of Exposure	Population Characteristics	DNA Damage	Reference/DOI
Teixeira	2010	Styrene	Portugal	Styrene in inhaled air Urinary excretion styrene metabolites, mandelic, and phenylglyoxylic acids (MAPGA)	106 (52 fibreglass workers, 54 controls)	• **Comet tail length** (PBMNC): exposed (49.20 ± 0.93) vs. controls (47.64 ± 0.64); **non-sig.**	[395] 10.1093/mutage/geq049
Tovalin **	2006	VOCs, PM2.5, ozone	Mexico	Personal occupational and non-occupational monitoring	55 city traffic exposure (28 outdoor workers, 27 indoor workers)	• **Comet tail length** (WBC): outdoor workers (median 46.80 [maximum 132.41]) vs. indoor workers (median 30.11 [maximum 51.47]); **sig.**	[104] 10.1136/oem.2005.019802
Xiong	2016	Benzene, toluene, ethylbenzene, and xylenes (BTEX)	China	Air sampling	252 gas station workers (200 refueling workers, 52 controls)	• **Comet tail moment** (lymphocytes): exposed (0.094 [0.045–0.215]) vs. controls (0.064 [0.027–0.113]); **sig.**	[396] 10.3390/ijerph13121212
Zhao	2017	Benzene, acetone, xylene, toluene, lead, isopropanol, and physical factors	China	Air sampling	722 workers in electronics factory (584 exposed, 138 controls)	• **% DNA tail** (peripheral blood): lead+high temp (12.06 ± 17.89) vs. isopropanol (20.15 ± 15.41) vs. controls (6.36); **sig.**	[397] 10.1016/j.mrfmmm.2017.07.005
Environmental exposure							
Avogbe **	2005	Benzene, ultrafine particles	Benin	Ambient UFP Urinary excretion of S-PMA	135 city traffic exposure (29 drivers, 37 roadside residents, 42 suburban, 27 rural)	• **% DNA tail** (PBMNC): drivers (6.09 ± 3.46) vs. roadside residents (6.32 ± 4.00) vs. suburban (5.42 ± 2.28) vs. rural (4.26 ± 1.76); **sig.**	[121] 10.1093/carcin/bgh353
Koppen **	2007	PAHs, VOCs (benzene and toluene)	Belgium	Outdoor ozone concentrations Urinary concentrations of PAH, t,t′-muconic acid, o-cresol, VOCs metabolites	200 adolescents air pollution	• **% DNA tail** (WBC): 1.16 ± 0.51 • **Correlation** DNA damage/o-cresol and OHpyrene; **sig.**	[138] 10.1002/jat.1174

Table 6. *Cont.*

Author	Year	Main Chemical Exposure	Country	Exposure Assessment or Biomarkers of Exposure	Population Characteristics	DNA Damage	Reference/DOI
Mukherjee **	2013	Particulate pollutants and benzene	India	Urinary trans, trans-muconic acid	105 (56 biomass users, 49 cleaner liquefied petroleum gas users)	• **% DNA tail** (sputum cells): biomass users (36.2 ± 9.4) vs. gas users (9.0 ± 4.1); **sig.** • **Comet tail** length (sputum cells): biomass users (44.2 ± 6.0) vs. gas users (32.3 ± 7.3); **sig.** • **Olive tail moment** (sputum cells): biomass users (6.2 ± 2.2) vs. gas users (1.2 ± 0.5); **sig.**	[144] 10.1002/jat.1748
Pelallo-Martínez **,ɣ	2014	Lead, benzene, toluene, PAHs	Mexico	Urinary and blood Pb, benzene, toluene, PAHs	97 children, air pollution (44 Allende, 37 Nuevo Mundo, 16 Lopez Mateos)	• **Olive tail moment** (WBC): Allende (8.3 [3.1–16.8]) vs. Nuevo Mundo (10.6 [5.6–22.9]) vs. Lopez Mateos (11.7 [7.4–15.9]); **sig.**	[149] 10.1007/s00244-014-9999-4
Sørensen	2003	Benzene	Denmark	Exposure benzene, toluene, MTBE 8-oxodG in blood Urinary ttMA, S-PMA	40 subjects, air pollution	• **Visual score** (lymphocytes): 13.0 (7.0–21.5) • **No correlation** comet/exposure	[398] 10.1016/S0048-9697(03)00054-8
Wilhelm **,ɣ	2007	PAH, benzene, heavy metals	Germany	Monitored ambient air quality data Urinary (PAH) metabolites, benzene metabolites	935 air pollution close to industrial settings (620 exposed children, 315 unexposed)	• **Comet tail moment** (lymphocytes): —percentile 50: exposed (1.99) vs. unexposed (1.32); **sig.** • **Comet tail moment**—percentile 90: exposed (6.69) vs. unexposed (1.89); **non-sig.**	[160] 10.1016/j.ijheh.2007.02.007
Zani **	2020	PM10, PM2.5, NO_2, CO, SO_2, benzene, O_3	Italy	Air monitoring by regional agency	152 children, air pollution	**Saliva leukocytes from sputum** • **Comet tail intensity:** 6.2 ± 4.3 • **Visual score**: 182.1 ± 30.9; **non-sig.**	[162] 10.3390/ijerph17093276

PBMNC—Peripheral blood mononuclear cells. WBC—Whole blood cells. ɣ Updated studies from the same author/group of authors. * We noted that the studies have most likely been conducted on partly overlapping samples of benzene-exposed workers in a printing company (4 out of 41 samples from the first study appear to have been included in the second). The references are counted as separate studies; ** studies also in air pollution table; ɣ studies also in heavy metals table.

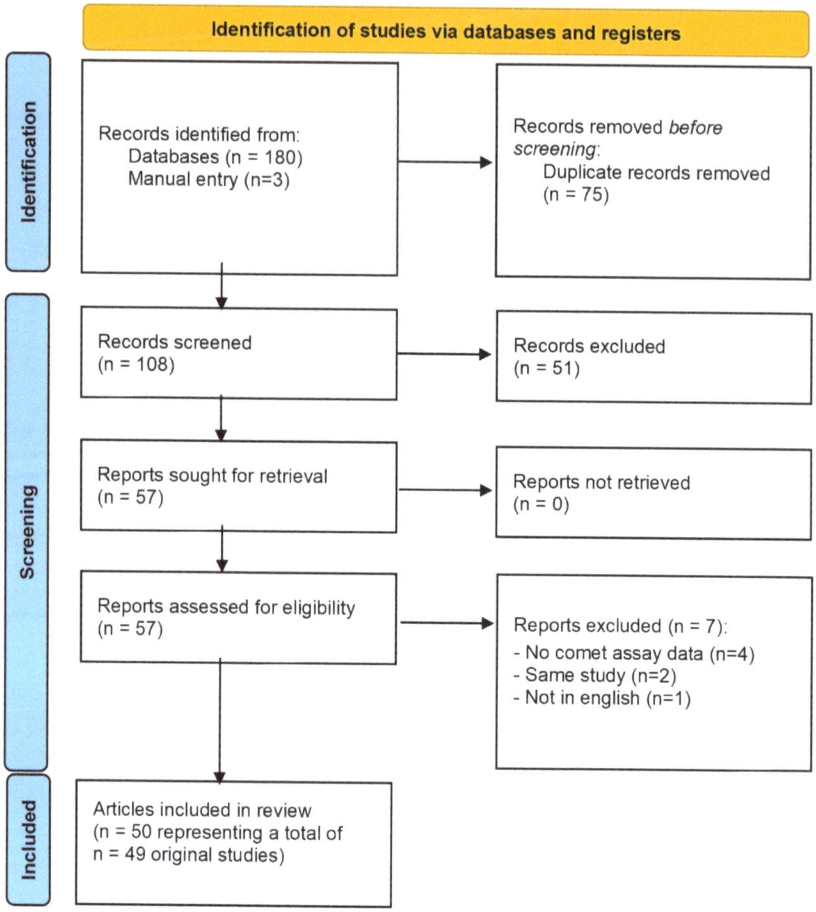

Figure 6. PRISMA flow diagram of systematic scoping review for solvents.

The studies mostly focused on occupational exposure to organic solvents, namely benzene, toluene, xylenes, ethylbenzene, styrene, perchloroethylene, and isopropyl alcohol. In many cases, subjects were exposed to mixtures of different organic solvents or mixtures of solvents and other toxicants such as heavy metals, PAHs, or pesticides. Around 40% of the studies (n = 20) evaluated workers in factories (plastics, polymers, shoemaking and others) [96,97,363,365,368,372–376,378–380,384,385,387,393–395,397], a quarter (n = 12; 24.0%) assessed gas station and petrochemical industry workers [70,369,371,377,381,387–392,396] and fewer studies addressed painters (n = 6; 12.0%) [96,364,366,382,383,386], dry cleaners (n = 2; 4.0%) [66,362], biomedical laboratory workers (n = 1; 2.0%) [385], sewage workers (n = 1; 2.0%) [41], and employees in biomass fuel burning (n = 1; 2.0%) [144]. Nine studies (18.0%) evaluated exposure to pollutants in adults [49,104,121,398] or in adolescents/children [138,149,160,162,387] and one (2.0%) in glue sniffers [367]. Around half of the studies were conducted in Europe (n = 23; 46.0%), one-third in Asia (n = 14; 28.0%) and around 22% in Southern America (n = 11); only two studies were performed in Africa (4.0%).

All the studies were observational, and most of them used a cross-sectional design comparing the exposed and non-exposed subjects. Only a few studies (n = 3; 6.1%) evaluated the correlation between DNA damage and exposure markers in the exposed subjects [138,394,398].

Overall, 43 studies (86.0%) used either environmental or biological monitoring of exposure or both. Studies with exposure evaluation by questionnaire (n = 7; 14.0%) [96,362, 364,370,380,382,383] were considered as limited regarding evidence. A significant increase in DNA damage in subjects exposed to solvents, or a positive correlation between DNA damage and exposure markers, was reported in 41 studies (82.0%) [of which 7 were limited based on the exposure evaluation], whereas in 8 studies (16.3%), the authors did not find any effect; in 1 paper (2.0%) a significant decrease in DNA damage was observed in the exposed subjects [374].

All of the studies reviewed took into consideration participants' age and sex matching or a correction for variables in their analysis (19 were restricted to male subjects, and 2 to female participants). In the majority of the included studies (n = 48, 96.0%), a smoking habit was considered as a confounding factor, or the study was conducted in non-smokers, with the exception of two studies [384,392] that did not consider smoking. Alcohol drinking was evaluated in 13 (26.0%) studies. With the exception of Azimi [362], statistical power calculations were not presented.

The parameters used to express the comet assay data (one or more in the same study) were as follows: % DNA in the tail was used in 19, tail moment in 13, tail length in 13, and visual scoring in 9 papers. The cells used for biomonitoring were mostly blood cells, with saliva leukocytes from sputum in two cases [144,162]. In one study, urine genotoxicity was assessed [41], and in another, buccal cells were used to monitor exposure in car painters [383], while two studies focused on sperm DNA in workers in plastic factories [384,385].

In summary, the synthetised evidence from 50 studies confirms the positive effect of solvent exposure (different types/mixtures) on DNA damage (both in adults and children/adolescents) measured by the comet assay in sentinel cells. However, further well-designed observational studies properly accounting for confounding variables are still needed.

4. Considerations

This broad systematic scoping review provides a critical assessment of the available evidence on the use of the comet assay in human biomonitoring, based on 334 different primary studies on the genotoxic effects from occupational or environmental exposures to six major groups of chemical substances (i.e., air pollutants, anaesthetics, antineoplastic drugs, heavy metals, pesticides, and solvents). In general, the information gathered in this scoping systematic review shows that the comet assay can be a good candidate to provide reliable information for health risk evaluations; and the volume of publications that applied this methodology contributes to its validation.

The comet assay has, in fact, become an important method in the field of bio-assaying to assess genetic damage in a great variety of cells in exposed populations. Historically, peripheral blood mononuclear cells (PBMNCs), mainly represented by lymphocytes, have been regarded as long-living sentinel cells [399], which are useful for detecting past exposures to genotoxic compounds and are widely used in human biomonitoring studies [400]. Lately, whole blood preparations containing all leukocytes have been increasingly used in spite of their lower cellular homogeneity, as they do not involve cell isolation procedures and can be readily and safely stored frozen [17]. Moreover, there is already a substantial number of studies of exfoliated buccal cells obtained by a minimally invasive method. The comet assay is recommended for monitoring populations chronically exposed to genotoxic agents, combined with the cytokinesis-blocked micronucleus assay [16,203], since the first identifies injuries resulting from a recent exposure (over the previous few weeks), which are still reparable, such as single- and double-strand DNA breaks, alkali labile lesions converted to strand breaks under alkaline conditions, and single-strand breaks associated with incomplete excision repair sites [12,18,401,402]. It is highly desirable that each laboratory should set up and implement standard operating procedures for experimental protocols, manipulation of samples, and analyses [12,18,401,402]. To facilitate this, a compendium of comet assay protocols for the analysis of different types of samples was recently published [14].

The results of this systematic scoping review indicate that, in general, for all the groups of chemicals included, for both occupational and environmental exposure, increased levels of DNA damage are seen in subjects exposed in comparison to the non-exposed subjects, with a majority of statistically significant results. There is great heterogeneity in the assessment of exposure-outcome association, with a preponderance of studies with a lack of exposure assessment and/or biomarkers of exposure and accountability of confounding variables scarcely considered, which fits with the underuse of exposure assessment tools [403].

Human biomonitoring provides additional information, which can contribute to a more accurate risk assessment at the individual and/or group level. With respect to occupational exposure and the biomonitoring of workers, the scenario is clearer, and three main goals can be drafted as follows: the first is an individual or collective exposure assessment, the second is health protection, and the ultimate objective is an occupational health risk assessment [404].

Biomonitoring tools provide information for several actions related to occupational health interventions, such as the following: determining if a specific exposure has occurred and if it implies a risk to workers' health; providing knowledge of exposure by all possible exposure routes; realising if health outcomes can be expected from exposure; helping to clarify the results from clinical testing in some circumstances; recognising the adequacy of control measures in place; helping to demonstrate the link between occupational exposure and a health effect [405]; and ultimately supporting health monitoring and surveillance programmes [406].

Emphasis should be given to monitoring populations which—at the environmental and/or occupational levels—are known to be exposed to hazardous substances, and to providing reliable health risk evaluations. This information can also be used to support regulations on environmental protection and/or to define limits in occupational settings. However, it is important to point out a critical issue in the application of any predictive biomarker in public health policies involving environmental and/or occupational exposures, namely, the meaning of the differing levels of predictive biomarkers at an individual level versus a group level. The latter (conservative approach) considers risk prediction to be valid only at a group level, allowing the effect of inter-individual variability and variability due to technical parameters being neglected [407]. The other (progressive approach) advocates that variability is a fundamental source of information, allowing the application of preventive measures in subsets of high-risk subjects. The other crucial aspect of predictive biomarkers is validation. A biomarker must be validated before it can be used for health risk assessments, especially as far as regulatory aspects are concerned.

The biomonitoring studies provide results on the associations between exposures and genotoxicity. There is an over-representation of studies with statistically significant increases in DNA damage in exposed subjects. Many studies use relatively simple statistical analyses such as ANOVA (or Student's *t*-test) or the corresponding non-parametric tests (i.e., Kruskal–Wallis and Mann–Whitney U tests). The smallest studies have roughly group sizes of 20–30 subjects, whereas the largest studies have more than a hundred subjects in each exposure group. A conservative estimate indicates that a group size of 40 subjects is necessary to obtain a statistically significant two-fold difference between two groups if the coefficient of variation in each group is 100% ($\alpha = 0.05$, $\beta = 0.80$, calculated in Stata version 15, StataCorp, College Station, TX, USA). Correction for confounding by multivariate analyses decreases the statistical power, implying that more subjects are required to obtain the same statistical significance as with a crude analysis (i.e., adjusted analyses decrease the effect size in cases of classical confounding). However, some studies in the database also make use of confounders in stratified analyses of genotoxicity, such as the genotoxic effects of exposure in the strata of non-smokers and smokers. Statistical planning before conducting studies on the interactions between host factors and exposures requires knowledge of the anticipated effects of both factors. In addition, it is important to consider both the intra- and inter-individual variations when assessing the statistical power of studies on comet assay endpoints. Inter-individual variation is relatively easy to assess as the difference between levels of DNA damage; coefficient of variation values range between 10% and 100% in different biomonitoring studies and larger studies typically have larger variations than small studies. The lower variation in small studies is most likely due to less effect of the between-day variation in the comet assay, which is an important contributor to the overall variation. The relatively large between-day variation in the comet assay also increases the uncertainty of the intra-individual variation assessments because it contributes to the overall variation if the samples are isolated and analysed on different days. The alternative—specimens are stored and analysed in the same batch—entails uncertainty about the stability of stored samples for the comet assay and/or whether, for instance, the freezing/thawing of samples affects DNA damage in case cryopreservation is used to store the samples. Given the current knowledge of the sources of variability in the comet assay, a conservative approach is that the magnitude of the intra- and inter-individual variations are similar, and both of these contributors are smaller than the between-day variation in the comet assay. Therefore, it may be relevant to use block designs when analysing samples in biomonitoring studies. This can be accomplished by analysing matched samples in the same comet assay experiment in biomonitoring studies where individual or group matching has been used in the study design.

Our study has some limitations. No quantitative analyses or further in-depth comparisons among studies were possible given the heterogeneity of data from the different study designs and the lack of studies properly reporting outcomes measurements and units. Moreover, most studies have a small number of subjects, rendering them insufficiently powerful to tease out the statistical effects of individual chemicals in complex mixtures, which is often the case in human biomonitoring studies. The absence of a core outcome set or standardised reporting of data [408] using the comet assay may contribute to selective bias and a loss of information and may impair evidence gathering on the effects of occupational or environmental exposures to different types of substances in different populations. Yet, although the results are only exploratory, a systematic and critical review process was followed in our study; the data summarised by means of tables support the development of further research in this field. It should be noted that the findings and conclusions of the studies were considered as presented by the authors, meaning that the results cannot be generalised to different scenarios/settings and geographical regions.

In summary, our findings may support further scientific, technological, and innovative development in this field, especially regarding the incorporation of the comet assay as a validated tool for human biomonitoring studies. The gathered evidence may also be used to monitor and reassess the value of this assay, as well as to assist in the development of guidelines.

Author Contributions: Conceptualization, C.L.; methodology, C.L., P.M., L.G. and F.S.T.; formal analysis, all authors; writing—original draft preparation, C.L.; writing each section, all authors; writing—review and editing, all authors; supervision, C.L. and F.S.T. All authors have read and agreed to the published version of the manuscript.

Funding: This research received no external funding.

Institutional Review Board Statement: Not applicable.

Informed Consent Statement: Not applicable.

Data Availability Statement: Data are available from the corresponding author by request.

Acknowledgments: This work was supported by the affiliated institutions, European Regional Development Fund project KK.01.1.1.02.0007 (Rec-IMI), the Croatian Science Foundation (HUMNap project #1192), the Horizon Europe (EDIAQI project #101057497), the European Union—Next Generation EU 533-03-23-0006 (BioMolTox), and the International Comet Assay Working Group (ICAWG).

Conflicts of Interest: The authors declare no conflicts of interest.

References

1. World Health Organization (WHO/HEP/ECH/EHD/22.01). Compendium of WHO and Other UN Guidance on Health and Environment Geneva: World Health Organization: Licence: CC BY-NC-SA 3.0 IGO. 2022. Available online: https://iris.who.int/bitstream/handle/10665/352844/WHO-HEP-ECH-EHD-22.01-eng.pdf?sequence=1 (accessed on 10 February 2024).
2. Bocato, M.Z.; Bianchi Ximenez, J.P.; Hoffmann, C.; Barbosa, F. An overview of the current progress, challenges, and prospects of human biomonitoring and exposome studies. *J. Toxicol. Environ. Health Part B Crit. Rev.* **2019**, *22*, 131–156. [CrossRef] [PubMed]
3. Rodriguez-Carrillo, A.; Mustieles, V.; Salamanca-Fernandez, E.; Olivas-Martinez, A.; Suarez, B.; Bajard, L.; Baken, K.; Blaha, L.; Bonefeld-Jorgensen, E.C.; Couderq, S.; et al. Implementation of effect biomarkers in human biomonitoring studies: A systematic approach synergizing toxicological and epidemiological knowledge. *Int. J. Hyg. Environ. Health* **2023**, *249*, 114140. [CrossRef] [PubMed]
4. DeCaprio, A. Biomarkers: Coming of age for environmental health and risk assessment. *Environ. Sci. Technol.* **1997**, *31*, 1837–1848. [CrossRef]
5. Sung, H.; Ferlay, J.; Siegel, R.; Laversanne, M.; Soerjomataram, I.; Jemal, A.; Bray, F. Global cancer statistics 2020: GLOBOCAN estimates of incidence and mortality worldwide for 36 cancers in 185 countries. *CA Cancer J. Clin.* **2021**, *71*, 209–249. [CrossRef] [PubMed]
6. Viegas, S.; Ladeira, C.; Costa-Veiga, A.; Perelman, J.; Gajski, G. Forgotten public health impacts of cancer—An overview. *Arh. Hig. Rada Toksikol.* **2017**, *68*, 287–297. [CrossRef] [PubMed]
7. Ladeira, C.; Smajdova, L. The use of genotoxicity biomarkers in molecular epidemiology: Applications in environmental, occupational and dietary studies. *AIMS Genet.* **2017**, *4*, 166–191. [CrossRef] [PubMed]
8. Tice, R.R.; Agurell, E.; Anderson, D.; Burlinson, B.; Hartmann, A.; Kobayashi, H.; Miyamae, Y.; Rojas, E.; Ryu, J.C.; Sasaki, Y.F. Single cell gel/comet assay: Guidelines for in vitro and in vivo genetic toxicology testing. *Environ. Mol. Mutagen.* **2000**, *35*, 206–221. [CrossRef]
9. Collins, A.R. The comet assay for DNA damage and repair: Principles, applications, and limitations. *Mol. Biotechnol.* **2004**, *26*, 249–261. [CrossRef] [PubMed]
10. Collins, A.R. Investigating oxidative DNA damage and its repair using the comet assay. *Mutat. Res.* **2009**, *681*, 24–32. [CrossRef] [PubMed]
11. Dusinska, M.; Collins, A.R. The comet assay in human biomonitoring: Gene-environment interactions. *Mutagenesis* **2008**, *23*, 191–205. [CrossRef] [PubMed]
12. Laffon, B.; Teixeira, J.P.; Silva, S.; Loureiro, J.; Torres, J.; Pasaro, E.; Mendez, J.; Mayan, O. Genotoxic effects in a population of nurses handling antineoplastic drugs, and relationship with genetic polymorphisms in DNA repair enzymes. *Am. J. Ind. Med.* **2005**, *48*, 128–136. [CrossRef] [PubMed]
13. Jiang, N.; Naz, S.; Ma, Y.; Ullah, Q.; Khan, M.Z.; Wang, J.; Lu, X.; Luosand, D.-Z.; Tabassum, S.; Chatha, A.M.M.; et al. An Overview of Comet Assay Application for Detecting DNA Damage in Aquatic Animals. *Agriculture* **2023**, *13*, 623. [CrossRef]
14. Collins, A.; Moller, P.; Gajski, G.; Vodenkova, S.; Abdulwahed, A.; Anderson, D.; Bankoglu, E.E.; Bonassi, S.; Boutet-Robinet, E.; Brunborg, G.; et al. Measuring DNA modifications with the comet assay: A compendium of protocols. *Nat. Protoc.* **2023**, *18*, 929–989. [CrossRef] [PubMed]
15. Moller, P.; Knudsen, L.E.; Loft, S.; Wallin, H. The comet assay as a rapid test in biomonitoring occupational exposure to DNA-damaging agents and effect of confounding factors. *Cancer Epidemiol. Biomark. Prev.* **2000**, *9*, 1005–1015.
16. Cavallo, D.; Ursini, C.L.; Rondinone, B.; Iavicoli, S. Evaluation of a suitable DNA damage biomarker for human biomonitoring of exposed workers. *Environ. Mol. Mutagen.* **2009**, *50*, 781–790. [CrossRef] [PubMed]
17. Ladeira, C.; Koppen, G.; Scavone, F.; Giovannelli, L. The comet assay for human biomonitoring: Effect of cryopreservation on DNA damage in different blood cell preparations. *Mutat. Res. Genet. Toxicol. Environ. Mutagen.* **2019**, *843*, 11–17. [CrossRef] [PubMed]
18. Valverde, M.; Rojas, E. Environmental and occupational biomonitoring using the Comet assay. *Mutat. Res.* **2009**, *681*, 93–109. [CrossRef] [PubMed]
19. Russo, C.; Acito, M.; Fatigoni, C.; Villarini, M.; Moretti, M. B-Comet Assay (Comet Assay on Buccal Cells) for the Evaluation of Primary DNA Damage in Human Biomonitoring Studies. *Int. J. Environ. Res. Public Health* **2020**, *17*, 9234. [CrossRef] [PubMed]
20. Gajski, G.; Zegura, B.; Ladeira, C.; Novak, M.; Sramkova, M.; Pourrut, B.; Del Bo, C.; Milic, M.; Gutzkow, K.B.; Costa, S.; et al. The comet assay in animal models: From bugs to whales—(Part 2 Vertebrates). Mutation research. *Rev. Mutat. Res.* **2019**, *781*, 130–164. [CrossRef]
21. Gajski, G.; Zegura, B.; Ladeira, C.; Pourrut, B.; Del Bo, C.; Novak, M.; Sramkova, M.; Milic, M.; Gutzkow, K.B.; Costa, S.; et al. The comet assay in animal models: From bugs to whales—(Part 1 Invertebrates). Mutation research. *Rev. Mutat. Res.* **2019**, *779*, 82–113. [CrossRef] [PubMed]
22. Moller, P.; Azqueta, A.; Boutet-Robinet, E.; Koppen, G.; Bonassi, S.; Milic, M.; Gajski, G.; Costa, S.; Teixeira, J.P.; Costa Pereira, C.; et al. Minimum Information for Reporting on the Comet Assay (MIRCA): Recommendations for describing comet assay procedures and results. *Nat. Protoc.* **2020**, *15*, 3817–3826. [CrossRef] [PubMed]

23. Gianfredi, V.; Nucci, D.; Fatigoni, C.; Salvatori, T.; Villarini, M.; Moretti, M. Extent of Primary DNA Damage Measured by the Comet Assay in Health Professionals Exposed to Antineoplastic Drugs: A Systematic Review and Meta-Analysis. *Int. J. Environ. Res. Public Health* **2020**, *17*, 523. [CrossRef] [PubMed]
24. Zare Sakhvidi, M.J.; Hajaghazadeh, M.; Mostaghaci, M.; Mehrparvar, A.H.; Zare Sakhvidi, F.; Naghshineh, E. Applicability of the comet assay in evaluation of DNA damage in healthcare providers' working with antineoplastic drugs: A systematic review and meta-analysis. *Int. J. Occup. Environ. Health* **2016**, *22*, 52–67. [CrossRef] [PubMed]
25. Moller, P.; Hemmingsen, J.G.; Jensen, D.M.; Danielsen, P.H.; Karottki, D.G.; Jantzen, K.; Roursgaard, M.; Cao, Y.; Kermanizadeh, A.; Klingberg, H.; et al. Applications of the comet assay in particle toxicology: Air pollution and engineered nanomaterials exposure. *Mutagenesis* **2015**, *30*, 67–83. [CrossRef] [PubMed]
26. Lee, R.F.; Steinert, S. Use of the single cell gel electrophoresis/comet assay for detecting DNA damage in aquatic (marine and freshwater) animals. *Mutat. Res.* **2003**, *544*, 43–64. [CrossRef] [PubMed]
27. Olive, P.L.; Banath, J.P.; Durand, R.E. Heterogeneity in radiation-induced DNA damage and repair in tumor and normal cells measured using the "comet" assay. *Radiat. Res.* **1990**, *122*, 86–94. [CrossRef] [PubMed]
28. Kumaravel, T.S.; Vilhar, B.; Faux, S.P.; Jha, A.N. Comet Assay measurements: A perspective. *Cell Biol. Toxicol.* **2009**, *25*, 53–64. [CrossRef]
29. Peters, M.D.J.; Marnie, C.; Tricco, A.C.; Pollock, D.; Munn, Z.; Alexander, L.; McInerney, P.; Godfrey, C.M.; Khalil, H. Updated methodological guidance for the conduct of scoping reviews. *JBI Evid. Synth.* **2020**, *18*, 2119–2126. [CrossRef] [PubMed]
30. Aromataris, E.; Munn, Z. Jonna Briggs Institute (JBI) Manual for Evidence Synthesis. 2020. Available online: https://synthesismanual.jbi.global (accessed on 12 April 2023).
31. Higgins, J.P.T.; Thomas, J.; Chandler, J.; Cumpston, M.; Li, T.; Page, M.J.; Welch, V.A. (Eds.) *Cochrane Handbook for Systematic Reviews of Interventions Version 6.0*; (Updated 2020); Cochrane: Hoboken, NJ, USA, 2020.
32. Page, M.J.; McKenzie, J.E.; Bossuyt, P.M.; Boutron, I.; Hoffmann, T.C.; Mulrow, C.D.; Shamseer, L.; Tetzlaff, J.M.; Akl, E.A.; Brennan, S.E.; et al. The PRISMA 2020 statement: An updated guideline for reporting systematic reviews. *BMJ* **2021**, *372*, n71. [CrossRef] [PubMed]
33. Tricco, A.C.; Lillie, E.; Zarin, W.; O'Brien, K.K.; Colquhoun, H.; Levac, D.; Moher, D.; Peters, M.D.J.; Horsley, T.; Weeks, L.; et al. PRISMA Extension for Scoping Reviews (PRISMA-ScR): Checklist and Explanation. *Ann. Intern. Med.* **2018**, *169*, 467–473. [CrossRef]
34. Boogaard, H.; Walker, K.; Cohen, A.J. Air pollution: The emergence of a major global health risk factor. *Int. Health* **2019**, *11*, 417–421. [CrossRef] [PubMed]
35. Dandotiya, B. Health Effects of Air Pollution in Urban Environment. In *Climate Change and Its Impact on Ecosystem Services and Biodiversity in Arid and Semi-Arid Zones*; IGI Global: Hershey, PA, USA, 2019. [CrossRef]
36. Manisalidis, I.; Stavropoulou, E.; Stavropoulos, A.; Bezirtzoglou, E. Environmental and Health Impacts of Air Pollution: A Review. *Front. Public Health* **2020**, *8*, 14. [CrossRef] [PubMed]
37. World Health Organization. WHO Global Urban Ambient Air Pollution Database. 2018. Available online: https://www.who.int/data/gho/data/themes/air-pollution/who-air-quality-database (accessed on 17 December 2023).
38. Smith, K.R. National Burden of Disease in India from Indoor Air Pollution. *Proc. Natl. Acad. Sci. USA* **2000**, *97*, 13286–13293. [CrossRef] [PubMed]
39. Andersen, M.H.G.; Saber, A.T.; Pedersen, J.E.; Pedersen, P.B.; Clausen, P.A.; Lohr, M.; Kermanizadeh, A.; Loft, S.; Ebbehoj, N.E.; Hansen, A.M.; et al. Assessment of polycyclic aromatic hydrocarbon exposure, lung function, systemic inflammation, and genotoxicity in peripheral blood mononuclear cells from firefighters before and after a work shift. *Environ. Mol. Mutagen.* **2018**, *59*, 539–548. [CrossRef] [PubMed]
40. Andersen, M.H.G.; Saber, A.T.; Frederiksen, M.; Clausen, P.A.; Sejbaek, C.S.; Hemmingsen, C.H.; Ebbehoj, N.E.; Catalan, J.; Aimonen, K.; Koivisto, J.; et al. Occupational exposure and markers of genetic damage, systemic inflammation and lung function: A Danish cross-sectional study among air force personnel. *Sci. Rep.* **2021**, *11*, 17998. [CrossRef] [PubMed]
41. Al Zabadi, H.; Ferrari, L.; Sari-Minodier, I.; Kerautret, M.A.; Tiberguent, A.; Paris, C.; Zmirou-Navier, D. Integrated exposure assessment of sewage workers to genotoxicants: An urinary biomarker approach and oxidative stress evaluation. *Environ. Health* **2011**, *10*, 23. [CrossRef] [PubMed]
42. Aydin, S.; Canpinar, H.; Undeger, U.; Guc, D.; Colakoglu, M.; Kars, A.; Basaran, N. Assessment of immunotoxicity and genotoxicity in workers exposed to low concentrations of formaldehyde. *Arch. Toxicol.* **2013**, *87*, 145–153. [CrossRef] [PubMed]
43. Bacaksiz, A.; Kayaalti, Z.; Soylemez, E.; Tutkun, E.; Soylemezoglu, T. Lymphocyte DNA damage in Turkish asphalt workers detected by the comet assay. *Int. J. Environ. Health Res.* **2014**, *24*, 11–17. [CrossRef] [PubMed]
44. Bagryantseva, Y.; Novotna, B.; Rossner, P., Jr.; Chvatalova, I.; Milcova, A.; Svecova, V.; Lnenickova, Z.; Solansky, I.; Sram, R.J. Oxidative damage to biological macromolecules in Prague bus drivers and garagemen: Impact of air pollution and genetic polymorphisms. *Toxicol. Lett.* **2010**, *199*, 60–68. [CrossRef] [PubMed]
45. Becit, M.; Çilekar, Ş.; Başaran, M.M.; Koca, H.B.; Çelik, S.; Dilsiz, S.A. Changes in genotoxicity, inflammatory and oxidative stress parameters of workers in marble processing plants. *Environ. Res.* **2021**, *197*, 111209. [CrossRef] [PubMed]
46. Barth, A.; Brucker, N.; Moro, A.M.; Nascimento, S.; Goethel, G.; Souto, C.; Fracasso, R.; Sauer, E.; Altknecht, L.; da Costa, B.; et al. Association between inflammation processes, DNA damage, and exposure to environmental pollutants. *Environ. Sci. Pollut. Res. Int.* **2017**, *24*, 353–362. [CrossRef] [PubMed]

47. Balamuralikrishnan, B.; Balachandar, V.; Subramaniam, M.D.; Alagumuthu, K.K.; Sureshkumar, S.; Arun, M.; Arun, S.; Padmavathi, K.; Razeena, A.H.; Gomathi, M.; et al. Assessment of genotoxic and humoral immune system alterations in silica exposed workers from pottery industries in South India. *Stoch. Environ. Res. Risk Assess.* **2014**, *28*, 1801–1814. [CrossRef]
48. Bruschweiler, E.D.; Wild, P.; Huynh, C.K.; Savova-Bianchi, D.; Danuser, B.; Hopf, N.B. DNA Damage among Wood Workers Assessed with the Comet Assay. *Environ. Health Insights* **2016**, *10*, 105–112. [CrossRef]
49. Carere, A.; Andreoli, C.; Galati, R.; Leopardi, P.; Marcon, F.; Rosati, M.V.; Rossi, S.; Tomei, F.; Verdina, A.; Zijno, A.; et al. Biomonitoring of exposure to urban air pollutants: Analysis of sister chromatid exchanges and DNA lesions in peripheral lymphocytes of traffic policemen. *Mutat. Res.* **2002**, *518*, 215–224. [CrossRef] [PubMed]
50. Cavallo, D.; Ursini, C.L.; Bavazzano, P.; Cassinelli, C.; Frattini, A.; Perniconi, B.; Di Francesco, A.; Ciervo, A.; Rondinone, B.; Iavicoli, S. Sister chromatid exchange and oxidative DNA damage in paving workers exposed to PAHs. *Ann. Occup. Hyg.* **2006**, *50*, 211–218. [CrossRef] [PubMed]
51. Cavallo, D.; Ursini, C.L.; Carelli, G.; Iavicoli, I.; Ciervo, A.; Perniconi, B.; Rondinone, B.; Gismondi, M.; Iavicoli, S. Occupational exposure in airport personnel: Characterization and evaluation of genotoxic and oxidative effects. *Toxicology* **2006**, *223*, 26–35. [CrossRef] [PubMed]
52. Cavallo, D.; Ursini, C.L.; Fresegna, A.M.; Ciervo, A.; Boccuni, F.; Ferrante, R.; Tombolini, F.; Maiello, R.; Chiarella, P.; Buresti, G.; et al. A follow-up study on workers involved in the graphene production process after the introduction of exposure mitigation measures: Evaluation of genotoxic and oxidative effects. *Nanotoxicology* **2022**, *16*, 776–790. [CrossRef] [PubMed]
53. Cebulska-Wasilewska, A.; Wiechec, A.; Panek, A.; Binkova, B.; Sram, R.J.; Farmer, P.B. Influence of environmental exposure to PAHs on the susceptibility of lymphocytes to DNA-damage induction and on their repair capacity. *Mutat. Res.* **2005**, *588*, 73–81. [CrossRef] [PubMed]
54. Cebulska-Wasilewska, A.; Pawlyk, I.; Panek, A.; Wiechec, A.; Kalina, I.; Popov, T.; Georgieva, T.; Farmer, P.B. Exposure to environmental polycyclic aromatic hydrocarbons: Influences on cellular susceptibility to DNA damage (sampling Kosice and Sofia). *Mutat. Res.* **2007**, *620*, 145–154. [CrossRef] [PubMed]
55. Cebulska-Wasilewska, A.; Binkova, B.; Sram, R.J.; Kalina, I.; Popov, T.; Farmer, P.B. Repair competence assay in studies of the influence of environmental exposure to c-PAHs on individual susceptibility to induction of DNA damage. *Mutat. Res.* **2007**, *620*, 155–164. [CrossRef]
56. Ceppi, M.; Smolkova, B.; Staruchova, M.; Kazimirova, A.; Barancokova, M.; Volkovova, K.; Collins, A.; Kocan, A.; Dzupinkova, Z.; Horska, A.; et al. Genotoxic effects of occupational exposure to glass fibres—A human biomonitoring study. *Mutat. Res. Genet. Toxicol. Environ. Mutagen.* **2023**, *885*, 503572. [CrossRef] [PubMed]
57. Chen, Y.; Bai, Y.; Yuan, J.; Chen, W.; Sun, J.; Wang, H.; Liang, H.; Guo, L.; Yang, X.; Tan, H.; et al. Association of polymorphisms in AhR, CYP1A1, GSTM1, and GSTT1 genes with levels of DNA damage in peripheral blood lymphocytes among coke-oven workers. *Cancer Epidemiol. Biomark. Prev.* **2006**, *15*, 1703–1707. [CrossRef]
58. Chen, H.L.; Chen, I.J.; Chia, T.P. Occupational exposure and DNA strand breakage of workers in bottom ash recovery and fly ash treatment plants. *J. Hazard. Mater.* **2010**, *174*, 23–27. [CrossRef] [PubMed]
59. Cheng, J.; Leng, S.; Li, H.; Huang, C.; Niu, Y.; Zhang, L.; Liang, X.; Lin, H.; Zheng, Y. Suboptimal DNA repair capacity predisposes coke-oven workers to accumulate more chromosomal damages in peripheral lymphocytes. *Cancer Epidemiol. Biomark. Prev. Publ. Am. Assoc. Cancer Res. Cosponsored Am. Soc. Prev. Oncol.* **2009**, *18*, 987–993. [CrossRef]
60. Chia, T.; Hsu, C.Y.; Chen, H.L. Oxidative damage of workers in secondary metal recovery plants affected by smoking status and joining the smelting work. *Ind. Health* **2008**, *46*, 174–182. [CrossRef] [PubMed]
61. Costa, S.; Coelho, P.; Costa, C.; Silva, S.; Mayan, O.; Santos, L.S.; Gaspar, J.; Teixeira, J.P. Genotoxic damage in pathology anatomy laboratory workers exposed to formaldehyde. *Toxicology* **2008**, *252*, 40–48. [CrossRef] [PubMed]
62. Costa, S.; Pina, C.; Coelho, P.; Costa, C.; Silva, S.; Porto, B.; Laffon, B.; Teixeira, J.P. Occupational exposure to formaldehyde: Genotoxic risk evaluation by comet assay and micronucleus test using human peripheral lymphocytes. *J. Toxicol. Environ. Health Part A* **2011**, *74*, 1040–1051. [CrossRef] [PubMed]
63. Costa, S.; Carvalho, S.; Costa, C.; Coelho, P.; Silva, S.; Santos, L.S.; Gaspar, J.F.; Porto, B.; Laffon, B.; Teixeira, J.P. Increased levels of chromosomal aberrations and DNA damage in a group of workers exposed to formaldehyde. *Mutagenesis* **2015**, *30*, 463–473. [CrossRef] [PubMed]
64. De Boeck, M.; Lardau, S.; Buchet, J.P.; Kirsch-Volders, M.; Lison, D. Absence of significant genotoxicity in lymphocytes and urine from workers exposed to moderate levels of cobalt-containing dust: A cross-sectional study. *Environ. Mol. Mutagen.* **2000**, *36*, 151–160. [CrossRef] [PubMed]
65. Duan, H.; Jia, X.; Zhai, Q.; Ma, L.; Wang, S.; Huang, C.; Wang, H.; Niu, Y.; Li, X.; Dai, Y.; et al. Long-term exposure to diesel engine exhaust induces primary DNA damage: A population-based study. *Occup. Environ. Med.* **2016**, *73*, 83–90. [CrossRef] [PubMed]
66. Everatt, R.; Slapsyte, G.; Mierauskiene, J.; Dedonyte, V.; Bakiene, L. Biomonitoring study of dry-cleaning workers using cytogenetic tests and the comet assay. *J. Occup. Environ. Hyg.* **2013**, *10*, 609–621. [CrossRef] [PubMed]
67. Galiotte, M.P.; Kohler, P.; Mussi, G.; Gattas, G.J. Assessment of occupational genotoxic risk among Brazilian hairdressers. *Ann. Occup. Hyg.* **2008**, *52*, 645–651. [CrossRef]
68. Giri, S.K.; Yadav, A.; Kumar, A.; Dev, K.; Gupta, R.; Aggarwal, N.; Seth, N.; Gautam, S.K. Association of GSTM1 and GSTT1 polymorphisms with DNA damage in coal-tar workers. *Sci. Total Environ.* **2011**, *409*, 4465–4469. [CrossRef] [PubMed]

69. Gomaa, M.S.; Elmesallamy, G.E.; Sameer, M.M. Evaluation of Genotoxic Effects of Formaldehyde in Adult Albino Rats and Its Implication In Case of Human Exposure. *Life Sci. J.* **2012**, *9*, 3085–3093.
70. Goethel, G.; Brucker, N.; Moro, A.M.; Charao, M.F.; Fracasso, R.; Barth, A.; Bubols, G.; Durgante, J.; Nascimento, S.; Baierle, M.; et al. Evaluation of genotoxicity in workers exposed to benzene and atmospheric pollutants. *Mutat. Res. Genet. Toxicol. Environ. Mutagen.* **2014**, *770*, 61–65. [CrossRef] [PubMed]
71. Rezaei Hachesu, V.; Naderyan Fe'li, S.; Kargar Shouroki, F.; Mehrparvar, A.H.; Zavar Reza, J.; Azimi, M.; Zare Sakhvidi, M.J. Carbon load in airway macrophages, DNA damage and lung function in taxi drivers exposed to traffic-related air pollution. *Environ. Sci. Pollut. Res. Int.* **2019**, *26*, 6868–6876. [CrossRef] [PubMed]
72. Huang, G.; Guo, H.; Wu, T. Genetic variations of CYP2B6 gene were associated with plasma BPDE-Alb adducts and DNA damage levels in coke oven workers. *Toxicol. Lett.* **2012**, *211*, 232–238. [CrossRef]
73. Jasso-Pineda, Y.; Díaz-Barriga, F.; Yáñez-Estrada, L.; Pérez-Vázquez, F.J.; Pérez-Maldonado, I.N. DNA damage in Mexican children living in high-risk contaminated scenarios. *Sci. Total Environ.* **2015**, *518–519*, 38–48. [CrossRef]
74. Jiang, S.; Yu, L.; Cheng, J.; Leng, S.; Dai, Y.; Zhang, Y.; Niu, Y.; Yan, H.; Qu, W.; Zhang, C.; et al. Genomic damages in peripheral blood lymphocytes and association with polymorphisms of three glutathione S-transferases in workers exposed to formaldehyde. *Mutat. Res.* **2010**, *695*, 9–15. [CrossRef] [PubMed]
75. Khanna, A.; Gautam, D.S.; Gokhale, M.; Jain, S.K. Tobacco dust induced genotoxicity as an occupational hazard in workers of bidi making cottage industry of central India. *Toxicol. Int.* **2014**, *21*, 18–23. [CrossRef]
76. Khisroon, M.; Khan, A.; Ayub, A.; Ullah, I.; Farooqi, J.; Ullah, A. DNA damage analysis concerning GSTM1 and GSTT1 gene polymorphism in gold jewellery workers from Peshawar Pakistan. *Biomarkers* **2020**, *25*, 483–489. [CrossRef] [PubMed]
77. Kianmehr, M.; Hajavi, J.; Gazeri, J. Assessment of DNA damage in blood lymphocytes of bakery workers by comet assay. *Toxicol. Ind. Health* **2017**, *33*, 726–735. [CrossRef] [PubMed]
78. Knudsen, L.E.; Gaskell, M.; Martin, E.A.; Poole, J.; Scheepers, P.T.; Jensen, A.; Autrup, H.; Farmer, P.B. Genotoxic damage in mine workers exposed to diesel exhaust, and the effects of glutathione transferase genotypes. *Mutat. Res.* **2005**, *583*, 120–132. [CrossRef] [PubMed]
79. Krieg, E.F., Jr.; Mathias, P.I.; Toennis, C.A.; Clark, J.C.; Marlow, K.L.; B'Hymer, C.; Singh, N.P.; Gibson, R.L.; Butler, M.A. Detection of DNA damage in workers exposed to JP-8 jet fuel. *Mutat. Res.* **2012**, *747*, 218–227. [CrossRef] [PubMed]
80. Kvitko, K.; Bandinelli, E.; Henriques, J.A.; Heuser, V.D.; Rohr, P.; da Silva, F.R.; Schneider, N.B.; Fernandes, S.; Ancines, C.; da Silva, J. Susceptibility to DNA damage in workers occupationally exposed to pesticides, to tannery chemicals and to coal dust during mining. *Genet. Mol. Biol.* **2012**, *35*, 1060–1068. [CrossRef] [PubMed]
81. Leng, S.; Cheng, J.; Pan, Z.; Huang, C.; Niu, Y.; Dai, Y.; Li, B.; He, F.; Zheng, Y. Associations between XRCC1 and ERCC2 polymorphisms and DNA damage in peripheral blood lymphocyte among coke oven workers. *Biomarkers* **2004**, *9*, 395–406. [CrossRef] [PubMed]
82. Leon-Mejia, G.; Espitia-Perez, L.; Hoyos-Giraldo, L.S.; Da Silva, J.; Hartmann, A.; Henriques, J.A.; Quintana, M. Assessment of DNA damage in coal open-cast mining workers using the cytokinesis-blocked micronucleus test and the comet assay. *Sci. Total Environ.* **2011**, *409*, 686–691. [CrossRef] [PubMed]
83. Leon-Mejia, G.; Luna-Rodriguez, I.; Trindade, C.; Oliveros-Ortiz, L.; Anaya-Romero, M.; Luna-Carrascal, J.; Navarro-Ojeda, N.; Ruiz-Benitez, M.; Franco-Valencia, K.; Da Silva, J.; et al. Cytotoxic and genotoxic effects in mechanics occupationally exposed to diesel engine exhaust. *Ecotoxicol. Environ. Saf.* **2019**, *171*, 264–273. [CrossRef] [PubMed]
84. Lin, D.; Guo, Y.; Yi, J.; Kuang, D.; Li, X.; Deng, H.; Huang, K.; Guan, L.; He, Y.; Zhang, X.; et al. Occupational exposure to formaldehyde and genetic damage in the peripheral blood lymphocytes of plywood workers. *J. Occup. Health* **2013**, *55*, 284–291. [CrossRef]
85. Marczynski, B.; Rihs, H.P.; Rossbach, B.; Holzer, J.; Angerer, J.; Scherenberg, M.; Hoffmann, G.; Bruning, T.; Wilhelm, M. Analysis of 8-oxo-7,8-dihydro-2′-deoxyguanosine and DNA strand breaks in white blood cells of occupationally exposed workers: Comparison with ambient monitoring, urinary metabolites and enzyme polymorphisms. *Carcinogenesis* **2002**, *23*, 273–281. [CrossRef] [PubMed]
86. Marczynski, B.; Raulf-Heimsoth, M.; Pesch, B.; Kendzia, B.; Kafferlein, H.U.; Vosshans, B.; Borowitzki, G.; Lee, E.H.; Bramer, R.; Bruning, T. Detection of DNA strand breaks by comet assay in sputum leucocytes of bitumen-exposed workers: A pilot study *Hum. Exp. Toxicol.* **2010**, *29*, 721–729. [CrossRef] [PubMed]
87. Marczynski, B.; Raulf-Heimsoth, M.; Spickenheuer, A.; Pesch, B.; Kendzia, B.; Mensing, T.; Engelhardt, B.; Lee, E.H.; Schindler, B.K.; Heinze, E.; et al. DNA adducts and strand breaks in workers exposed to vapours and aerosols of bitumen: Associations between exposure and effect. *Arch. Toxicol.* **2011**, *85* (Suppl. 1), S53–S64. [CrossRef]
88. Moretti, M.; Dell'Omo, M.; Villarini, M.; Pastorelli, R.; Muzi, G.; Airoldi, L.; Pasquini, R. Primary DNA damage and genetic polymorphisms for CYP1A1, EPHX and GSTM1 in workers at a graphite electrode manufacturing plant. *BMC Public Health* **2007**, *7*, 270. [CrossRef] [PubMed]
89. Novotna, B.; Topinka, J.; Solansky, I.; Chvatalova, I.; Lnenickova, Z.; Sram, R.J. Impact of air pollution and genotype variability on DNA damage in Prague policemen. *Toxicol. Lett.* **2007**, *172*, 37–47. [CrossRef] [PubMed]
90. Oh, E.; Im, H.; Kang, H.S.; Jung, W.; Won, N.H.; Lee, E.; Sul, D. Comparison of immunnological and genotoxicological parameters in automobile emission inspectors exposed to polycyclic aromatic hydrocarbons. *Environ. Toxicol. Pharmacol.* **2006**, *21*, 108–117. [CrossRef] [PubMed]

91. Peteffi, G.P.; da Silva, L.B.; Antunes, M.V.; Wilhelm, C.; Valandro, E.T.; Glaeser, J.; Kaefer, D.; Linden, R. Evaluation of genotoxicity in workers exposed to low levels of formaldehyde in a furniture manufacturing facility. *Toxicol. Ind. Health* **2016**, *32*, 1763–1773. [CrossRef] [PubMed]
92. Peteffi, G.P.; Antunes, M.V.; Carrer, C.; Valandro, E.T.; Santos, S.; Glaeser, J.; Mattos, L.; da Silva, L.B.; Linden, R. Environmental and biological monitoring of occupational formaldehyde exposure resulting from the use of products for hair straightening. *Environ. Sci. Pollut. Res. Int.* **2016**, *23*, 908–917. [CrossRef] [PubMed]
93. Recio-Vega, R.; Olivas-Calderon, E.; Michel-Ramirez, G.; Martinez-Salinas, R.I.; Gallegos-Arreola, M.P.; Ocampo-Gomez, G.L.; Perez-Morales, R. Associations between sperm quality, DNA damage, and CYP1A1, GSTT1 and GSTM1 polymorphisms with 1-hydroxypyrene urinary levels in men occupationally exposed to polycyclic aromatic hydrocarbons. *Int. Arch. Occup. Environ. Health* **2018**, *91*, 725–734. [CrossRef] [PubMed]
94. Rekhadevi, P.V.; Mahboob, M.; Rahman, M.F.; Grover, P. Genetic damage in wood dust-exposed workers. *Mutagenesis* **2009**, *24*, 59–65. [CrossRef] [PubMed]
95. Rohr, P.; Kvitko, K.; da Silva, F.R.; Menezes, A.P.; Porto, C.; Sarmento, M.; Decker, N.; Reyes, J.M.; Allgayer Mda, C.; Furtado, T.C.; et al. Genetic and oxidative damage of peripheral blood lymphocytes in workers with occupational exposure to coal. *Mutat. Res. Genet. Toxicol. Environ. Mutagen.* **2013**, *758*, 23–28. [CrossRef] [PubMed]
96. Sardas, S.; Omurtag, G.Z.; Tozan, A.; Gul, H.; Beyoglu, D. Evaluation of DNA damage in construction-site workers occupationally exposed to welding fumes and solvent-based paints in Turkey. *Toxicol. Ind. Health* **2010**, *26*, 601–608. [CrossRef] [PubMed]
97. Scheepers, P.T.; Coggon, D.; Knudsen, L.E.; Anzion, R.; Autrup, H.; Bogovski, S.; Bos, R.P.; Dahmann, D.; Farmer, P.; Martin, E.A.; et al. BIOMarkers for occupational diesel exhaust exposure monitoring (BIOMODEM)—A study in underground mining. *Toxicol. Lett.* **2002**, *134*, 305–317. [CrossRef]
98. Sellappa, S.; Prathyumnan, S.; Balachandar, V. DNA damage induction and repair inhibition among building construction workers in South India. *Asian Pac. J. Cancer Prev.* **2010**, *11*, 875–880. [PubMed]
99. Sellappa, S.; Mani, B.; Keyan, K.S. Cytogenetic biomonitoring of road paving workers occupationally exposed to polycyclic aromatic hydrocarbons. *Asian Pac. J. Cancer Prev.* **2011**, *12*, 713–717.
100. Shen, M.; Bin, P.; Li, H.; Zhang, X.; Sun, X.; Duan, H.; Niu, Y.; Meng, T.; Dai, Y.; Gao, W.; et al. Increased levels of etheno-DNA adducts and genotoxicity biomarkers of long-term exposure to pure diesel engine exhaust. *Sci. Total Environ.* **2016**, *543*, 267–273. [CrossRef] [PubMed]
101. Siwinska, E.; Mielzynska, D.; Kapka, L. Association between urinary 1-hydroxypyrene and genotoxic effects in coke oven workers. *Occup. Environ. Med.* **2004**, *61*, e10. [CrossRef]
102. Sul, D.; Oh, E.; Im, H.; Yang, M.; Kim, C.W.; Lee, E. DNA damage in T- and B-lymphocytes and granulocytes in emission inspection and incineration workers exposed to polycyclic aromatic hydrocarbons. *Mutat. Res.* **2003**, *538*, 109–119. [CrossRef] [PubMed]
103. Toraason, M.; Lynch, D.W.; DeBord, D.G.; Singh, N.; Krieg, E.; Butler, M.A.; Toennis, C.A.; Nemhauser, J.B. DNA damage in leukocytes of workers occupationally exposed to 1-bromopropane. *Mutat. Res.* **2006**, *603*, 1–14. [CrossRef] [PubMed]
104. Tovalin, H.; Valverde, M.; Morandi, M.T.; Blanco, S.; Whitehead, L.; Rojas, E. DNA damage in outdoor workers occupationally exposed to environmental air pollutants. *Occup. Environ. Med.* **2006**, *63*, 230–236. [CrossRef]
105. Ullah, I.; Zahid, M.; Jawad, M.; Arsh, A. Assessment of DNA damage and oxidative stress among traffic conductors and coal miners. *Pak. J. Med. Sci.* **2021**, *37*, 499–502. [CrossRef]
106. van Delft, J.H.M.; Steenwinkel, M.J.S.; van Asten, J.G.; de Vogel, N.; Bruijntjes-Rozier, T.C.; Schouten, T.; Cramers, P.; Maas, L.; van Herwijnen, M.H.; van Schooten, F.J.; et al. Biological monitoring the exposure to polycyclic aromatic hydrocarbons of coke oven workers in relation to smoking and genetic polymorphisms for GSTM1 and GSTT1. *Ann. Occup. Hyg.* **2001**, *45*, 395–408. [CrossRef]
107. Villarini, M.; Moretti, M.; Fatigoni, C.; Agea, E.; Dominici, L.; Mattioli, A.; Volpi, R.; Pasquini, R. Evaluation of primary DNA damage, cytogenetic biomarkers and genetic polymorphisms for CYP1A1 and GSTM1 in road tunnel construction workers. *J. Toxicol. Environ. Health Part A* **2008**, *71*, 1430–1439. [CrossRef] [PubMed]
108. Vital, N.; Antunes, S.; Louro, H.; Vaz, F.; Simoes, T.; Penque, D.; Silva, M.J. Environmental Tobacco Smoke in Occupational Settings: Effect and Susceptibility Biomarkers in Workers From Lisbon Restaurants and Bars. *Front. Public Health* **2021**, *9*, 674142. [CrossRef] [PubMed]
109. Wang, H.; Chen, W.; Zheng, H.; Guo, L.; Liang, H.; Yang, X.; Bai, Y.; Sun, J.; Su, Y.; Chen, Y.; et al. Association between plasma BPDE-Alb adduct concentrations and DNA damage of peripheral blood lymphocytes among coke oven workers. *Occup. Environ. Med.* **2007**, *64*, 753–758. [CrossRef] [PubMed]
110. Wang, F.; He, Y.; Guo, H.; Li, J.; Yang, Y.; Wu, Z.; Zheng, H.; Wu, T. Genetic variants of nucleotide excision repair genes are associated with DNA damage in coke oven workers. *Cancer Epidemiol. Biomark. Prev.* **2010**, *19*, 211–218. [CrossRef]
111. Wang, J.; Luo, X.; Xu, B.; Wei, J.; Zhang, Z.; Zhu, H. Elevated oxidative damage in kitchen workers in Chinese restaurants. *J. Occup. Health* **2011**, *53*, 327–333. [CrossRef] [PubMed]
112. Wultsch, G.; Mišík, M.; Nersesyan, A.; Knasmueller, S. Genotoxic effects of occupational exposure measured in lymphocytes of waste-incinerator workers. *Mutat. Res.* **2011**, *720*, 3–7. [CrossRef] [PubMed]
113. Yang, X.; Zheng, J.; Bai, Y.; Tian, F.; Yuan, J.; Sun, J.; Liang, H.; Guo, L.; Tan, H.; Chen, W.; et al. Using lymphocyte and plasma Hsp70 as biomarkers for assessing coke oven exposure among steel workers. *Environ. Health Perspect.* **2007**, *115*, 1573–1577. [CrossRef] [PubMed]

114. Yu, J.; Liu, M.; Fang, Q.; Zhang, X. Polycyclic aromatic hydrocarbons, long non-coding RNA expression, and DNA damage in coke oven workers. *Environ. Sci. Pollut. Res. Int.* **2022**, *29*, 57277–57286. [CrossRef] [PubMed]
115. Zhang, Z.; Xing, X.; Jiang, S.; Qiu, C.; Mo, Z.; Chen, S.; Chen, L.; Wang, Q.; Xiao, Y.; Dong, G.; et al. Global H3K79 di-methylation mediates DNA damage response to PAH exposure in Chinese coke oven workers. *Environ. Pollut.* **2021**, *268*, 115956. [CrossRef]
116. Zendehdel, R.; Jouni, F.J.; Hajipour, B.; Panjali, Z.; Kheiri, H.; Vahabi, M. DNA damage in workers exposed to formaldehyde at concentrations below occupational exposure limits. *Toxicol. Environ. Chem.* **2017**, *99*, 1409–1417. [CrossRef]
117. Zendehdel, R.; Vahabi, M.; Sedghi, R. Estimation of formaldehyde occupational exposure limit based on genetic damage in some Iranian exposed workers using benchmark dose method. *Environ. Sci. Pollut. Res. Int.* **2018**, *25*, 31183–31189. [CrossRef] [PubMed]
118. Zendehdel, R.; Abdolmaleki, P.; Jouni, F.J.; Mazinani, M. Genetic variation and risk of DNA damage in peripheral blood lymphocytes of Iranian formaldehyde-exposed workers. *Hum. Exp. Toxicol.* **2018**, *37*, 690–696. [CrossRef] [PubMed]
119. Alvarado-Cruz, I.; Sanchez-Guerra, M.; Hernandez-Cadena, L.; De Vizcaya-Ruiz, A.; Mugica, V.; Pelallo-Martinez, N.A.; Solis-Heredia, M.J.; Byun, H.M.; Baccarelli, A.; Quintanilla-Vega, B. Increased methylation of repetitive elements and DNA repair genes is associated with higher DNA oxidation in children in an urbanized, industrial environment. *Mutat. Res. Genet. Toxicol. Environ. Mutagen.* **2017**, *813*, 27–36. [CrossRef] [PubMed]
120. Andersen, M.H.G.; Frederiksen, M.; Saber, A.T.; Wils, R.S.; Fonseca, A.S.; Koponen, I.K.; Johannesson, S.; Roursgaard, M.; Loft, S.; Moller, P.; et al. Health effects of exposure to diesel exhaust in diesel-powered trains. *Part. Fibre Toxicol.* **2019**, *16*, 21. [CrossRef] [PubMed]
121. Avogbe, P.H.; Ayi-Fanou, L.; Autrup, H.; Loft, S.; Fayomi, B.; Sanni, A.; Vinzents, P.; Moller, P. Ultrafine particulate matter and high-level benzene urban air pollution in relation to oxidative DNA damage. *Carcinogenesis* **2005**, *26*, 613–620. [CrossRef] [PubMed]
122. Beyoglu, D.; Ozkozaci, T.; Akici, N.; Omurtag, G.Z.; Akici, A.; Ceran, O.; Sardas, S. Assessment of DNA damage in children exposed to indoor tobacco smoke. *Int. J. Hyg. Environ. Health* **2010**, *213*, 40–43. [CrossRef] [PubMed]
123. Cetkovic, T.; Haveric, A.; Behmen, S.; Hadzic Omanovic, M.; Caluk Klacar, L.; Dzaferspahic, A.; Durmisevic, I.; Mehanovic, M.; Haveric, S. A pilot biomonitoring study of air pollution in the urban area of Sarajevo, Bosnia and Herzegovina: Genotoxicity assessment in buccal cells. *Mutagenesis* **2023**, *38*, 33–42. [CrossRef] [PubMed]
124. Cho, J.A.; Oh, E.; Lee, E.; Sul, D. Effects of hair dyeing on DNA damage in human lymphocytes. *J. Occup. Health* **2003**, *45*, 376–381. [CrossRef] [PubMed]
125. Chu, M.; Sun, C.; Chen, W.; Jin, G.; Gong, J.; Zhu, M.; Yuan, J.; Dai, J.; Wang, M.; Pan, Y.; et al. Personal exposure to PM2.5, genetic variants and DNA damage: A multi-center population-based study in Chinese. *Toxicol. Lett.* **2015**, *235*, 172–178. [CrossRef] [PubMed]
126. Coronas, M.V.; Pereira, T.S.; Rocha, J.A.; Lemos, A.T.; Fachel, J.M.; Salvadori, D.M.; Vargas, V.M. Genetic biomonitoring of an urban population exposed to mutagenic airborne pollutants. *Environ. Int.* **2009**, *35*, 1023–1029. [CrossRef] [PubMed]
127. Coronas, M.V.; Rocha, J.A.; Salvadori, D.M.; Vargas, V.M. Evaluation of area contaminated by wood treatment activities: Genetic markers in the environment and in the child population. *Chemosphere* **2016**, *144*, 1207–1215. [CrossRef]
128. Danielsen, P.H.; Brauner, E.V.; Barregard, L.; Sallsten, G.; Wallin, M.; Olinski, R.; Rozalski, R.; Moller, P.; Loft, S. Oxidatively damaged DNA and its repair after experimental exposure to wood smoke in healthy humans. *Mutat. Res.* **2008**, *642*, 37–42. [CrossRef] [PubMed]
129. Silva da Silva, C.; Rossato, J.M.; Vaz Rocha, J.A.; Vargas, V.M. Characterization of an area of reference for inhalable particulate matter (PM2.5) associated with genetic biomonitoring in children. *Mutat. Res. Genet. Toxicol. Environ. Mutagen.* **2015**, *778*, 44–55. [CrossRef] [PubMed]
130. Forchhammer, L.; Moller, P.; Riddervold, I.S.; Bonlokke, J.; Massling, A.; Sigsgaard, T.; Loft, S. Controlled human wood smoke exposure: Oxidative stress, inflammation and microvascular function. *Part. Fibre Toxicol.* **2012**, *9*, 7. [CrossRef] [PubMed]
131. Gamboa, R.T.; Gamboa, A.R.; Bravo, A.H.; Ostrosky, W.P. Genotoxicity in child populations exposed to polycyclic aromatic hydrocarbons (PAHs) in the air from Tabasco, Mexico. *Int. J. Environ. Res. Public Health* **2008**, *5*, 349–355. [CrossRef] [PubMed]
132. Gong, J.; Zhu, M.; Chu, M.; Sun, C.; Chen, W.; Jin, G.; Yuan, J.; Dai, J.; Wang, M.; Pan, Y.; et al. Genetic variants in SMARC genes are associated with DNA damage levels in Chinese population. *Toxicol. Lett.* **2014**, *229*, 327–332. [CrossRef] [PubMed]
133. Han, X.; Zhou, N.; Cui, Z.; Ma, M.; Li, L.; Cai, M.; Li, Y.; Lin, H.; Li, Y.; Ao, L.; et al. Association between urinary polycyclic aromatic hydrocarbon metabolites and sperm DNA damage: A population study in Chongqing, China. *Environ. Health Perspect.* **2011**, *119*, 652–657. [CrossRef] [PubMed]
134. Hemmingsen, J.G.; Moller, P.; Jantzen, K.; Jonsson, B.A.; Albin, M.; Wierzbicka, A.; Gudmundsson, A.; Loft, S.; Rissler, J. Controlled exposure to diesel exhaust and traffic noise—Effects on oxidative stress and activation in mononuclear blood cells. *Mutat. Res.* **2015**, *775*, 66–71. [CrossRef] [PubMed]
135. Hisamuddin, N.H.; Jalaludin, J.; Abu Bakar, S.; Latif, M.T. The Influence of Environmental Polycyclic Aromatic Hydrocarbons (PAHs) Exposure on DNA Damage among School Children in Urban Traffic Area, Malaysia. *Int. J. Environ. Res. Public Health* **2022**, *19*, 2193. [CrossRef] [PubMed]
136. Ismail, I.N.; Alaludin, J.; Bakar, S.A.; Hisamuddin, N.H.; Suhaimi, N.F. Association of Traffic-Related Air Pollution (TRAP) with DNA Damage and Respiratory Health Symptoms among Primary School Children in Selangor. *Asian J. Atmos. Environ.* **2019**, *13*, 106–116. [CrossRef]

137. Jensen, A.; Karottki, D.G.; Christensen, J.M.; Bonlokke, J.H.; Sigsgaard, T.; Glasius, M.; Loft, S.; Moller, P. Biomarkers of oxidative stress and inflammation after wood smoke exposure in a reconstructed Viking Age house. *Environ. Mol. Mutagen.* **2014**, *55*, 652–661. [CrossRef]
138. Koppen, G.; Verheyen, G.; Maes, A.; Van Gorp, U.; Schoeters, G.; Hond, E.D.; Staessen, J.; Nawrot, T.; Roels, H.A.; Vlietinck, R.; et al. A battery of DNA effect biomarkers to evaluate environmental exposure of Flemish adolescents. *J. Appl. Toxicol. JAT* **2007**, *27*, 238–246. [CrossRef] [PubMed]
139. Koppen, G.; Franken, C.; Den Hond, E.; Plusquin, M.; Reimann, B.; Leermakers, M.; Covaci, A.; Nawrot, T.; Van Larebeke, N.; Schoeters, G.; et al. Pooled analysis of genotoxicity markers in relation to exposure in the Flemish Environment and Health Studies (FLEHS) between 1999 and 2018. *Environ. Res.* **2020**, *190*, 110002. [CrossRef] [PubMed]
140. Lemos, A.T.; Lemos, C.T.; Coronas, M.V.; Rocha, J.R.D.; Vargas, V.M.F. Integrated study of genotoxicity biomarkers in schoolchildren and inhalable particles in areas under petrochemical influence. *Environ. Res.* **2020**, *188*, 109443. [CrossRef] [PubMed]
141. Leon-Mejia, G.; Vargas, J.E.; Quintana-Sosa, M.; Rueda, R.A.; Perez, J.P.; Miranda-Guevara, A.; Moreno, O.F.; Trindade, C.; Acosta-Hoyos, A.; Dias, J.; et al. Exposure to coal mining can lead to imbalanced levels of inorganic elements and DNA damage in individuals living near open-pit mining sites. *Environ. Res.* **2023**, *227*, 115773. [CrossRef] [PubMed]
142. Mondal, N.K.; Mukherjee, B.; Das, D.; Ray, M.R. Micronucleus formation, DNA damage and repair in premenopausal women chronically exposed to high level of indoor air pollution from biomass fuel use in rural India. *Mutat. Res.* **2010**, *697*, 47–54. [CrossRef] [PubMed]
143. Mondal, N.K.; Bhattacharya, P.; Ray, M.R. Assessment of DNA damage by comet assay and fast halo assay in buccal epithelial cells of Indian women chronically exposed to biomass smoke. *Int. J. Hyg. Environ. Health* **2011**, *214*, 311–318. [CrossRef] [PubMed]
144. Mukherjee, B.; Dutta, A.; Roychoudhury, S.; Ray, M.R. Chronic inhalation of biomass smoke is associated with DNA damage in airway cells: Involvement of particulate pollutants and benzene. *J. Appl. Toxicol. JAT* **2013**, *33*, 281–289. [CrossRef] [PubMed]
145. Mukherjee, B.; Bindhani, B.; Saha, H.; Ray, M.R. Increased oxidative DNA damage and decreased expression of base excision repair proteins in airway epithelial cells of women who cook with biomass fuels. *Environ. Toxicol. Pharmacol.* **2014**, *38*, 341–352. [CrossRef] [PubMed]
146. Nagiah, S.; Phulukdaree, A.; Naidoo, D.; Ramcharan, K.; Naidoo, R.N.; Moodley, D.; Chuturgoon, A. Oxidative stress and air pollution exposure during pregnancy: A molecular assessment. *Hum. Exp. Toxicol.* **2015**, *34*, 838–847. [CrossRef] [PubMed]
147. Pacini, S.; Giovannelli, L.; Gulisano, M.; Peruzzi, B.; Polli, G.; Boddi, V.; Ruggiero, M.; Bozzo, C.; Stomeo, F.; Fenu, G.; et al. Association between atmospheric ozone levels and damage to human nasal mucosa in Florence, Italy. *Environ. Mol. Mutagen.* **2003**, *42*, 127–135. [CrossRef] [PubMed]
148. Pandey, A.K.; Bajpayee, M.; Parmar, D.; Rastogi, S.K.; Mathur, N.; Seth, P.K.; Dhawan, A. DNA damage in lymphocytes of rural Indian women exposed to biomass fuel smoke as assessed by the Comet assay. *Environ. Mol. Mutagen.* **2005**, *45*, 435–441. [CrossRef] [PubMed]
149. Pelallo-Martínez, N.A.; Batres-Esquivel, L.; Carrizales-Yáñez, L.; Díaz-Barriga, F.M. Genotoxic and hematological effects in children exposed to a chemical mixture in a petrochemical area in Mexico. *Arch. Environ. Contam. Toxicol.* **2014**, *67*, 1–8. [CrossRef] [PubMed]
150. Pereira, T.S.; Beltrami, L.S.; Rocha, J.A.; Broto, F.P.; Comellas, L.R.; Salvadori, D.M.; Vargas, V.M. Toxicogenetic monitoring in urban cities exposed to different airborne contaminants. *Ecotoxicol. Environ. Saf.* **2013**, *90*, 174–182. [CrossRef] [PubMed]
151. Perez-Cadahia, B.; Laffon, B.; Pasaro, E.; Mendez, J. Genetic damage induced by accidental environmental pollutants. *Sci. World J.* **2006**, *6*, 1221–1237. [CrossRef] [PubMed]
152. Piperakis, S.M.; Petrakou, E.; Tsilimigaki, S. Effects of air pollution and smoking on DNA damage of human lymphocytes. *Environ. Mol. Mutagen.* **2000**, *36*, 243–249. [CrossRef] [PubMed]
153. Rojas, E.; Valverde, M.; Lopez, M.C.; Naufal, I.; Sanchez, I.; Bizarro, P.; Lopez, I.; Fortoul, T.I.; Ostrosky-Wegman, P. Evaluation of DNA damage in exfoliated tear duct epithelial cells from individuals exposed to air pollution assessed by single cell gel electrophoresis assay. *Mutat. Res.* **2000**, *468*, 11–17. [CrossRef]
154. Sanchez-Guerra, M.; Pelallo-Martinez, N.; Diaz-Barriga, F.; Rothenberg, S.J.; Hernandez-Cadena, L.; Faugeron, S.; Oropeza-Hernandez, L.F.; Guaderrama-Diaz, M.; Quintanilla-Vega, B. Environmental polycyclic aromatic hydrocarbon (PAH) exposure and DNA damage in Mexican children. *Mutat. Res.* **2012**, *742*, 66–71. [CrossRef] [PubMed]
155. Shermatov, K.; Zeyrek, D.; Yildirim, F.; Kilic, M.; Cebi, N.; Kocyigit, A. DNA damage in children exposed to secondhand cigarette smoke and its association with oxidative stress. *Indian Pediatr.* **2012**, *49*, 958–962. [CrossRef] [PubMed]
156. Sopian, N.A.; Jalaludin, J.; Abu Bakar, S.; Hamedon, T.R.; Latif, M.T. Exposure to Particulate PAHs on Potential Genotoxicity and Cancer Risk among School Children Living Near the Petrochemical Industry. *Int. J. Environ. Res. Public Health* **2021**, *18*, 2575. [CrossRef] [PubMed]
157. Torres-Dosal, A.; Perez-Maldonado, I.N.; Jasso-Pineda, Y.; Martinez Salinas, R.I.; Alegria-Torres, J.A.; Diaz-Barriga, F. Indoor air pollution in a Mexican indigenous community: Evaluation of risk reduction program using biomarkers of exposure and effect. *Sci. Total Environ.* **2008**, *390*, 362–368. [CrossRef] [PubMed]
158. Verschaeve, L.; Koppen, G.; Gorp, U.V.; Schoeters, G.; Jacobs, G.; Zwijzen, C. Seasonal variations in spontaneous levels of DNA damage; implication in the risk assessment of environmental chemicals. *J. Appl. Toxicol. JAT* **2007**, *27*, 612–620. [CrossRef] [PubMed]
159. Vinzents, P.S.; Moller, P.; Sorensen, M.; Knudsen, L.E.; Hertel, O.; Jensen, F.P.; Schibye, B.; Loft, S. Personal exposure to ultrafine particles and oxidative DNA damage. *Environ. Health Perspect.* **2005**, *113*, 1485–1490. [CrossRef]

160. Wilhelm, M.; Eberwein, G.; Holzer, J.; Gladtke, D.; Angerer, J.; Marczynski, B.; Behrendt, H.; Ring, J.; Sugiri, D.; Ranft, U. Influence of industrial sources on children's health—Hot spot studies in North Rhine Westphalia, Germany. *Int. J. Hyg. Environ. Health* **2007**, *210*, 591–599. [CrossRef]
161. Wu, F.Y.; Wu, H.D.; Yang, H.L.; Kuo, H.W.; Ying, J.C.; Lin, C.J.; Yang, C.C.; Lin, L.Y.; Chiu, T.H.; Lai, J.S. Associations among genetic susceptibility, DNA damage, and pregnancy outcomes of expectant mothers exposed to environmental tobacco smoke. *Sci. Total Environ.* **2007**, *386*, 124–133. [CrossRef] [PubMed]
162. Zani, C.; Ceretti, E.; Zerbini, I.; Viola, G.C.V.; Donato, F.; Gelatti, U.; Feretti, D. Comet Test in Saliva Leukocytes of Pre-School Children Exposed to Air Pollution in North Italy: The Respira Study. *Int. J. Environ. Res. Public Health* **2020**, *17*, 3276. [CrossRef] [PubMed]
163. Zani Zani, C.; Ceretti, E.; Feretti, D.; Villarini, M.; Moretti, M.; Verani, M.; De Donno, A.; Bonetta, S.; Buschini, A.; Bonetti, A.; et al. Winter Air Pollution and Genotoxic Effects in Children Living in a Highly Polluted Urban Area. *Atmosphere* **2021**, *12*, 1191. [CrossRef]
164. Zeller, J.; Neuss, S.; Mueller, J.U.; Kuhner, S.; Holzmann, K.; Hogel, J.; Klingmann, C.; Bruckner, T.; Triebig, G.; Speit, G. Assessment of genotoxic effects and changes in gene expression in humans exposed to formaldehyde by inhalation under controlled conditions. *Mutagenesis* **2011**, *26*, 555–561. [CrossRef]
165. Eftimova, B.; Sholjakova, M.; Mirakovski, D.; Hadzi-Nikolova, M. Health Effects Associated With Exposure to Anesthetic Gas Nitrous Oxide-N(2)O in Clinical Hospital—Shtip Personel. *Open Access Maced. J. Med. Sci.* **2017**, *5*, 800–804. [CrossRef] [PubMed]
166. Fodale, V.; Mondello, S.; Aloisi, C.; Schifilliti, D.; Santamaria, L. Genotoxic effects of anesthetic agents. *Expert. Opin. Drug Saf.* **2008**, *7*, 447–458. [CrossRef] [PubMed]
167. Schifilliti, D.; Mondello, S.; D'Arrigo, M.G.; Chille, G.; Fodale, V. Genotoxic effects of anesthetic agents: An update. *Expert. Opin. Drug Saf.* **2011**, *10*, 891–899. [CrossRef] [PubMed]
168. Kiani, F.; Jorfi, S.; Soltani, F.; Ghanbari, S.; Rezaee, R.; Mohammadi, M.J. Exposure to anesthetic gases in the operating rooms and assessment of non-carcinogenic risk among health care workers. *Toxicol. Rep.* **2023**, *11*, 1–8. [CrossRef] [PubMed]
169. Yilmaz, S.; Calbayram, N.C. Exposure to anesthetic gases among operating room personnel and risk of genotoxicity: A systematic review of the human biomonitoring studies. *J. Clin. Anesth.* **2016**, *35*, 326–331. [CrossRef] [PubMed]
170. Garcia-Alvarez, J.M.; Escribano-Sanchez, G.; Osuna, E.; Molina-Rodriguez, A.; Diaz-Agea, J.L.; Garcia-Sanchez, A. Occupational Exposure to Inhalational Anesthetics and Teratogenic Effects: A Systematic Review. *Healthcare* **2023**, *11*, 883. [CrossRef] [PubMed]
171. Nagella, A.B.; Ravishankar, M.; Hemanth Kumar, V.R. Anaesthesia practice and reproductive outcomes: Facts unveiled. *Indian J. Anaesth.* **2016**, *60*, 225. [CrossRef] [PubMed]
172. Nilsson, R.; Bjordal, C.; Andersson, M.; Bjordal, J.; Nyberg, A.; Welin, B.; Willman, A. Health risks and occupational exposure to volatile anaesthetics—A review with a systematic approach. *J. Clin. Nurs.* **2005**, *14*, 173–186. [CrossRef] [PubMed]
173. Aun, A.G.; Golim, M.A.; Nogueira, F.R.; Souza, K.M.; Arruda, N.M.; Braz, J.R.C.; Braz, L.G.; Braz, M.G. Monitoring early cell damage in physicians who are occupationally exposed to inhalational anesthetics. *Mutat. Res.* **2018**, *812*, 5–9. [CrossRef] [PubMed]
174. Baysal, Z.; Cengiz, M.; Ozgonul, A.; Cakir, M.; Celik, H.; Kocyigit, A. Oxidative status and DNA damage in operating room personnel. *Clin. Biochem.* **2009**, *42*, 189–193. [CrossRef] [PubMed]
175. Chandrasekhar, M.; Rekhadevi, P.V.; Sailaja, N.; Rahman, M.F.; Reddy, J.P.; Mahboob, M.; Grover, P. Evaluation of genetic damage in operating room personnel exposed to anaesthetic gases. *Mutagenesis* **2006**, *21*, 249–254. [CrossRef]
176. El-Ebiary, A.A.; Abuelfadl, A.A.; Sarhan, N.I.; Othman, M.M. Assessment of genotoxicity risk in operation room personnel by the alkaline comet assay. *Hum. Exp. Toxicol.* **2013**, *32*, 563–570. [CrossRef] [PubMed]
177. Figueiredo, D.B.S.; Aun, A.G.; Souza, K.M.; Nishimoto, I.H.; Silva, M.A.P.; de Carvalho, L.R.; Braz, L.G.; Braz, M.G. High anesthetic (isoflurane) indoor pollution is associated with genetic instability, cytotoxicity, and proliferative alterations in professionals working in a veterinary hospital. *Environ. Sci. Pollut. Res. Int.* **2022**, *29*, 71774–71784. [CrossRef] [PubMed]
178. Izdes, S.; Sardas, S.; Kadioglu, E.; Kaymak, C.; Ozcagli, E. Assessment of genotoxic damage in nurses occupationally exposed to anaesthetic gases or antineoplastic drugs by the comet assay. *J. Occup. Health* **2009**, *51*, 283–286. [CrossRef] [PubMed]
179. Izdes, S.; Sardas, S.; Kadioglu, E.; Karakaya, A.E. DNA damage, glutathione, and total antioxidant capacity in anesthesia nurses. *Arch. Environ. Occup. Health* **2010**, *65*, 211–217. [CrossRef]
180. Khisroon, M.; Humayun, M.; Khan, A.; Farooqi, J.; Humayun; Khan, J. Polymorphism in GSTM1 and GSTT1 genes influence DNA damage in personnel occupationally exposed to volatile anaesthetics (VA), from Peshawar, Pakistan. *Occup. Environ. Med.* **2020**, *77*, 769–774. [CrossRef] [PubMed]
181. Rozgaj, R.; Kasuba, V.; Brozovic, G.; Jazbec, A. Genotoxic effects of anaesthetics in operating theatre personnel evaluated by the comet assay and micronucleus test. *Int. J. Hyg. Environ. Health* **2009**, *212*, 11–17. [CrossRef] [PubMed]
182. Sardas, S.; Izdes, S.; Ozcagli, E.; Kanbak, O.; Kadioglu, E. The role of antioxidant supplementation in occupational exposure to waste anaesthetic gases. *Int. Arch. Occup. Environ. Health* **2006**, *80*, 154–159. [CrossRef] [PubMed]
183. Souza, K.M.; Braz, L.G.; Nogueira, F.R.; Souza, M.B.; Bincoleto, L.F.; Aun, A.G.; Corrente, J.E.; Carvalho, L.R.; Braz, J.R.C.; Braz, M.G. Occupational exposure to anesthetics leads to genomic instability, cytotoxicity and proliferative changes. *Mutat. Res.* **2016**, *791–792*, 42–48. [CrossRef] [PubMed]
184. Szyfter, K.; Szulc, R.; Mikstacki, A.; Stachecki, I.; Rydzanicz, M.; Jaloszynski, P. Genotoxicity of inhalation anaesthetics: DNA lesions generated by sevoflurane in vitro and in vivo. *J. Appl. Genet.* **2004**, *45*, 369–374. [PubMed]
185. Szyfter, K.; Stachecki, I.; Kostrzewska-Poczekaj, M.; Szaumkessel, M.; Szyfter-Harris, J.; Sobczynski, P. Exposure to volatile anaesthetics is not followed by a massive induction of single-strand DNA breaks in operation theatre personnel. *J. Appl. Genet.* **2016**, *57*, 343–348. [CrossRef] [PubMed]

186. Wronska-Nofer, T.; Palus, J.; Krajewski, W.; Jajte, J.; Kucharska, M.; Stetkiewicz, J.; Wasowicz, W.; Rydzynski, K. DNA damage induced by nitrous oxide: Study in medical personnel of operating rooms. *Mutat. Res.* **2009**, *666*, 39–43. [CrossRef] [PubMed]
187. Wronska-Nofer, T.; Nofer, J.R.; Jajte, J.; Dziubaltowska, E.; Szymczak, W.; Krajewski, W.; Wasowicz, W.; Rydzynski, K. Oxidative DNA damage and oxidative stress in subjects occupationally exposed to nitrous oxide (N(2)O). *Mutat Res.* **2012**, *731*, 58–63. [CrossRef] [PubMed]
188. Connor, T.H.; McDiarmid, M.A. Preventing occupational exposures to antineoplastic drugs in health care settings. *CA Cancer J. Clin.* **2006**, *56*, 354–365. [CrossRef]
189. Grosse, Y.; Baan, R.; Straif, K.; Secretan, B.; El Ghissassi, F.; Bouvard, V.; Benbrahim-Tallaa, L.; Guha, N.; Galichet, L.; Cogliano, V.; et al. A review of human carcinogens—Part A: Pharmaceuticals. *Lancet Oncol.* **2009**, *10*, 13–14. [CrossRef] [PubMed]
190. CDC. The National Institute for Occupational Safety and Health (NIOSH). Available online: http://www.cdc.gov/niosh/2004 (accessed on 8 May 2023).
191. Kopjar, N.; Garaj-Vrhovac, V.; Kasuba, V.; Rozgaj, R.; Ramic, S.; Pavlica, V.; Zeljezic, D. Assessment of genotoxic risks in Croatian health care workers occupationally exposed to cytotoxic drugs: A multi-biomarker approach. *Int. J. Hyg. Environ. Health* **2009**, *212*, 414–431. [CrossRef]
192. Mahboob, M.; Rahman, M.F.; Rekhadevi, P.V.; Sailaja, N.; Balasubramanyam, A.; Prabhakar, P.V.; Singh, S.P.; Reddy, U.A.; Rao, G.S.; Grover, P. Monitoring of oxidative stress in nurses occupationally exposed to antineoplastic drugs. *Toxicol. Int.* **2012**, *19*, 20–24. [CrossRef]
193. Ziegler, E.; Mason, H.J.; Baxter, P.J. Occupational exposure to cytotoxic drugs in two UK oncology wards. *Occup. Environ. Med.* **2002**, *59*, 608–612. [CrossRef] [PubMed]
194. Valanis, B.G.; Vollmer, W.M.; Labuhn, K.T.; Glass, A.G. Association of antineoplastic drug handling with acute adverse effects in pharmacy personnel. *Am. J. Hosp. Pharm.* **1993**, *50*, 455–462. [CrossRef] [PubMed]
195. Fransman, W.; Kager, H.; Meijster, T.; Heederik, D.; Kromhout, H.; Portengen, L.; Blaauboer, B.J. Leukemia from dermal exposure to cyclophosphamide among nurses in The Netherlands: Quantitative assessment of the risk. *Ann. Occup. Hyg.* **2014**, *58*, 271–282. [CrossRef] [PubMed]
196. Kopjar, N.; Garaj-Vrhovac, V. Application of the alkaline comet assay in human biomonitoring for genotoxicity: A study on Croatian medical personnel handling antineoplastic drugs. *Mutagenesis* **2001**, *16*, 71–78. [CrossRef] [PubMed]
197. Skov, T.; Maarup, B.; Olsen, J.; Rorth, M.; Winthereik, H.; Lynge, E. Leukaemia and reproductive outcome among nurses handling antineoplastic drugs. *Br. J. Ind. Med.* **1992**, *49*, 855–861. [CrossRef] [PubMed]
198. Gunnarsdottir, H.K.; Aspelund, T.; Karlsson, T.; Rafnsson, V.V. Occupational Risk Factors for Breast Cancer among Nurses. *Int. J. Occup. Environ. Health* **1997**, *3*, 254–258. [CrossRef] [PubMed]
199. Ratner, P.A.; Spinelli, J.J.; Beking, K.; Lorenzi, M.; Chow, Y.; Teschke, K.; Le, N.D.; Gallagher, R.P.; Dimich-Ward, H. Cancer incidence and adverse pregnancy outcome in registered nurses potentially exposed to antineoplastic drugs. *BMC Nurs.* **2010**, *9*, 15. [CrossRef] [PubMed]
200. Maluf, S.W.; Erdtmann, B. Follow-up study of the genetic damage in lymphocytes of pharmacists and nurses handling antineoplastic drugs evaluated by cytokinesis-block micronuclei analysis and single cell gel electrophoresis assay. *Mutat. Res.* **2000**, *471*, 21–27. [CrossRef] [PubMed]
201. Ursini, C.L.; Cavallo, D.; Colombi, A.; Giglio, M.; Marinaccio, A.; Iavicoli, S. Evaluation of early DNA damage in healthcare workers handling antineoplastic drugs. *Int. Arch. Occup. Environ. Health* **2006**, *80*, 134–140. [CrossRef]
202. Yoshida, J.; Kosaka, H.; Tomioka, K.; Kumagai, S. Genotoxic risks to nurses from contamination of the work environment with antineoplastic drugs in Japan. *J. Occup. Health* **2006**, *48*, 517–522. [CrossRef] [PubMed]
203. Rekhadevi, P.V.; Sailaja, N.; Chandrasekhar, M.; Mahboob, M.; Rahman, M.F.; Grover, P. Genotoxicity assessment in oncology nurses handling anti-neoplastic drugs. *Mutagenesis* **2007**, *22*, 395–401. [CrossRef] [PubMed]
204. Cornetta, T.; Padua, L.; Testa, A.; Ievoli, E.; Festa, F.; Tranfo, G.; Baccelliere, L.; Cozzi, R. Molecular biomonitoring of a population of nurses handling antineoplastic drugs. *Mutat. Res.* **2008**, *638*, 75–82. [CrossRef] [PubMed]
205. Rombaldi, F.; Cassini, C.; Salvador, M.; Saffi, J.; Erdtmann, B. Occupational risk assessment of genotoxicity and oxidative stress in workers handling anti-neoplastic drugs during a working week. *Mutagenesis* **2009**, *24*, 143–148. [CrossRef] [PubMed]
206. Sasaki, M.; Dakeishi, M.; Hoshi, S.; Ishii, N.; Murata, K. Assessment of DNA damage in Japanese nurses handling antineoplastic drugs by the comet assay. *J. Occup. Health* **2008**, *50*, 7–12. [CrossRef] [PubMed]
207. Connor, T.H.; DeBord, D.G.; Pretty, J.R.; Oliver, M.S.; Roth, T.S.; Lees, P.S.; Krieg, E.F., Jr.; Rogers, B.; Escalante, C.P.; Toennis, C.A.; et al. Evaluation of antineoplastic drug exposure of health care workers at three university-based US cancer centers. *J. Occup. Environ. Med.* **2010**, *52*, 1019–1027. [CrossRef] [PubMed]
208. Villarini, M.; Dominici, L.; Piccinini, R.; Fatigoni, C.; Ambrogi, M.; Curti, G.; Morucci, P.; Muzi, G.; Monarca, S.; Moretti, M. Assessment of primary, oxidative and excision repaired DNA damage in hospital personnel handling antineoplastic drugs. *Mutagenesis* **2011**, *26*, 359–369. [CrossRef] [PubMed]
209. Buschini, A.; Villarini, M.; Feretti, D.; Mussi, F.; Dominici, L.; Zerbini, I.; Moretti, M.; Ceretti, E.; Bonfiglioli, R.; Carrieri, M.; et al. Multicentre study for the evaluation of mutagenic/carcinogenic risk in nurses exposed to antineoplastic drugs: Assessment of DNA damage. *Occup. Environ. Med.* **2013**, *70*, 789–794. [CrossRef] [PubMed]
210. Ladeira, C.; Viegas, V.; Pádua, M.; Carolino, E.; Gomes, M.C.; Brito, M. Relation between DNA damage measured by comet assay and OGG1 Ser326Cys polymorphism in antineoplastic drugs biomonitoring. *AIMS Genet.* **2015**, *2*, 204–2018. [CrossRef]

211. Oltulu, C.; Yesil Devecioglu, T.; Akinci, M.; Akgun Olmez, S.G.; Obeidin, S.V.; Beceren, A. Evaluation of Genotoxicity Risk in Health Care Workers Exposed to Antineoplastic Drugs. *Clin. Exp. Health Sci.* **2019**, *9*, 166–170. [CrossRef]
212. Aristizabal-Pachon, A.F.; Castillo, W.O. Genotoxic evaluation of occupational exposure to antineoplastic drugs. *Toxicol. Res.* **2020**, *36*, 29–36. [CrossRef] [PubMed]
213. Huang, X.; Gao, C.; Cai, W.; Tao, Y.; Zhong, X.; Liu, H.; Hong, X.; Ding, X.; Lu, H.; Lai, W.; et al. Effect of occupational exposure to antineoplastic drugs on DNA damage in nurses: A cross-sectional study. *Occup. Environ. Med.* **2022**, *79*, 253–258. [CrossRef] [PubMed]
214. Hongping, D.; Jianlin, L.; Meibian, Z.; Wei, W.; Lifen, J.; Shijie, C.; Wei, Z.; Baohong, W.; Jiliang, H. Detecting the cytogenetic effects in workers occupationally exposed to vincristine with four genetic tests. *Mutat. Res.* **2006**, *599*, 152–159. [CrossRef] [PubMed]
215. Omrane, F.; Gargouri, I.; Khadhraoui, M.; Elleuch, B.; Zmirou-Navier, D. Risk assessment of occupational exposure to heavy metal mixtures: A study protocol. *BMC Public Health* **2018**, *18*, 314. [CrossRef]
216. Wiwanitkit, V. Minor heavy metal: A review on occupational and environmental intoxication. *Indian. J. Occup. Environ. Med.* **2008**, *12*, 116–121. [CrossRef] [PubMed]
217. Tchounwou, P.B.; Yedjou, C.G.; Patlolla, A.K.; Sutton, D.J. Heavy metal toxicity and the environment. *Exp. Suppl.* **2012**, *101*, 133–164. [CrossRef] [PubMed]
218. Aksu, İ.; Anlar, H.G.; Taner, G.; Bacanlı, M.; İritaş, S.; Tutkun, E.; Basaran, N. Assessment of DNA damage in welders using comet and micronucleus assays. *Mutat. Res. Genet. Toxicol. Environ. Mutagen.* **2019**, *843*, 40–45. [CrossRef] [PubMed]
219. Balachandar, V.; Arun, M.; Mohana Devi, S.; Velmurugan, P.; Manikantan, P.; Karthick Kumar, A.; Sasikala, K.; Venkatesan, C. Evaluation of the genetic alterations in direct and indirect exposures of hexavalent chromium [Cr(VI)] in leather tanning industry workers North Arcot District, South India. *Int. Arch. Occup. Environ. Health* **2010**, *83*, 791–801. [CrossRef] [PubMed]
220. Batra, J.; Thakur, A.; Deepak, J.; Shrawan, K. Lead Induced Oxidative DNA Damage among the Occupationally Exposed Workers: A Case-Control Study. *J. Clin. Diagn. Res.* **2020**, *14*, 12–16. [CrossRef]
221. Cavallo, D.; Iavicoli, I.; Setini, A.; Marinaccio, A.; Perniconi, B.; Carelli, G.; Iavicoli, S. Genotoxic risk and oxidative DNA damage in workers exposed to antimony trioxide. *Environ. Mol. Mutagen.* **2002**, *40*, 184–189. [CrossRef] [PubMed]
222. Chinde, S.; Kumari, M.; Devi, K.R.; Murty, U.S.; Rahman, M.F.; Kumari, S.I.; Mahboob, M.; Grover, P. Assessment of genotoxic effects of lead in occupationally exposed workers. *Environ. Sci. Pollut. Res. Int.* **2014**, *21*, 11469–11480. [CrossRef] [PubMed]
223. Coelho, P.; García-Lestón, J.; Costa, S.; Costa, C.; Silva, S.; Dall'Armi, V.; Zoffoli, R.; Bonassi, S.; de Lima, J.P.; Gaspar, J.F.; et al. Genotoxic effect of exposure to metal(loid)s. A molecular epidemiology survey of populations living and working in Panasqueira mine area, Portugal. *Environ. Int.* **2013**, *60*, 163–170. [CrossRef] [PubMed]
224. Danadevi, K.; Rozati, R.; Saleha Banu, B.; Hanumanth Rao, P.; Grover, P. DNA damage in workers exposed to lead using comet assay. *Toxicology* **2003**, *187*, 183–193. [CrossRef]
225. Danadevi, K.; Rozati, R.; Banu, B.S.; Grover, P. Genotoxic evaluation of welders occupationally exposed to chromium and nickel using the Comet and micronucleus assays. *Mutagenesis* **2004**, *19*, 35–41. [CrossRef] [PubMed]
226. De Olivera, J.V.; Boufleur, L.A.; Dos Santos, C.E.; Dias, J.F.; Squeff, C.H.; Silva, G.R.; Ianistcki, M.; Benvegnu, V.C.; Da Silva, J. Occupational genotoxicity among copper smelters. *Toxicol. Ind. Health* **2012**, *28*, 789–795. [CrossRef] [PubMed]
227. Restrepo, H.G.; Sicard, D.; Torres, M.M. DNA damage and repair in cells of lead exposed people. *Am. J. Ind. Med.* **2000**, *38*, 330–334. [CrossRef] [PubMed]
228. Fracasso, M.E.; Perbellini, L.; Soldà, S.; Talamini, G.; Franceschetti, P. Lead induced DNA strand breaks in lymphocytes of exposed workers: Role of reactive oxygen species and protein kinase C. *Mutat. Res.* **2002**, *515*, 159–169. [CrossRef] [PubMed]
229. Gambelunghe, A.; Piccinini, R.; Ambrogi, M.; Villarini, M.; Moretti, M.; Marchetti, C.; Abbritti, G.; Muzi, G. Primary DNA damage in chrome-plating workers. *Toxicology* **2003**, *188*, 187–195. [CrossRef] [PubMed]
230. García-Lestón, J.; Roma-Torres, J.; Vilares, M.; Pinto, R.; Cunha, L.M.; Prista, J.; Teixeira, J.P.; Mayan, O.; Pásaro, E.; Méndez, J.; et al. Biomonitoring of a population of Portuguese workers exposed to lead. *Mutat. Res.* **2011**, *721*, 81–88. [CrossRef] [PubMed]
231. Grover, P.; Rekhadevi, P.V.; Danadevi, K.; Vuyyuri, S.B.; Mahboob, M.; Rahman, M.F. Genotoxicity evaluation in workers occupationally exposed to lead. *Int. J. Hyg. Environ. Health* **2010**, *213*, 99–106. [CrossRef] [PubMed]
232. Hernández-Franco, P.; Maldonado-Vega, M.; Calderón-Salinas, J.V.; Rojas, E.; Valverde, M. Role of Apel in Impaired DNA Repair Capacity in Battery Recycling Plant Workers Exposed to Lead. *Int. J. Environ. Res. Public Health* **2022**, *19*, 7961. [CrossRef] [PubMed]
233. Iarmarcovai, G.; Sari-Minodier, I.; Chaspoul, F.; Botta, C.; De Méo, M.; Orsière, T.; Bergé-Lefranc, J.L.; Gallice, P.; Botta, A. Risk assessment of welders using analysis of eight metals by ICP-MS in blood and urine and DNA damage evaluation by the comet and micronucleus assays; influence of XRCC1 and XRCC3 polymorphisms. *Mutagenesis* **2005**, *20*, 425–432. [CrossRef] [PubMed]
234. Kašuba, V.; Rozgaj, R.; Milić, M.; Zelježić, D.; Kopjar, N.; Pizent, A.; Kljaković-Gašpić, Z.; Jazbec, A. Evaluation of genotoxic effects of lead in pottery-glaze workers using micronucleus assay, alkaline comet assay and DNA diffusion assay. *Int. Arch. Occup. Environ. Health* **2012**, *85*, 807–818. [CrossRef] [PubMed]
235. Kašuba, V.; Milić, M.; Želježić, D.; Mladinić, M.; Pizent, A.; Kljaković-Gašpić, Z.; Balija, M.; Jukić, I. Biomonitoring findings for occupational lead exposure in battery and ceramic tile workers using biochemical markers, alkaline comet assay, and micronucleus test coupled with fluorescence in situ hybridisation. *Arh. Hig. Rada Toksikol.* **2020**, *71*, 339–352. [CrossRef] [PubMed]
236. Kayaaltı, Z.; Yavuz, İ.; Söylemez, E.; Bacaksız, A.; Tutkun, E.; Sayal, A.; Söylemezoğlu, T. Evaluation of DNA damage using 3 comet assay parameters in workers occupationally exposed to lead. *Arch. Environ. Occup. Health* **2015**, *70*, 120–125. [CrossRef]
237. Khisroon, M.; Khan, A.; Shah, A.A.; Ullah, I.; Farooqi, J.; Ullah, A. Scalp Hair Metal Analysis Concerning DNA Damage in Welders of Peshawar Khyber Pakhtunkhwa Pakistan. *Biol. Trace Elem. Res.* **2021**, *199*, 1649–1656. [CrossRef]

238. Liu, N.; Guan, Y.; Xue, L.; Yu, Y.; Xiao, J.; Chang, Z.; Li, Q.; Bai, Y.; Li, B.; Guan, W. Assessment of DNA/Chromosome Damage in the Peripheral Blood Lymphocytes of Workers Exposed to Indium Compounds. *Toxicol. Sci.* **2017**, *157*, 41–49. [CrossRef] [PubMed]
239. Zhang, M.; Chen, Z.; Chen, Q.; Zou, H.; Lou, J.; He, J. Investigating DNA damage in tannery workers occupationally exposed to trivalent chromium using comet assay. *Mutat. Res.* **2008**, *654*, 45–51. [CrossRef] [PubMed]
240. Minozzo, R.; Deimling, L.I.; Santos-Mello, R. Cytokinesis-blocked micronucleus cytome and comet assays in peripheral blood lymphocytes of workers exposed to lead considering folate and vitamin B12 status. *Mutat. Res.* **2010**, *697*, 24–32. [CrossRef] [PubMed]
241. Muller, C.D.; Garcia, S.C.; Brucker, N.; Goethel, G.; Sauer, E.; Lacerda, L.M.; Oliveira, E.; Trombini, T.L.; Machado, A.B.; Pressotto, A.; et al. Occupational risk assessment of exposure to metals in chrome plating workers. *Drug Chem. Toxicol.* **2022**, *45*, 560–567. [CrossRef] [PubMed]
242. Olewińska, E.; Kasperczyk, A.; Kapka, L.; Kozłowska, A.; Pawlas, N.; Dobrakowski, M.; Birkner, E.; Kasperczyk, S. Level of DNA damage in lead-exposed workers. *Ann. Agric. Environ. Med.* **2010**, *17*, 231–236.
243. Palus, J.; Rydzynski, K.; Dziubaltowska, E.; Wyszynska, K.; Natarajan, A.T.; Nilsson, R. Genotoxic effects of occupational exposure to lead and cadmium. *Mutat. Res.* **2003**, *540*, 19–28. [CrossRef]
244. Palus, J.; Lewinska, D.; Dziubaltowska, E.; Stepnik, M.; Beck, J.; Rydzynski, K.; Nilsson, R. DNA damage in leukocytes of workers occupationally exposed to arsenic in copper smelters. *Environ. Mol. Mutagen.* **2005**, *46*, 81–87. [CrossRef] [PubMed]
245. Pandeh, M.; Fathi, S.; Zare Sakhvidi, M.J.; Zavar Reza, J.; Sedghian, L. Oxidative stress and early DNA damage in workers exposed to iron-rich metal fumes. *Environ. Sci. Pollut. Res. Int.* **2017**, *24*, 9645–9650. [CrossRef] [PubMed]
246. Pawlas, N.; Olewińska, E.; Markiewicz-Górka, I.; Kozłowska, A.; Januszewska, L.; Lundh, T.; Januszewska, E.; Pawlas, K. Oxidative damage of DNA in subjects occupationally exposed to lead. *Adv. Clin. Exp. Med.* **2017**, *26*, 939–945. [CrossRef] [PubMed]
247. Pérez-Cadahía, B.; Méndez, J.; Pásaro, E.; Lafuente, A.; Cabaleiro, T.; Laffon, B. Biomonitoring of human exposure to prestige oil: Effects on DNA and endocrine parameters. *Environ. Health Insights* **2008**, *2*, 83–92. [CrossRef] [PubMed]
248. Rashid, S.; Arshad, M.; Siddiqa, M.; Ahmad, R. Evaluation of DNA damage in traffic police wardens of Pakistan due to cadmium and zinc. *Sci. Total Environ.* **2018**, *630*, 1360–1364. [CrossRef] [PubMed]
249. Singh, Z.; Chadha, P. Assessment of DNA damage as an index of genetic toxicity in welding microenvironments among iron-based industries. *Toxicol. Ind. Health* **2016**, *32*, 1817–1824. [CrossRef] [PubMed]
250. Wang, Y.; Sun, X.; Fang, L.; Li, K.; Yang, P.; Du, L.; Ji, K.; Wang, J.; Liu, Q.; Xu, C.; et al. Genomic instability in adult men involved in processing electronic waste in Northern China. *Environ. Int.* **2018**, *117*, 69–81. [CrossRef] [PubMed]
251. Wani, A.L.; Ahmad, A.; Shadab, G.G.; Usmani, J.A. Possible role of zinc in diminishing lead-related occupational stress-a zinc nutrition concern. *Environ. Sci. Pollut. Res. Int.* **2017**, *24*, 8682–8691. [CrossRef] [PubMed]
252. Vuyyuri, S.B.; Ishaq, M.; Kuppala, D.; Grover, P.; Ahuja, Y.R. Evaluation of micronucleus frequencies and DNA damage in glass workers exposed to arsenic. *Environ. Mol. Mutagen.* **2006**, *47*, 562–570. [CrossRef] [PubMed]
253. Zhang, X.H.; Zhang, X.; Wang, X.C.; Jin, L.F.; Yang, Z.P.; Jiang, C.X.; Chen, Q.; Ren, X.B.; Cao, J.Z.; Wang, Q.; et al. Chronic occupational exposure to hexavalent chromium causes DNA damage in electroplating workers. *BMC Public Health* **2011**, *11*, 224. [CrossRef] [PubMed]
254. Chen, Z.; Lou, J.; Chen, S.; Zheng, W.; Wu, W.; Jin, L.; Deng, H.; He, J. Evaluating the genotoxic effects of workers exposed to lead using micronucleus assay, comet assay and TCR gene mutation test. *Toxicology* **2006**, *223*, 219–226. [CrossRef]
255. Andrew, A.S.; Burgess, J.L.; Meza, M.M.; Demidenko, E.; Waugh, M.G.; Hamilton, J.W.; Karagas, M.R. Arsenic exposure is associated with decreased DNA repair in vitro and in individuals exposed to drinking water arsenic. *Environ. Health Perspect.* **2006**, *114*, 1193–1198. [CrossRef]
256. Banerjee, M.; Sarma, N.; Biswas, R.; Roy, J.; Mukherjee, A.; Giri, A.K. DNA repair deficiency leads to susceptibility to develop arsenic-induced premalignant skin lesions. *Int. J. Cancer* **2008**, *123*, 283–287. [CrossRef] [PubMed]
257. Basu, A.; Som, A.; Ghoshal, S.; Mondal, L.; Chaubey, R.C.; Bhilwade, H.N.; Rahman, M.M.; Giri, A.K. Assessment of DNA damage in peripheral blood lymphocytes of individuals susceptible to arsenic induced toxicity in West Bengal, India. *Toxicol. Lett.* **2005**, *159*, 100–112. [CrossRef] [PubMed]
258. Cruz-Esquivel, Á.; Marrugo-Negrete, J.; Calao-Ramos, C. Genetic damage in human populations at mining sites in the upper basin of the San Jorge River, Colombia. *Environ. Sci. Pollut. Res. Int.* **2019**, *26*, 10961–10971. [CrossRef] [PubMed]
259. David, M.; Turi, N.; Ain, Q.U.; Rahman, H.; Jahan, S. Evaluation of environmental effects of heavy metals on biochemical profile and oxidative stress among children at brick kiln sites. *Arch. Environ. Occup. Health* **2021**, *76*, 441–449. [CrossRef] [PubMed]
260. Franken, C.; Koppen, G.; Lambrechts, N.; Govarts, E.; Bruckers, L.; Den Hond, E.; Loots, I.; Nelen, V.; Sioen, I.; Nawrot, T.S.; et al. Environmental exposure to human carcinogens in teenagers and the association with DNA damage. *Environ. Res.* **2017**, *152*, 165–174. [CrossRef] [PubMed]
261. Jasso-Pineda, Y.; Díaz-Barriga, F.; Calderón, J.; Yáñez, L.; Carrizales, L.; Pérez-Maldonado, I.N. DNA damage and decreased DNA repair in peripheral blood mononuclear cells in individuals exposed to arsenic and lead in a mining site. *Biol. Trace Elem. Res.* **2012**, *146*, 141–149. [CrossRef] [PubMed]
262. Jasso-Pineda, Y.; Espinosa-Reyes, G.; Gonzalez-Mille, D.; Razo-Soto, I.; Carrizales, L.; Torres-Dosal, A.; Mejia-Saavedra, J.; Monroy, M.; Ize, A.I.; Yarto, M.; et al. An integrated health risk assessment approach to the study of mining sites contaminated with arsenic and lead. *Integr. Environ. Assess. Manag.* **2007**, *3*, 344–350. [CrossRef] [PubMed]
263. Khan, F.H.; Ambreen, K.; Fatima, G.; Kumar, S. Assessment of health risks with reference to oxidative stress and DNA damage in chromium exposed population. *Sci. Total Environ.* **2012**, *430*, 68–74. [CrossRef] [PubMed]

264. Lourenço, J.; Pereira, R.; Pinto, F.; Caetano, T.; Silva, A.; Carvalheiro, T.; Guimarães, A.; Gonçalves, F.; Paiva, A.; Mendo, S. Biomonitoring a human population inhabiting nearby a deactivated uranium mine. *Toxicology* **2013**, *305*, 89–98. [CrossRef] [PubMed]
265. Mendez-Gomez, J.; Garcia-Vargas, G.G.; Lopez-Carrillo, L.; Calderon-Aranda, E.S.; Gomez, A.; Vera, E.; Valverde, M.; Cebrian, M.E.; Rojas, E. Genotoxic effects of environmental exposure to arsenic and lead on children in region Lagunera, Mexico. *Ann. N. Y. Acad. Sci.* **2008**, *1140*, 358–367. [CrossRef] [PubMed]
266. Sampayo-Reyes, A.; Hernández, A.; El-Yamani, N.; López-Campos, C.; Mayet-Machado, E.; Rincón-Castañeda, C.B.; Limones-Aguilar Mde, L.; López-Campos, J.E.; de León, M.B.; González-Hernández, S.; et al. Arsenic induces DNA damage in environmentally exposed Mexican children and adults. Influence of GSTO1 and AS3MT polymorphisms. *Toxicol. Sci.* **2010**, *117*, 63–71. [CrossRef]
267. Staessen, J.A.; Nawrot, T.; Hond, E.D.; Thijs, L.; Fagard, R.; Hoppenbrouwers, K.; Koppen, G.; Nelen, V.; Schoeters, G.; Vanderschueren, D.; et al. Renal function, cytogenetic measurements, and sexual development in adolescents in relation to environmental pollutants: A feasibility study of biomarkers. *Lancet* **2001**, *357*, 1660–1669. [CrossRef] [PubMed]
268. Wu, W.T.; Liou, S.H.; Lin, K.J.; Liu, T.E.; Liu, S.H.; Chen, C.Y.; Sung, F.C.; Wu, T.N. Changing blood lead levels and DNA damage (comet assay) among immigrant women in Taiwan. *Sci. Total Environ.* **2009**, *407*, 5931–5936. [CrossRef] [PubMed]
269. Yáñez, L.; García-Nieto, E.; Rojas, E.; Carrizales, L.; Mejía, J.; Calderón, J.; Razo, I.; Díaz-Barriga, F. DNA damage in blood cells from children exposed to arsenic and lead in a mining area. *Environ. Res.* **2003**, *93*, 231–240. [CrossRef] [PubMed]
270. Duffus, J.H. "Heavy metals" a meaningless term? (IUPAC Technical Report). *Pure Appl. Chem.* **2002**, *74*, 793–807. [CrossRef]
271. Kanno, T.; Nakamura, K.; Ikai, H.; Kikuchi, K.; Sasaki, K.; Niwano, Y. Literature review of the role of hydroxyl radicals in chemically-induced mutagenicity and carcinogenicity for the risk assessment of a disinfection system utilizing photolysis of hydrogen peroxide. *J. Clin. Biochem. Nutr.* **2012**, *51*, 9–14. [CrossRef] [PubMed]
272. United Nations Environment Programme. Synthesis Report on the Environmental and Health Impacts of Pesticides and Fertilizers and Ways to Minimize Them. 2022. Available online: https://wedocs.unep.org/xmlui/bitstream/handle/20.500.11822/38409/pesticides.pdf (accessed on 10 June 2023).
273. El-Nahhal, Y.; El-Nahhal, I. Cardiotoxicity of some pesticides and their amelioration. *Environ. Sci. Pollut. Res. Int.* **2021**, *28*, 44726–44754. [CrossRef] [PubMed]
274. Gilden, R.C.; Huffling, K.; Sattler, B. Pesticides and health risks. *J. Obstet. Gynecol. Neonatal Nurs. JOGNN* **2010**, *39*, 103–110. [CrossRef] [PubMed]
275. Nicolopoulou-Stamati, P.; Maipas, S.; Kotampasi, C.; Stamatis, P.; Hens, L. Chemical Pesticides and Human Health: The Urgent Need for a New Concept in Agriculture. *Front. Public Health* **2016**, *4*, 148. [CrossRef]
276. Food and Agriculture Organization. FAO Stat [PESTICIDES]. License: CC BY-NC-SA 3.0 IGO. 2023. Available online: https://www.fao.org/faostat/en/#data/RP (accessed on 14 August 2023).
277. Fenner, K.; Canonica, S.; Wackett, L.P.; Elsner, M. Evaluating pesticide degradation in the environment: Blind spots and emerging opportunities. *Science* **2013**, *341*, 752–758. [CrossRef] [PubMed]
278. Kim, K.H.; Kabir, E.; Jahan, S.A. Exposure to pesticides and the associated human health effects. *Sci. Total Environ.* **2017**, *575*, 525–535. [CrossRef]
279. Carvalho, F. Pesticides, environment, and food safety. *Food Energy Secur.* **2017**, *6*, 48–60. [CrossRef]
280. Alavanja, M.C.; Ross, M.K.; Bonner, M.R. Increased cancer burden among pesticide applicators and others due to pesticide exposure. *CA Cancer J. Clin.* **2013**, *63*, 120–142. [CrossRef] [PubMed]
281. Pascale, A.; Laborde, A. Impact of pesticide exposure in childhood. *Rev. Environ. Health* **2020**, *35*, 221–227. [CrossRef] [PubMed]
282. Bonner, M.R.; Freeman, L.E.; Hoppin, J.A.; Koutros, S.; Sandler, D.P.; Lynch, C.F.; Hines, C.J.; Thomas, K.; Blair, A.; Alavanja, M.C. Occupational Exposure to Pesticides and the Incidence of Lung Cancer in the Agricultural Health Study. *Environ. Health Perspect.* **2017**, *125*, 544–551. [CrossRef] [PubMed]
283. Hofmann, J.N.; Beane Freeman, L.E.; Murata, K.; Andreotti, G.; Shearer, J.J.; Thoren, K.; Ramanathan, L.; Parks, C.G.; Koutros, S.; Lerro, C.C.; et al. Lifetime Pesticide Use and Monoclonal Gammopathy of Undetermined Significance in a Prospective Cohort of Male Farmers. *Environ. Health Perspect.* **2021**, *129*, 17003. [CrossRef] [PubMed]
284. Lee, G.H.; Choi, K.C. Adverse effects of pesticides on the functions of immune system. Comparative biochemistry and physiology. *Toxicol. Pharmacol. CBP* **2020**, *235*, 108789. [CrossRef]
285. Lerro, C.C.; Beane Freeman, L.E.; DellaValle, C.T.; Andreotti, G.; Hofmann, J.N.; Koutros, S.; Parks, C.G.; Shrestha, S.; Alavanja, M.C.R.; Blair, A.; et al. Pesticide exposure and incident thyroid cancer among male pesticide applicators in agricultural health study. *Environ. Int.* **2021**, *146*, 106187. [CrossRef] [PubMed]
286. Sule, R.O.; Condon, L.; Gomes, A.V. A Common Feature of Pesticides: Oxidative Stress-The Role of Oxidative Stress in Pesticide-Induced Toxicity. *Oxidative Med. Cell. Longev.* **2022**, *2022*, 5563759. [CrossRef] [PubMed]
287. Burns, C.J.; Juberg, D.R. Cancer and occupational exposure to pesticides: An umbrella review. *Int. Arch. Occup. Environ. Health* **2021**, *94*, 945–957. [CrossRef]
288. Abhishek, S.; Kaur, N.; Kaur, S.; Lata, M.; Sharma, J.K.; Sharma, A. Association of GSTM1 and GSTT1 gene deletions with susceptibility to DNA damage in the pesticide-exposed workers of Punjab. *Rejuvenation Res.* **2010**, *13*, 281–284. [CrossRef] [PubMed]
289. Aiassa, D.E.; Manas, F.J.; Gentile, N.E.; Bosch, B.; Salinero, M.C.; Gorla, N.B.M. Evaluation of genetic damage in pesticides applicators from the province of Cordoba, Argentina. *Environ. Sci. Pollut. Res. Int.* **2019**, *26*, 20981–20988. [CrossRef] [PubMed]

290. Ali, T.; Ismail, M.; Asad, F.; Ashraf, A.; Waheed, U.; Khan, Q.M. Pesticide genotoxicity in cotton picking women in Pakistan evaluated using comet assay. *Drug Chem. Toxicol.* **2018**, *41*, 213–220. [CrossRef] [PubMed]
291. Alves, J.S.; da Silva, F.R.; da Silva, G.F.; Salvador, M.; Kvitko, K.; Rohr, P.; dos Santos, C.E.; Dias, J.F.; Henriques, J.A.; da Silva, J. Investigation of potential biomarkers for the early diagnosis of cellular stability after the exposure of agricultural workers to pesticides. *An. Da Acad. Bras. De. Cienc.* **2016**, *88*, 349–360. [CrossRef] [PubMed]
292. Arshad, M.; Siddiqa, M.; Rashid, S.; Hashmi, I.; Awan, M.A.; Ali, M.A. Biomonitoring of Toxic Effects of Pesticides in Occupationally Exposed Individuals. *Saf. Health Work.* **2016**, *7*, 156–160. [CrossRef]
293. Benedetti, D.; Nunes, E.; Sarmento, M.; Porto, C.; Dos Santos, C.E.; Dias, J.F.; da Silva, J. Genetic damage in soybean workers exposed to pesticides: Evaluation with the comet and buccal micronucleus cytome assays. *Mutat. Res.* **2013**, *752*, 28–33. [CrossRef] [PubMed]
294. Bhalli, J.A.; Khan, Q.M.; Nasim, A. DNA damage in Pakistani pesticide-manufacturing workers assayed using the Comet assay. *Environ. Mol. Mutagen.* **2006**, *47*, 587–593. [CrossRef]
295. Bhalli, J.A.; Ali, T.; Asi, M.R.; Khalid, Z.M.; Ceppi, M.; Khan, Q.M. DNA damage in Pakistani agricultural workers exposed to mixture of pesticides. *Environ. Mol. Mutagen.* **2009**, *50*, 37–45. [CrossRef]
296. Bian, Q.; Xu, L.C.; Wang, S.L.; Xia, Y.K.; Tan, L.F.; Chen, J.F.; Song, L.; Chang, H.C.; Wang, X.R. Study on the relation between occupational fenvalerate exposure and spermatozoa DNA damage of pesticide factory workers. *Occup. Environ. Med.* **2004**, *61*, 999–1005. [CrossRef] [PubMed]
297. Carbajal-Lopez, Y.; Gomez-Arroyo, S.; Villalobos-Pietrini, R.; Calderon-Segura, M.E.; Martinez-Arroyo, A. Biomonitoring of agricultural workers exposed to pesticide mixtures in Guerrero state, Mexico, with comet assay and micronucleus test. *Environ. Sci. Pollut. Res. Int.* **2016**, *23*, 2513–2520. [CrossRef] [PubMed]
298. Cayir, A.; Coskun, M.; Coskun, M.; Cobanoglu, H. Comet assay for assessment of DNA damage in greenhouse workers exposed to pesticides. *Biomarkers* **2019**, *24*, 592–599. [CrossRef] [PubMed]
299. Chen, C.C.; Huang, C.H.; Wu, M.T.; Chou, C.H.; Huang, C.C.; Tseng, T.Y.; Chang, F.Y.; Li, Y.T.; Tsai, C.C.; Wang, T.S.; et al. Multidrug resistance 1 gene variants, pesticide exposure, and increased risk of DNA damage. *Biomed. Res. Int.* **2014**, *2014*, 965729. [CrossRef]
300. Costa, C.; Garcia-Leston, J.; Costa, S.; Coelho, P.; Silva, S.; Pingarilho, M.; Valdiglesias, V.; Mattei, F.; Dall'Armi, V.; Bonassi, S.; et al. Is organic farming safer to farmers' health? A comparison between organic and traditional farming. *Toxicol. Lett.* **2014**, *230*, 166–176. [CrossRef] [PubMed]
301. da Silva, J.; Moraes, C.R.; Heuser, V.D.; Andrade, V.M.; Silva, F.R.; Kvitko, K.; Emmel, V.; Rohr, P.; Bordin, D.L.; Andreazza, A.C.; et al. Evaluation of genetic damage in a Brazilian population occupationally exposed to pesticides and its correlation with polymorphisms in metabolizing genes. *Mutagenesis* **2008**, *23*, 415–422. [CrossRef] [PubMed]
302. Da Silva, F.R.; Da Silva, J.; Allgayer Mda, C.; Simon, C.F.; Dias, J.F.; dos Santos, C.E.; Salvador, M.; Branco, C.; Schneider, N.B.; Kahl, V.; et al. Genotoxic biomonitoring of tobacco farmers: Biomarkers of exposure, of early biological effects and of susceptibility. *J. Hazard. Mater.* **2012**, *225–226*, 81–90. [CrossRef] [PubMed]
303. Da Silva, F.R.; Kvitko, K.; Rohr, P.; Abreu, M.B.; Thiesen, F.V.; Da Silva, J. Genotoxic assessment in tobacco farmers at different crop times. *Sci. Total Environ.* **2014**, *490*, 334–341. [CrossRef] [PubMed]
304. Dalberto, D.; Alves, J.; Garcia, A.L.H.; de Souza, M.R.; Abella, A.P.; Thiesen, F.V.; Salvador, M.; Santos Branco, C.D.; Marroni, N.; Bona, S.; et al. Exposure in the tobacco fields: Genetic damage and oxidative stress in tobacco farmers occupationally exposed during harvest and grading seasons. *Mutat. Res. Genet. Toxicol. Environ. Mutagen.* **2022**, *878*, 503485. [CrossRef] [PubMed]
305. Dhananjayan, V.; Ravichandran, B.; Panjakumar, K.; Kalaiselvi, K.; Rajasekar, K.; Mala, A.; Avinash, G.; Shridhar, K.; Manju, A.; Wilson, R. Assessment of genotoxicity and cholinesterase activity among women workers occupationally exposed to pesticides in tea garden. *Mutat. Res. Genet. Toxicol. Environ. Mutagen.* **2019**, *841*, 1–7. [CrossRef]
306. Dutta, S.; Bahadur, M. Comet assay genotoxicity evaluation of occupationally exposed tea-garden workers in northern West Bengal, India. *Mutat. Res. Genet. Toxicol. Environ. Mutagen.* **2019**, *844*, 1–9. [CrossRef] [PubMed]
307. Franco, F.C.; Alves, A.A.; Godoy, F.R.; Avelar, J.B.; Rodrigues, D.D.; Pedroso, T.M.; da Cruz, A.D.; Nomura, F.; de Melo, E.S.D. Evaluating genotoxic risks in Brazilian public health agents occupationally exposed to pesticides: A multi-biomarker approach. *Environ. Sci. Pollut. Res. Int.* **2016**, *23*, 19723–19734. [CrossRef] [PubMed]
308. Garaj-Vrhovac, V.; Zeljezic, D. Evaluation of DNA damage in workers occupationally exposed to pesticides using single-cell gel electrophoresis (SCGE) assay. *Pestic. Genotoxicity Reveal. By Comet. Assay. Mutat. Res.* **2000**, *469*, 279–285. [CrossRef]
309. Garaj-Vrhovac, V.; Zeljezic, D. Cytogenetic monitoring of croatian population occupationally exposed to a complex mixture of pesticides. *Toxicology* **2001**, *165*, 153–162. [CrossRef] [PubMed]
310. Garaj-Vrhovac, V.; Zeljezic, D. Assessment of genome damage in a population of Croatian workers employed in pesticide production by chromosomal aberration analysis, micronucleus assay and Comet assay. *J. Appl. Toxicol. JAT* **2002**, *22*, 249–255. [CrossRef] [PubMed]
311. Godoy, F.R.; Nunes, H.F.; Alves, A.A.; Carvalho, W.F.; Franco, F.C.; Pereira, R.R.; da Cruz, A.S.; da Silva, C.C.; Bastos, R.P.; de Melo, E.S.D. Increased DNA damage is not associated to polymorphisms in OGGI DNA repair gene, CYP2E1 detoxification gene, and biochemical and hematological findings in soybeans farmers from Central Brazil. *Environ. Sci. Pollut. Res. Int.* **2019**, *26*, 26553–26562. [CrossRef] [PubMed]

312. Grover, P.; Danadevi, K.; Mahboob, M.; Rozati, R.; Banu, B.S.; Rahman, M.F. Evaluation of genetic damage in workers employed in pesticide production utilizing the Comet assay. *Mutagenesis* **2003**, *18*, 201–205. [CrossRef] [PubMed]
313. Kahl, V.F.S.; da Silva, F.R.; Alves, J.D.S.; da Silva, G.F.; Picinini, J.; Dhillon, V.S.; Fenech, M.; de Souza, M.R.; Dias, J.F.; de Souza, C.T.; et al. Role of PON1, SOD2, OGG1, XRCC1, and XRCC4 polymorphisms on modulation of DNA damage in workers occupationally exposed to pesticides. *Ecotoxicol. Environ. Saf.* **2018**, *159*, 164–171. [CrossRef] [PubMed]
314. Kasiotis, K.M.; Kyriakopoulou, K.; Emmanouil, C.; Tsantila, N.; Liesivuori, J.; Souki, H.; Manakis, S.; Machera, K. Monitoring of systemic exposure to plant protection products and DNA damage in orchard workers. *Toxicol. Lett.* **2012**, *210*, 182–188. [CrossRef]
315. Kaur, R.; Kaur, S.; Lata, M. Evaluation of DNA damage in agricultural workers exposed to pesticides using single cell gel electrophoresis (comet) assay. *Indian. J. Hum. Genet.* **2011**, *17*, 179–187. [CrossRef] [PubMed]
316. Kaur, K.; Kaur, R. Polymorphisms in XPC and XPD genes modulate DNA damage in pesticide-exposed agricultural workers of Punjab, North-West India. *Mol. Biol. Rep.* **2020**, *47*, 5253–5262. [CrossRef] [PubMed]
317. Kaur, K.; Kaur, R. Impact of single nucleotide polymorphisms in the OGG1 and XRCC1 genes on modulation of DNA damage in pesticide-exposed agricultural workers in Punjab, North-West India. *Biomarkers* **2020**, *25*, 498–505. [CrossRef] [PubMed]
318. Kaur, K.; Kaur, R. Modulation of DNA damage by XPF, XPG and ERCC1 gene polymorphisms in pesticide-exposed agricultural workers of Punjab, North-West India. *Mutat. Res. Genet. Toxicol. Environ. Mutagen.* **2021**, *861–862*, 503302. [CrossRef]
319. Khayat, C.B.; Costa, E.O.; Goncalves, M.W.; da Cruz e Cunha, D.M.; da Cruz, A.S.; de Araujo Melo, C.O.; Bastos, R.P.; da Cruz, A.D.; de Melo e Silva, D. Assessment of DNA damage in Brazilian workers occupationally exposed to pesticides: A study from Central Brazil. *Environ. Sci. Pollut. Res. Int.* **2013**, *20*, 7334–7340. [CrossRef] [PubMed]
320. Lebailly, P.; Devaux, A.; Pottier, D.; De Meo, M.; Andre, V.; Baldi, I.; Severin, F.; Bernaud, J.; Durand, B.; Henry-Amar, M.; et al. Urine mutagenicity and lymphocyte DNA damage in fruit growers occupationally exposed to the fungicide captan. *Occup. Environ. Med.* **2003**, *60*, 910–917. [CrossRef] [PubMed]
321. Liu, Y.J.; Huang, P.L.; Chang, Y.F.; Chen, Y.H.; Chiou, Y.H.; Xu, Z.L.; Wong, R.H. GSTP1 genetic polymorphism is associated with a higher risk of DNA damage in pesticide-exposed fruit growers. *Cancer Epidemiol. Biomark. Prev.* **2006**, *15*, 659–666. [CrossRef] [PubMed]
322. Muniz, J.F.; McCauley, L.; Scherer, J.; Lasarev, M.; Koshy, M.; Kow, Y.W.; Nazar-Stewart, V.; Kisby, G.E. Biomarkers of oxidative stress and DNA damage in agricultural workers: A pilot study. *Toxicol. Appl. Pharmacol.* **2008**, *227*, 97–107. [CrossRef] [PubMed]
323. Naravaneni, R.; Jamil, K. Determination of AChE levels and genotoxic effects in farmers occupationally exposed to pesticides. *Hum. Exp. Toxicol.* **2007**, *26*, 723–731. [CrossRef]
324. Paiva, J.C.; Cabral, I.O.; Soares, B.M.; Sombra, C.M.; Ferreira, J.R.; Moraes, M.O.; Cavalcanti, B.C.; Pessoa, C. Biomonitoring of rural workers exposed to a complex mixture of pesticides in the municipalities of Tiangua and Ubajara (Ceara state, Brazil): Genotoxic and cytogenetic studies. *Environ. Mol. Mutagen.* **2011**, *52*, 492–501. [CrossRef] [PubMed]
325. Paz-y-Mino, C.; Arevalo, M.; Sanchez, M.E.; Leone, P.E. Chromosome and DNA damage analysis in individuals occupationally exposed to pesticides with relation to genetic polymorphism for CYP 1A1 gene in Ecuador. *Mutat. Res.* **2004**, *562*, 77–89. [CrossRef]
326. Prabha & Pooja Chadha. Risk Assessment of Occupational Exposure to Pesticides among Pesticide Distributors of Punjab (India) Using Single Cell Gel Electrophoresis. *Int. J. Hum. Genet.* **2015**, *15*, 149–155. [CrossRef]
327. Ramos, J.S.A.; Pedroso, T.M.A.; Godoy, F.R.; Batista, R.E.; de Almeida, F.B.; Francelin, C.; Ribeiro, F.L.; Parise, M.R.; de Melo, E.S.D. Multi-biomarker responses to pesticides in an agricultural population from Central Brazil. *Sci. Total Environ.* **2021**, *754*, 141893. [CrossRef] [PubMed]
328. Remor, A.P.; Totti, C.C.; Moreira, D.A.; Dutra, G.P.; Heuser, V.D.; Boeira, J.M. Occupational exposure of farm workers to pesticides: Biochemical parameters and evaluation of genotoxicity. *Environ. Int.* **2009**, *35*, 273–278. [CrossRef] [PubMed]
329. Rohr, P.; da Silva, J.; Erdtmann, B.; Saffi, J.; Guecheva, T.N.; Antonio Pegas Henriques, J.; Kvitko, K. BER gene polymorphisms (OGG1 Ser326Cys and XRCC1 Arg194Trp) and modulation of DNA damage due to pesticides exposure. *Environ. Mol. Mutagen.* **2011**, *52*, 20–27. [CrossRef] [PubMed]
330. Saad-Hussein, A.; Noshy, M.; Taha, M.; El-Shorbagy, H.; Shahy, E.; Abdel-Shafy, E.A. GSTP1 and XRCC1 polymorphisms and DNA damage in agricultural workers exposed to pesticides. *Mutat. Res. Genet. Toxicol. Environ. Mutagen.* **2017**, *819*, 20–25. [CrossRef] [PubMed]
331. Saad-Hussein, A.; Beshir, S.; Taha, M.M.; Shahy, E.M.; Shaheen, W.; Abdel-Shafy, E.A.; Thabet, E. Early prediction of liver carcinogenicity due to occupational exposure to pesticides. *Mutat. Res. Genet. Toxicol. Environ. Mutagen.* **2019**, *838*, 46–53. [CrossRef] [PubMed]
332. Sapbamrer, R.; Khacha-Ananda, S.; Sittitoon, N.; Wunnapuk, K.; Seesen, M.; Sidthilaw, S.; Chittrakul, J.; Suwannakul, B. A longitudinal follow-up study of oxidative stress and DNA damage among farmers exposed to pesticide mixtures. *Environ. Sci. Pollut. Res. Int.* **2019**, *26*, 13185–13194. [CrossRef]
333. Simoniello, M.F.; Kleinsorge, E.C.; Scagnetti, J.A.; Grigolato, R.A.; Poletta, G.L.; Carballo, M.A. DNA damage in workers occupationally exposed to pesticide mixtures. *J. Appl. Toxicol. JAT* **2008**, *28*, 957–965. [CrossRef] [PubMed]
334. Simoniello, M.F.; Kleinsorge, E.C.; Scagnetti, J.A.; Mastandrea, C.; Grigolato, R.A.; Paonessa, A.M.; Carballo, M.A. Biomarkers of cellular reaction to pesticide exposure in a rural population. *Biomarkers* **2010**, *15*, 52–60. [CrossRef] [PubMed]
335. Singh, S.; Kumar, V.; Thakur, S.; Banerjee, B.D.; Chandna, S.; Rautela, R.S.; Grover, S.S.; Rawat, D.S.; Pasha, S.T.; Jain, S.K.; et al. DNA damage and cholinesterase activity in occupational workers exposed to pesticides. *Environ. Toxicol. Pharmacol.* **2011**, *31*, 278–285. [CrossRef] [PubMed]

336. Singh, S.; Kumar, V.; Singh, P.; Thakur, S.; Banerjee, B.D.; Rautela, R.S.; Grover, S.S.; Rawat, D.S.; Pasha, S.T.; Jain, S.K.; et al. Genetic polymorphisms of GSTM1, GSTT1 and GSTP1 and susceptibility to DNA damage in workers occupationally exposed to organophosphate pesticides. *Mutat. Res.* **2011**, *725*, 36–42. [CrossRef]
337. Singh, S.; Kumar, V.; Singh, P.; Banerjee, B.D.; Rautela, R.S.; Grover, S.S.; Rawat, D.S.; Pasha, S.T.; Jain, S.K.; Rai, A. Influence of CYP2C9, GSTM1, GSTT1 and NAT2 genetic polymorphisms on DNA damage in workers occupationally exposed to organophosphate pesticides. *Mutat. Res.* **2012**, *741*, 101–108. [CrossRef]
338. Singh, S.; Kumar, V.; Vashisht, K.; Singh, P.; Banerjee, B.D.; Rautela, R.S.; Grover, S.S.; Rawat, D.S.; Pasha, S.T.; Jain, S.K.; et al. Role of genetic polymorphisms of CYP1A1, CYP3A5, CYP2C9, CYP2D6, and PON1 in the modulation of DNA damage in workers occupationally exposed to organophosphate pesticides. *Toxicol. Appl. Pharmacol.* **2011**, *257*, 84–92. [CrossRef]
339. Valencia-Quintana, R.; Lopez-Duran, R.M.; Milic, M.; Bonassi, S.; Ochoa-Ocana, M.A.; Uriostegui-Acosta, M.O.; Perez-Flores, G.A.; Gomez-Olivares, J.L.; Sanchez-Alarcon, J. Assessment of Cytogenetic Damage and Cholinesterases' Activity in Workers Occupationally Exposed to Pesticides in Zamora-Jacona, Michoacan, Mexico. *Int. J. Environ. Res. Public Health* **2021**, *18*, 6269. [CrossRef]
340. Varona-Uribe, M.E.; Torres-Rey, C.H.; Diaz-Criollo, S.; Palma-Parra, R.M.; Narvaez, D.M.; Carmona, S.P.; Briceno, L.; Idrovo, A.J. Exposure to pesticide mixtures and DNA damage among rice field workers. *Arch. Environ. Occup. Health* **2016**, *71*, 3–9. [CrossRef] [PubMed]
341. Perumalla Venkata, R.; Rahman, M.F.; Mahboob, M.; Indu Kumari, S.; Chinde, S.; M, B.; Dumala, N.; Grover, P. Assessment of genotoxicity in female agricultural workers exposed to pesticides. *Biomark. Biochem. Indic. Expo. Response Susceptibility Chem.* **2017**, *22*, 446–454. [CrossRef] [PubMed]
342. Wilhelm, C.M.; Calsing, A.K.; da Silva, L.B. Assessment of DNA damage in floriculturists in southern Brazil. *Environ. Sci. Pollut. Res. Int.* **2015**, *22*, 8182–8189. [CrossRef] [PubMed]
343. Wong, R.H.; Chang, S.Y.; Ho, S.W.; Huang, P.L.; Liu, Y.J.; Chen, Y.C.; Yeh, Y.H.; Lee, H.S. Polymorphisms in metabolic GSTP1 and DNA-repair XRCC1 genes with an increased risk of DNA damage in pesticide-exposed fruit growers. *Mutat. Res.* **2008**, *654*, 168–175. [CrossRef] [PubMed]
344. Yadav, A.; Sehrawat, G. Evaluation of Genetic Damage in Farmers Exposed to Pesticide Mixtures. *Int. J. Hum. Genet.* **2011**, *11*, 105–109. [CrossRef]
345. Zepeda-Arce, R.; Rojas-Garcia, A.E.; Benitez-Trinidad, A.; Herrera-Moreno, J.F.; Medina-Diaz, I.M.; Barron-Vivanco, B.S.; Villegas, G.P.; Hernandez-Ochoa, I.; Solis Heredia, M.J.; Bernal-Hernandez, Y.Y. Oxidative stress and genetic damage among workers exposed primarily to organophosphate and pyrethroid pesticides. *Environ. Toxicol.* **2017**, *32*, 1754–1764. [CrossRef]
346. Zeljezic, D.; Garaj-Vrhovac, V. Chromosomal aberration and single cell gel electrophoresis (Comet) assay in the longitudinal risk assessment of occupational exposure to pesticides. *Mutagenesis* **2001**, *16*, 359–363. [CrossRef] [PubMed]
347. Alvarado-Hernandez, D.L.; Montero-Montoya, R.; Serrano-Garcia, L.; Arellano-Aguilar, O.; Jasso-Pineda, Y.; Yanez-Estrada, L. Assessment of exposure to organochlorine pesticides and levels of DNA damage in mother-infant pairs of an agrarian community. *Environ. Mol. Mutagen.* **2013**, *54*, 99–111. [CrossRef] [PubMed]
348. Dwivedi, N.; Mahdi, A.A.; Deo, S.; Ahmad, M.K.; Kumar, D. Assessment of genotoxicity and oxidative stress in pregnant women contaminated to organochlorine pesticides and its correlation with pregnancy outcome. *Environ. Res.* **2022**, *204*, 112010. [CrossRef] [PubMed]
349. How, V.; Hashim, Z.; Ismail, P.; Md Said, S.; Omar, D.; Bahri Mohd Tamrin, S. Exploring cancer development in adulthood: Cholinesterase depression and genotoxic effect from chronic exposure to organophosphate pesticides among rural farm children. *J. Agromed.* **2014**, *19*, 35–43. [CrossRef] [PubMed]
350. Kapka-Skrzypczak, L.; Czajka, M.; Sawicki, K.; Matysiak-Kucharek, M.; Gabelova, A.; Sramkova, M.; Bartyzel-Lechforowicz, H.; Kruszewski, M. Assessment of DNA damage in Polish children environmentally exposed to pesticides. *Mutat. Res. Genet. Toxicol. Environ. Mutagen.* **2019**, *843*, 52–56. [CrossRef] [PubMed]
351. Leite, S.B.; Franco de Diana, D.M.; Segovia Abreu, J.A.; Avalos, D.S.; Denis, M.A.; Ovelar, C.C.; Samaniego Royg, M.J.; Thielmann Arbo, B.A.; Corvalan, R. DNA damage induced by exposure to pesticides in children of rural areas in Paraguay. *Indian. J. Med. Res.* **2019**, *150*, 290–296. [CrossRef] [PubMed]
352. Sutris, J.M.; How, V.; Sumeri, S.A.; Muhammad, M.; Sardi, D.; Mohd Mokhtar, M.T.; Muhammad, H.; Ghazi, H.F.; Isa, Z.M. Genotoxicity following Organophosphate Pesticides Exposure among Orang Asli Children Living in an Agricultural Island in Kuala Langat, Selangor, Malaysia. *Int. J. Occup. Environ. Med.* **2016**, *7*, 42–51. [CrossRef] [PubMed]
353. Azqueta, A.; Ladeira, C.; Giovannelli, L.; Boutet-Robinet, E.; Bonassi, S.; Neri, M.; Gajski, G.; Duthie, S.; Del Bo, C.; Riso, P.; et al. Application of the comet assay in human biomonitoring: An hCOMET perspective. Mutation research. *Rev. Mutat. Res.* **2020**, *783*, 108288. [CrossRef] [PubMed]
354. Durrani, T.; Clapp, R.; Harrison, R.; Shusterman, D. Solvent-based paint and varnish removers: A focused toxicologic review of existing and alternative constituents. *J. Appl. Toxicol. JAT* **2020**, *40*, 1325–1341. [CrossRef] [PubMed]
355. Brauner, C.; Joveleviths, D.; Alvares-da-Silva, M.R.; Marroni, N.; Bona, S.; Schemitt, E.; Nardi, R. Exposure to organic solvents and hepatotoxicity. *J. Environ. Sci. Health Part A* **2020**, *55*, 1173–1178. [CrossRef] [PubMed]
356. Sainio, M.A., Sr. Neurotoxicity of solvents. *Handb. Clin. Neurol.* **2015**, *131*, 93–110. [CrossRef] [PubMed]

357. Sit, G.; Letellier, N.; Iwatsubo, Y.; Goldberg, M.; Leynaert, B.; Nadif, R.; Ribet, C.; Roche, N.; Roquelaure, Y.; Varraso, R.; et al. Occupational Exposures to Organic Solvents and Asthma Symptoms in the CONSTANCES Cohort. *Int. J. Environ. Res. Public Health* **2021**, *18*, 9258. [CrossRef] [PubMed]
358. da Silva, J. DNA damage induced by occupational and environmental exposure to miscellaneous chemicals. *Mutat. Res. Rev. Mutat. Res.* **2016**, *770*, 170–182. [CrossRef] [PubMed]
359. International International Agency for Research on Cancer (IARC). Occupational Exposure as a Painter—IARC Monographs on the Evaluation of the Carcinogenic Risks of Chemicals to Humans, Chemical Agents and Related Occupations. 2018, pp. 509–539. Available online: https://monographs.iarc.who.int/wp-content/uploads/2018/06/mono100F-35.pdf (accessed on 10 June 2023).
360. International Agency for Research on Cancer (IARC). Boot and Shoe Manufacture and Repair—IARC Monographs on the Evaluation of the Carcinogenic Risks of Chemicals to Humans, Chemical Agents and Related Occupations. 1987, pp. 232–235. Available online: http://monographs.iarc.fr/ENG/Monographs/suppl7/Suppl7-96.pdf (accessed on 10 June 2023).
361. Onyije, F.M.; Hosseini, B.; Togawa, K.; Schuz, J.; Olsson, A. Cancer Incidence and Mortality among Petroleum Industry Workers and Residents Living in Oil Producing Communities: A Systematic Review and Meta-Analysis. *Int. J. Environ. Res. Public Health* **2021**, *18*, 4343. [CrossRef] [PubMed]
362. Azimi, M.; Bahrami, M.R.; Rezaei Hachesu, V.; Zavar Reza, J.; Mihanpour, H.; Zare Sakhvidi, M.J.; Mostaghaci, M. Primary DNA Damage in Dry Cleaners with Perchlorethylene Exposure. *Int. J. Occup. Environ. Med.* **2017**, *8*, 224–231. [CrossRef] [PubMed]
363. Buschini, A.; De Palma, G.; Poli, P.; Martino, A.; Rossi, C.; Mozzoni, P.; Scotti, E.; Buzio, L.; Bergamaschi, E.; Mutti, A. Genetic polymorphism of drug-metabolizing enzymes and styrene-induced DNA damage. *Environ. Mol. Mutagen.* **2003**, *41*, 243–252. [CrossRef] [PubMed]
364. Cassini, C.; Calloni, C.; Bortolini, G.; Garcia, S.C.; Dornelles, M.A.; Henriques, J.A.; Erdtmann, B.; Salvador, M. Occupational risk assessment of oxidative stress and genotoxicity in workers exposed to paints during a working week. *Int. J. Occup. Med. Environ. Health* **2011**, *24*, 308–319. [CrossRef] [PubMed]
365. Cavallo, D.; Tranfo, G.; Ursini, C.L.; Fresegna, A.M.; Ciervo, A.; Maiello, R.; Paci, E.; Pigini, D.; Gherardi, M.; Gatto, M.P.; et al. Biomarkers of early genotoxicity and oxidative stress for occupational risk assessment of exposure to styrene in the fibreglass reinforced plastic industry. *Toxicol. Lett.* **2018**, *298*, 53–59. [CrossRef] [PubMed]
366. Cavallo, D.; Ursini, C.L.; Fresegna, A.M.; Ciervo, A.; Maiello, R.; Buresti, G.; Paci, E.; Pigini, D.; Gherardi, M.; Carbonari, D.; et al. Occupational Exposure in Industrial Painters: Sensitive and Noninvasive Biomarkers to Evaluate Early Cytotoxicity, Genotoxicity and Oxidative Stress. *Int. J. Environ. Res. Public Health* **2021**, *18*, 4645. [CrossRef] [PubMed]
367. Cok, I.; Sardas, S.; Kadioglu, E.; Ozcagli, E. Assessment of DNA damage in glue sniffers by use of the alkaline comet assay. *Mutat. Res.* **2004**, *557*, 131–136. [CrossRef]
368. Costa, C.; Costa, S.; Silva, S.; Coelho, P.; Botelho, M.; Gaspar, J.; Rueff, J.; Laffon, B.; Teixeira, J.P. DNA damage and susceptibility assessment in industrial workers exposed to styrene. *J. Toxicol. Environ. Health Part A* **2012**, *75*, 735–746. [CrossRef]
369. Costa-Amaral, I.C.; Carvalho, L.V.B.; Santos, M.V.C.; Valente, D.; Pereira, A.C.; Figueiredo, V.O.; Souza, J.M.; Castro, V.S.; Trancoso, M.F.; Fonseca, A.S.A.; et al. Environmental Assessment and Evaluation of Oxidative Stress and Genotoxicity Biomarkers Related to Chronic Occupational Exposure to Benzene. *Int. J. Environ. Res. Public Health* **2019**, *16*, 2240. [CrossRef] [PubMed]
370. de Aquino, T.; Zenkner, F.F.; Ellwanger, J.H.; Pra, D.; Rieger, A. DNA damage and cytotoxicity in pathology laboratory technicians exposed to organic solvents. *An. Da Acad. Bras. De Cienc.* **2016**, *88*, 227–236. [CrossRef]
371. Fracasso, M.E.; Doria, D.; Bartolucci, G.B.; Carrieri, M.; Lovreglio, P.; Ballini, A.; Soleo, L.; Tranfo, G.; Manno, M. Low air levels of benzene: Correlation between biomarkers of exposure and genotoxic effects. *Toxicol. Lett.* **2010**, *192*, 22–28. [CrossRef] [PubMed]
372. Fracasso, M.E.; Doria, D.; Carrieri, M.; Bartolucci, G.B.; Quintavalle, S.; De Rosa, E. DNA single- and double-strand breaks by alkaline- and immuno-comet assay in lymphocytes of workers exposed to styrene. *Toxicol. Lett.* **2009**, *185*, 9–15. [CrossRef] [PubMed]
373. Godderis, L.; De Boeck, M.; Haufroid, V.; Emmery, M.; Mateuca, R.; Gardinal, S.; Kirsch-Volders, M.; Veulemans, H.; Lison, D. Influence of genetic polymorphisms on biomarkers of exposure and genotoxic effects in styrene-exposed workers. *Environ. Mol. Mutagen.* **2004**, *44*, 293–303. [CrossRef] [PubMed]
374. Hanova, M.; Stetina, R.; Vodickova, L.; Vaclavikova, R.; Hlavac, P.; Smerhovsky, Z.; Naccarati, A.; Polakova, V.; Soucek, P.; Kuricova, M.; et al. Modulation of DNA repair capacity and mRNA expression levels of XRCC1, hOGG1 and XPC genes in styrene-exposed workers. *Toxicol. Appl. Pharmacol.* **2010**, *248*, 194–200. [CrossRef]
375. Heuser, V.D.; de Andrade, V.M.; da Silva, J.; Erdtmann, B. Comparison of genetic damage in Brazilian footwear-workers exposed to solvent-based or water-based adhesive. *Mutat. Res.* **2005**, *583*, 85–94. [CrossRef] [PubMed]
376. Heuser, V.D.; Erdtmann, B.; Kvitko, K.; Rohr, P.; da Silva, J. Evaluation of genetic damage in Brazilian footwear-workers: Biomarkers of exposure, effect, and susceptibility. *Toxicology* **2007**, *232*, 235–247. [CrossRef] [PubMed]
377. Keretetse, G.S.; Laubscher, P.J.; Du Plessis, J.L.; Pretorius, P.J.; Van Der Westhuizen, F.H.; Van Deventer, E.; Van Dyk, E.; Eloff, F.C.; Van Aarde, M.N.; Du Plessis, L.H. DNA damage and repair detected by the comet assay in lymphocytes of african petrol attendants: A pilot study. *Ann. Occup. Hyg.* **2008**, *52*, 653–662. [CrossRef]
378. Ladeira, C.; Gajski, G.; Meneses, M.; Geric, M.; Viegas, S. The genotoxicity of an organic solvent mixture: A human biomonitoring study and translation of a real-scenario exposure to in vitro. *Regul. Toxicol. Pharmacol. RTP* **2020**, *116*, 104726. [CrossRef] [PubMed]
379. Laffon, B.; Pasaro, E.; Mendez, J. Evaluation of genotoxic effects in a group of workers exposed to low levels of styrene. *Toxicology* **2002**, *171*, 175–186. [CrossRef] [PubMed]

380. Lam, T.H.; Zhu, C.Q.; Jiang, C.Q. Lymphocyte DNA damage in elevator manufacturing workers in Guangzhou, China. *Mutat. Res.* **2002**, *515*, 147–157. [CrossRef] [PubMed]
381. Li, J.; Zhang, X.; He, Z.; Sun, Q.; Qin, F.; Huang, Z.; Zhang, X.; Sun, X.; Liu, L.; Chen, L.; et al. MGMT hypomethylation is associated with DNA damage in workers exposed to low-dose benzene. *Biomarkers* **2017**, *22*, 470–475. [CrossRef]
382. Londono-Velasco, E.; Martinez-Perafan, F.; Carvajal-Varona, S.; Garcia-Vallejo, F.; Hoyos-Giraldo, L.S. Assessment of DNA damage in car spray painters exposed to organic solvents by the high-throughput comet assay. *Toxicol. Mech. Methods* **2016**, *26*, 238–242. [CrossRef] [PubMed]
383. Martino-Roth, M.G.; Viegas, J.; Roth, D.M. Occupational genotoxicity risk evaluation through the comet assay and the micronucleus test. *Genet. Mol. Res. GMR* **2003**, *2*, 410–417. [PubMed]
384. Migliore, L.; Colognato, R.; Naccarati, A.; Bergamaschi, E. Relationship between genotoxicity biomarkers in somatic and germ cells: Findings from a biomonitoring study. *Mutagenesis* **2006**, *21*, 149–152. [CrossRef]
385. Migliore, L.; Naccarati, A.; Zanello, A.; Scarpato, R.; Bramanti, L.; Mariani, M. Assessment of sperm DNA integrity in workers exposed to styrene. *Hum. Reprod.* **2002**, *17*, 2912–2918. [CrossRef] [PubMed]
386. Moro, A.M.; Brucker, N.; Charao, M.; Bulcao, R.; Freitas, F.; Baierle, M.; Nascimento, S.; Valentini, J.; Cassini, C.; Salvador, M.; et al. Evaluation of genotoxicity and oxidative damage in painters exposed to low levels of toluene. *Mutat. Res.* **2012**, *746*, 42–48. [CrossRef] [PubMed]
387. Navasumrit, P.; Chanvaivit, S.; Intarasunanont, P.; Arayasiri, M.; Lauhareungpanya, N.; Parnlob, V.; Settachan, D.; Ruchirawat, M. Environmental and occupational exposure to benzene in Thailand. *Chemico-Biol. Interact.* **2005**, *153–154*, 75–83. [CrossRef]
388. Pandey, A.K.; Bajpayee, M.; Parmar, D.; Kumar, R.; Rastogi, S.K.; Mathur, N.; Thorning, P.; de Matas, M.; Shao, Q.; Anderson, D.; et al. Multipronged evaluation of genotoxicity in Indian petrol-pump workers. *Environ. Mol. Mutagen.* **2008**, *49*, 695–707. [CrossRef]
389. Poca, K.S.D.; Giardini, I.; Silva, P.V.B.; Geraldino, B.R.; Bellomo, A.; Alves, J.A.; Conde, T.R.; Zamith, H.; Otero, U.B.; Ferraris, F.K.; et al. Gasoline-station workers in Brazil: Benzene exposure; Genotoxic and immunotoxic effects. *Mutat. Res. Genet. Toxicol. Environ. Mutagen.* **2021**, *865*, 503322. [CrossRef] [PubMed]
390. Rekhadevi, P.V.; Rahman, M.F.; Mahboob, M.; Grover, P. Genotoxicity in filling station attendants exposed to petroleum hydrocarbons. *Ann. Occup. Hyg.* **2010**, *54*, 944–954. [CrossRef]
391. Roma-Torres, J.; Teixeira, J.P.; Silva, S.; Laffon, B.; Cunha, L.M.; Mendez, J.; Mayan, O. Evaluation of genotoxicity in a group of workers from a petroleum refinery aromatics plant. *Mutat. Res.* **2006**, *604*, 19–27. [CrossRef] [PubMed]
392. Sakhvidi, M.J.Z.; Zarei, A.; Hachesu, V.R.; Zolfaghari, A. Evaluating the relationship between the respiratory exposure to the benzene with the primary damages of deoxyribonucleic acid and total antioxidant capacity in one of the oil companies in Iran. *Environ. Sci. Pollut. Res. Int.* **2022**, *29*, 48340–48346. [CrossRef] [PubMed]
393. Sul, D.; Lee, D.; Im, H.; Oh, E.; Kim, J.; Lee, E. Single strand DNA breaks in T- and B-lymphocytes and granulocytes in workers exposed to benzene. *Toxicol. Lett.* **2002**, *134*, 87–95. [CrossRef] [PubMed]
394. Sul, D.; Lee, E.; Lee, M.Y.; Oh, E.; Im, H.; Lee, J.; Jung, W.W.; Won, N.; Kang, H.S.; Kim, E.M.; et al. DNA damage in lymphocytes of benzene exposed workers correlates with trans,trans-muconic acids and breath benzene levels. *Mutat. Res.* **2005**, *582*, 61–70. [CrossRef] [PubMed]
395. Teixeira, J.P.; Gaspar, J.; Coelho, P.; Costa, C.; Pinho-Silva, S.; Costa, S.; Da Silva, S.; Laffon, B.; Pasaro, E.; Rueff, J.; et al. Cytogenetic and DNA damage on workers exposed to styrene. *Mutagenesis* **2010**, *25*, 617–621. [CrossRef] [PubMed]
396. Xiong, F.; Li, Q.; Zhou, B.; Huang, J.; Liang, G.; Zhang, L.; Ma, S.; Qing, L.; Liang, L.; Su, J.; et al. Oxidative Stress and Genotoxicity of Long-Term Occupational Exposure to Low Levels of BTEX in Gas Station Workers. *Int. J. Environ. Res. Public Health* **2016**, *13*, 1212. [CrossRef]
397. Zhao, Z.; Xing, X.; Ou, X.; Liu, X.; Zhou, R.; Zhang, H.; Yang, L.; Zhuang, Z.; Su, X.; Lu, Y.; et al. DNA damage levels in electronics workers in Southern China: A micro-whole blood comet assay. *Mutat. Res.* **2017**, *803–805*, 17–21. [CrossRef] [PubMed]
398. Sorensen, M.; Skov, H.; Autrup, H.; Hertel, O.; Loft, S. Urban benzene exposure and oxidative DNA damage: Influence of genetic polymorphisms in metabolism genes. *Sci. Total Environ.* **2003**, *309*, 69–80. [CrossRef] [PubMed]
399. Salama, S.A.; Serrana, M.; Au, W.W. Biomonitoring using accessible human cells for exposure and health risk assessment. *Mutat. Res.* **1999**, *436*, 99–112. [CrossRef]
400. Collins, A.; Dusinska, M.; Franklin, M.; Somorovska, M.; Petrovska, H.; Duthie, S.; Fillion, L.; Panayiotidis, M.; Raslova, K.; Vaughan, N. Comet assay in human biomonitoring studies: Reliability, validation, and applications. *Environ. Mol. Mutagen.* **1997**, *30*, 139–146. [CrossRef]
401. Villarini, M.; Dominici, L.; Fatigoni, C.; Muzi, G.; Monarca, S.; Moretti, M. Biological effect monitoring in peripheral blood lymphocytes from subjects occupationally exposed to antineoplastic drugs: Assessment of micronuclei frequency. *J. Occup. Health* **2012**, *54*, 405–415. [CrossRef] [PubMed]
402. Ladeira, C.; Viegas, S.; Padua, M.; Gomes, M.; Carolino, E.; Gomes, M.C.; Brito, M. Assessment of genotoxic effects in nurses handling cytostatic drugs. *J. Toxicol. Environ. Health Part A* **2014**, *77*, 879–887. [CrossRef] [PubMed]
403. Viegas, S.; Zare Jeddi, M.B.; Hopf, N.; Bessems, J.; Palmen, N.S.; Galea, K.; Jones, K.; Kujath, P.; Duca, R.C.; Verhagen, H.; et al. Biomonitoring as an Underused Exposure Assessment Tool in Occupational Safety and Health Context-Challenges and Way Forward. *Int. J. Environ. Res. Public Health* **2020**, *17*, 5884. [CrossRef] [PubMed]
404. Manno, M.; Viau, C.; in collaboration, w.; Cocker, J.; Colosio, C.; Lowry, L.; Mutti, A.; Nordberg, M.; Wang, S. Biomonitoring for occupational health risk assessment (BOHRA). *Toxicol. Lett.* **2010**, *192*, 3–16. [CrossRef] [PubMed]

405. Ladeira, C.; Viegas, S. Human biomonitoring: An overview on biomarkers and their application in occupational and environmental health. *Biomonitoring* **2016**, 3, 15–24. [CrossRef]
406. Decker, J.A.; DeBord, D.G.; Bernard, B.; Dotson, G.S.; Halpin, J.; Hines, C.J.; Kiefer, M.; Myers, K.; Page, E.; Schulte, P.; et al. Recommendations for biomonitoring of emergency responders: Focus on occupational health investigations and occupational health research. *Mil. Med.* **2013**, *178*, 68–75. [CrossRef] [PubMed]
407. Bonassi, S.; Au, W.W. Biomarkers in molecular epidemiology studies for health risk prediction. *Mutat. Res.* **2002**, *511*, 73–86. [CrossRef] [PubMed]
408. Kirkham, J.J.; Gargon, E.; Clarke, M.; Williamson, P.R. Can a core outcome set improve the quality of systematic reviews?—A survey of the Co-ordinating Editors of Cochrane Review Groups. *Trials* **2013**, *14*, 21. [CrossRef]

toxics

Review

Hand Sanitizer Gels: Classification, Challenges, and the Future of Multipurpose Hand Hygiene Products

Yilei Ma [1,†], Jia Yi [1,†], Jiahui Ma [1], Haiyang Yu [1], Li Luo [2], Wei Wu [3], Libo Jin [1], Qinsi Yang [4], Ting Lou [5], Da Sun [1,*] and Min Cao [6,*]

[1] Institute of Life Sciences & Biomedical Collaborative Innovation Center of Zhejiang Province, Wenzhou University, Wenzhou 325035, China
[2] Affiliated Dongguan Hospital, Southern Medical University, Dongguan 523059, China
[3] Bioengineering College of Chongqing University, Chongqing 400044, China
[4] Wenzhou Institute, University of Chinese Academy of Sciences, Wenzhou 325000, China
[5] Yiwu Center for Disease Control and Prevention, Yiwu 322000, China; louting22@163.com
[6] The Quzhou Affiliated Hospital of Wenzhou Medical University, Quzhou People's Hospital, Quzhou 324000, China
* Correspondence: sunday@wzu.edu.cn (D.S.); dzzhb@qzhospital.com (M.C.)
† These authors contributed equally to this work.

Abstract: Hand hygiene is a crucial measure in the prevention and control of infections, and there is a growing awareness among individuals who are making a conscious effort to maintain hand cleanliness. With the advent of the SARS-CoV-2 outbreak, the demand for hand hygiene products has also gradually shifted towards those with antimicrobial properties. Among these products, hand sanitizer gels (HSGs) have gained considerable popularity as an efficient method of hand cleaning, due to their rapid drying and sustained antimicrobial efficacy. Concurrently, there has been a growing interest in novel HSGs that offer additional functions such as skin whitening, moisturizing, and anti-inflammatory effects. These novel HSGs effectively address concerns associated with the ingestion of antimicrobial ingredients and demonstrate reduced skin irritation, thereby alleviating hand dermatological issues. This review provides an extensive overview of the application scenarios, classification, and challenges associated with HSGs while emphasizing the emergence of novel components with biological functions, aiming to contribute to the advancement of hand hygiene practices and offer novel insights for the development of novel HSGs with outstanding antimicrobial properties with other multiple biological functions and desirable biosafety profiles.

Keywords: hand sanitizer gels; antimicrobial properties; biological functions; hand hygiene

Citation: Ma, Y.; Yi, J.; Ma, J.; Yu, H.; Luo, L.; Wu, W.; Jin, L.; Yang, Q.; Lou, T.; Sun, D.; et al. Hand Sanitizer Gels: Classification, Challenges, and the Future of Multipurpose Hand Hygiene Products. *Toxics* **2023**, *11*, 687. https://doi.org/10.3390/toxics11080687

Academic Editor: Carina Ladeira

Received: 4 July 2023
Revised: 2 August 2023
Accepted: 8 August 2023
Published: 10 August 2023

1. Introduction

Hand hygiene plays a crucial role in preventing infections and controlling the spread of diseases, as hands serve as a major medium for the transmission of pathogenic microorganisms [1,2]. Notably, over half of respiratory viruses and enterovirus infections, along with bacterial infections, are transmitted through hand-to-mouth contact among people living in close contact [3–5]. Therefore, hand hygiene is crucial in preventing infections and controlling the spread of epidemics, which can ultimately safeguard human life and health [6,7]. In response to the global COVID-19 pandemic, the demand for hand cleaning products with antimicrobial properties has surged [8,9]. Instant hand sanitizers (IHSs) have gained significant popularity due to their effectiveness in reducing hand pathogens, along with their convenience, accessibility, and wide consumer acceptance [10,11]. Compared to traditional soaps and wipes, IHSs are considered more convenient and efficient in reducing bacterial counts [12]. Consequently, using IHSs as an alternative to disposable wipes before meals is considered more convenient and efficient [13]. As the importance of hand hygiene continues to be emphasized, promoting the use of IHSs as an integral part of everyday routines can significantly contribute to reducing the transmission of infectious diseases and improving overall public health.

IHSs typically contain active functional ingredients, as well as inactive carrier ingredients, such as excipients, humectants, fragrances, and colorants [14,15]. They are available in various forms, including gels, liquids, sprays, and foams [16]. Each formulation has its own characteristics and considerations: (I) The low viscosity of the liquid form makes it more challenging to use effectively [17]; (II) The spray formulation of IHSs offers broader coverage, allowing for convenient application to larger surfaces. However, it requires a valve mechanism and poses a safety risk due to the flammability of ethanol. Extra precautions must be taken to avoid accidents and ensure proper storage [18]; (III) The high humidity of the foam formulation can reduce its antimicrobial properties and prolong the drying time, leading to discomfort [19]. In comparison, hand sanitizer gels (HSGs) are a popular choice due to their rapid drying time and continuous antimicrobial efficacy [20]. Additionally, HSGs form a protective layer at the application site, providing a longer period of protection on the skin. It is essential to rub HSGs thoroughly to ensure complete coverage of the hand skin (see Table 1) [21,22]. It is important to consider factors such as ease of use, safety, antimicrobial efficacy, and user comfort when selecting the most suitable IHSs for personal hygiene practices. This review will take HSGs as the main character to introduce their research progress in the field of hand hygiene.

Table 1. Different formulations of IHS and feature comparison.

The Formulations of IHS	Inactive Ingredients	Characteristics	Ref.
Liquid formulations	Humectant, fragrance, colorant.	Widely available, but with low viscosity and hard to dispense.	[17]
Spray formulations	Valve actuation, humectant, fragrance, colorant.	Higher flammability risk at room temperature.	[18]
Foam formulations	Foaming agent, humectant, fragrance, colorant.	Longer drying time, difficult to eliminate the feeling of dissimilarity and more expensive than gel.	[19,23]
Gel formulations	Emollients, thickeners, neutralizers, chelators, fragrances, and dyes or colorants.	With better antimicrobial action and fast drying time, formation of a protective layer on the application site.	[20,22]

Since the onset of the pandemic, many countries have emphasized the implementation of non-pharmaceutical preventive measures, prominently including the intensive use of HSGs [24]. Additionally, the critical role of hand sanitizers in reducing the transmission of infectious diseases such as COVID-19 has led to a surge in global demand for HSGs since 2020, resulting in a remarkable 600-fold increase in market production [25,26]. Consequently, there has been a significant decrease in the incidence of infectious diseases in recent times [27]. This surge in market sales of HSGs has spurred the refinement of their applications, resulting in the development of a wide range of novel products with specific biological functions. These products have been designed to meet the varying needs of different consumer groups.

Herein, we focus on the research progress of HSGs in hand hygiene, highlighting the application scenarios, challenges, and possible solutions, as well as the outlook for the future application of biological functions. By making informed decisions, individuals can enhance their hand hygiene routine and contribute to the prevention of infections and the control of disease transmission.

2. Application Scenarios for HSGs

Hand hygiene products that are widely accepted for use should have good sensory characteristics, which include a pH value similar to that of the skin, as well as appropriate viscosity and spreadability [22], as depicted in Figure 1a. Additionally, it is important not to develop an allergic reaction to the skin. The majority of hand hygiene products contain high levels of alcohol, which can cause skin irritation and dryness, as well as harm the environment [28]. Therefore, it is essential to find hand hygiene products that are safe and environmentally friendly. Products with poor sensory characteristics can reduce the frequency with which it is necessary to wash your hands [29]. The components of HSGs are biocompatible with each other at an excellent level, and the most abundant component is the thickening agent, mainly using carbomer, which is extensively used for skin and eye wound healing [30,31]. Simultaneously, carbomer has high viscosity at low concentrations and has the advantages of a wide viscosity range, great flowability, compatibility with many active ingredients, high transparency, good thermal stability, and high consumer acceptance [32,33].

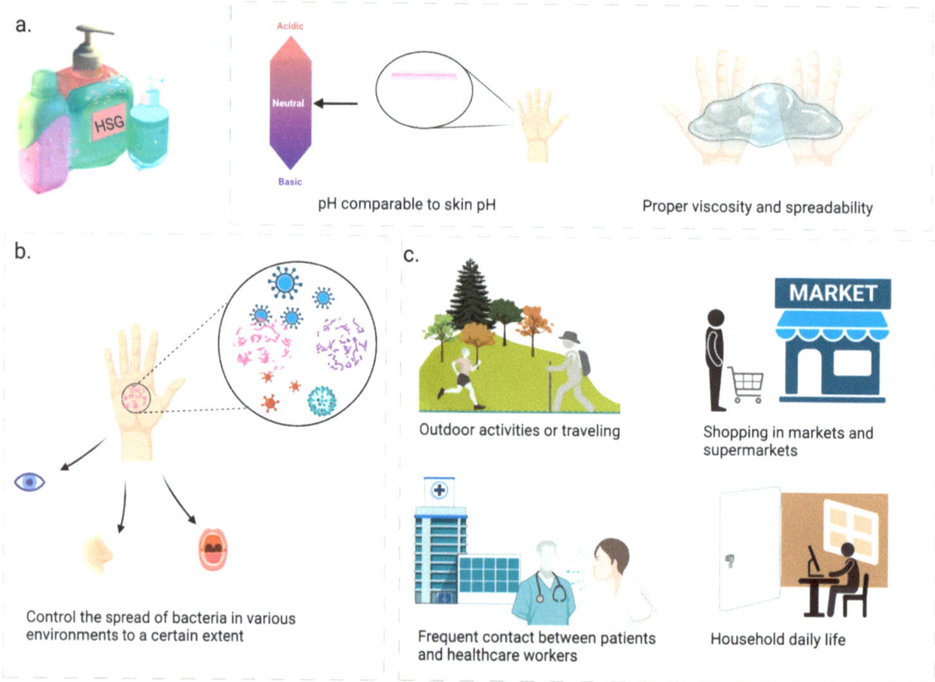

Figure 1. Hand sanitizer gels (HSGs) possess favorable sensory characteristics, exert a certain degree of control over bacterial transmission, and can be applied in various scenarios. (**a**) Hand hygiene products with good sensory properties, including pH similar to skin, as well as appropriate viscosity and spreadability; (**b**) HSGs can control the spread of bacteria in various environments to a certain extent and significantly reduce the transfer of microorganisms, thus effectively reducing respiratory and gastrointestinal infections; (**c**) The versatility of HSGs extends to multiple scenarios, embracing outdoor activities, supermarket shopping, health centers, domestic routines and other scenarios.

HSGs can be utilized in a wide variety of situations, owing to its effectiveness in minimizing respiratory and gastrointestinal infections based on its ability to control to a certain extent the spread of bacteria in a range of environments and to significantly reduce the transfer of microorganisms [34–36], as illustrated in Figure 1b,c. (I) When engaging in outdoor activities or traveling, the compact size of the HSGs allows for convenient and hygienic hand cleaning without the necessity of additional water sources and towels, thus avoiding infections and contagious diseases caused by the presence of pollutants and bacteria [37], making HSGs suitable for travel, camping, and other activities [10]; (II) during shopping in markets and supermarkets, customers frequently touch surfaces such as cart handles and shelves, resulting in bacterial contamination and transmission through hand-to-mouth and other hand-to-face touching behaviors [37,38]. In pursuit of creating a safe, healthy, and hygienic environment, HSGs offer a convenient alternative to hand soaps and other cleaning methods. They effectively eliminate hand bacteria, ensuring personal cleanliness [39]; (III) frequent contact between patients and healthcare workers can lead to the spread of bacteria in health centers such as hospitals, clinics, and pharmacies, where HSGs are more effective, act faster and can be made available at the point of patient care [40]. Studies have also shown that HSGs are generally better tolerated by the skin than soap and water [41]; (IV) family members use mobile phones, keyboards and other electronic devices daily. It is possible to spread bacteria to the hands by touching these devices [42]. In addition to furniture, doorknobs, and other common items, family members can also spread bacteria from one item to another. Therefore, using HSGs is a quick, efficient, and convenient way to help maintain good hand hygiene and prevent the spread of bacteria [34,43].

3. Classification of HSGs

HSGs can be classified into two categories based on their active ingredients: Alcohol-based HSGs (ABHSGs) and Non-alcohol-based HSGs (NABHSGs). Both types are effective in inhibiting microorganisms on the hands and no significant difference in efficacy has been observed between them [44]. Nevertheless, the mechanisms of action for these two HSGs types differ [14].

3.1. ABHSGs

ABHSGs exert rapid antimicrobial effects by denaturing and deactivating microorganisms on the hands through their interaction with membrane proteins, thus acting as effective antimicrobial agents [1]. However, the alcohol-based components present in ABHSGs can often adversely affect the lipid structure of epidermal keratinocytes, resulting in side effects such as hand dryness and cracking [45].

Typically, commercially available ABHSGs contain primary ingredients such as ethanol (ranging from 60% to 95% concentration), isopropanol, or a combination of both [29,46]. Additional ingredients like hydrogen peroxide, water, and notably, glycerol as emollients, can be incorporated into the product to reduce irritation and minimize the occurrence of contact dermatitis [47]. ABHSGs have been demonstrated to eliminate most microorganisms present on the hands, and the alcohol group within the product acts as a potent antimicrobial agent against a broad spectrum of bacteria, viruses, and fungi, including methicillin-resistant *Staphylococcus aureus*, *E. coli*, H1N1, and the Ebola virus [48–50]. The World Health Organization has confirmed that corresponding concentrations of these alcohols effectively inactivate SARS-CoV-2 [51]. Ethanol is particularly effective against non-enveloped viruses, making it the primary active ingredient in most HSG products [52,53]. The antimicrobial activity of ABHSGs largely depends on the percentage concentration of alcohol-based substances [54]; if the content falls below 60% v/v, product quality may be compromised [16]. In response to the COVID-19 pandemic, the FDA has recommended increasing the concentration of isopropanol to 91% v/v and ensuring that the ethanol concentration does not exceed 94.9% v/v [46].

ABHSGs typically achieve their full antimicrobial activity within 15 s [55]. In accordance with EN1500 guidelines, approximately 3 mL of ABHSGs should be applied to the hands for approximately 30 s to ensure their effectiveness against microorganisms [29,56]. Nonetheless, factors such as the amount of ABHSGs applied, the duration of application, and hand size can influence the reduction of microbial load, the extent of hand surface coverage, and the drying time [57,58]. A significant correlation has been observed between hand area and the number of detected CFUs [59]. Larger hand surfaces necessitate a greater amount of ABHSGs to guarantee complete coverage, which, in turn, prolongs the required duration of hand friction. This could decrease the acceptability, practicality, and ultimately, the sterilization efficacy of ABHSGs [60].

3.2. NABHSGs

The efficacy of NABHSGs in reducing the incidence and transmission of bacterial infections has been demonstrated [61]. The antimicrobial function of NABHSGs is primarily mediated through the presence of cationic surfactants, such as quaternary ammonium compounds (QACs) [62]. These compounds exhibit a broad spectrum of antimicrobial properties due to their hydrophilic and hydrophobic components, specifically the positively charged ammonium cations in their hydrophilic part [37,63]. The presence of both hydrophilic and hydrophobic components enhances their surface activity and the ability to form liposomes, thereby increasing their antibacterial capacity [64,65]. Furthermore, ingredients such as aloe vera, glycerol, and vitamin E can be incorporated into NABHSGs to reduce skin irritation and enhance their moisturizing effects [22].

Benzalkonium chloride (BAK), a representative of QACs, exhibits excellent sterilization performance within 30 s and has been widely employed in medical institutions and cleaning supplies as a non-alcohol-based functional antimicrobial agent for NABHSGs [66,67]. Compared to alcohol, QACs offer several advantages for hand disinfection: (1) they are non-toxic, less irritating to the skin, and non-flammable [68]; (2) the combination of ABHSGs and NABHSGs (primarily referring to benzalkonium chloride) generates a synergistic effect that enhances the antimicrobial range and efficacy of the alcohol group while also providing an immediate and long-lasting antimicrobial effect [69]. Therefore, the identification of various QACs components with antimicrobial activity and good biocompatibility could potentially address future challenges associated with alcohol demand and side effects.

4. Application Challenges of HSGs

4.1. Antimicrobial Functional Ingredients

Since the onset of the COVID-19 pandemic, there has been a significant surge in sales of ABHSGs [70]. To meet the challenge of increased demand and raw material shortages, many countries have permitted manufacturers and medical institutions to develop and produce hand sanitizers independently. Unfortunately, this has led to a situation where some manufacturers fail to adhere to proper quality control measures or use denatured alcohol, resulting in a notable increase in substandard hand sanitizer products in the market [47,71–73]. Furthermore, the fermentation and distillation processes employed by manufacturers, along with the impact of production equipment and the environment, may introduce impurities such as benzene and acetaldehyde into hand sanitizers, exacerbating public health concerns [74,75]. Consequently, it is recommended that users critically evaluate the antimicrobial functional ingredients in hand sanitizer formulas and carefully review product labels to avoid potential allergens based on their personal sensitivities before use [76], as shown in Figure 2.

Common antimicrobial components found in regular hand sanitizer products (such as ABHSGs) can cause corneal epithelial cell destruction at high concentrations, leading to eye diseases. Prolonged contact with hands may also increase the risk of local adverse reactions, such as contact dermatitis and atopic dermatitis [77]. In recent years, clinical cases involving adverse effects from ABHSGs have been on the rise, particularly among children [78,79]. Children tend to use hand sanitizers more frequently than adults [80], and face an increased risk of ingestion and exposure problems after using informal ABHSGs, underscoring the importance of adult supervision during use [78,81]. Additionally, healthcare workers who apply ABHSGs before handling newborns may heighten the risk of neonatal exposure to ethanol vapor, potentially causing neuronal damage and leading to neurodevelopmental delay and behavioral issues. However, the impact of such low-dose ethanol exposure on neonatal brain development is currently unclear [82,83]. Furthermore, isopropanol ingestion can result in common health problems. Accidental inhalation or dermal exposure may cause poisoning, although it usually does not lead to serious health hazards [84]. Repeated exposure of microbes to disinfectants, antibiotics, and other genotoxic chemicals leads to the development of resistance, becoming a significant global concern, particularly burdening healthcare professionals [85]. Triclosan serves as a prime example, having been used as an antimicrobial component in NABHSGs over the past decades, but evidence indicates its potential for environmental impact and antibiotic resistance [86]. Therefore, excessive and prolonged use of HSGs also entails potential risks. Furthermore, the impact of ethanol, isopropanol, and other ingredients on the environment has long been a topic of concern [87]. These compounds can volatilize or seep into soil and groundwater, significantly affecting aquatic organisms [88]. Thus, while addressing potential issues associated with the long-term use of HSGs, it is crucial to emphasize effective management of good hygiene practices.

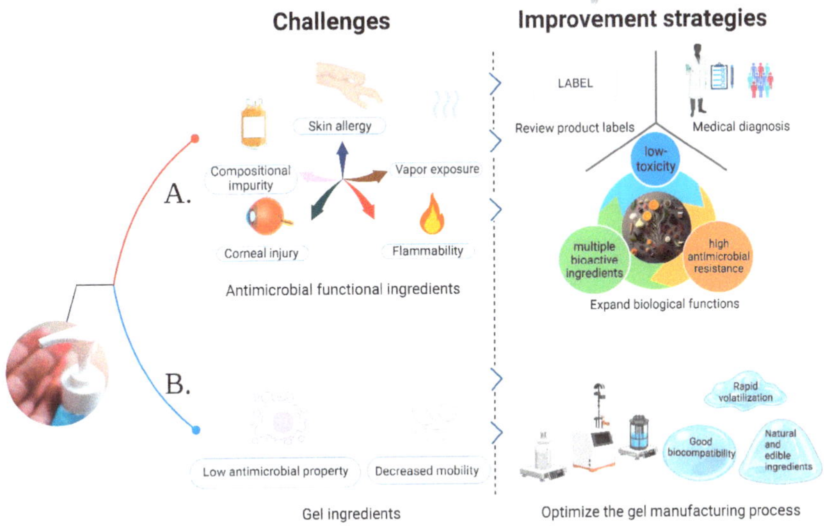

Figure 2. Application challenges and the corresponding improvement strategies of hand sanitizer gels (HSGs). (**A**) The challenges of antimicrobial functional ingredients at present and the corresponding improvement measures are put forward; (**B**) The current challenges associated with gel ingredients are identified, along with proposed measures for their improvement.

To maintain hand hygiene, particularly among children and healthcare workers, frequent hand washing is crucial. Therefore, it is important to select an emollient that effectively strengthens the skin barrier. Research indicates that BAK, an antiseptic ingredient, provides immediate and long-term antimicrobial effects, and the addition of emollients typically does not diminish its effectiveness [89]. Furthermore, BAK tends to be less irritating to the skin and rarely causes allergic reactions [63]. In cases where adverse reactions, such as hand rashes, arise from using ABHSGs, it is recommended to either switch to hand sanitizers containing BAK or seek medical treatment [90]. It is important to note that ethanol and isopropanol, commonly found in ABHSGs, are volatile and combustible substances that could cause fires when used near flames or exposed to high temperatures [16,18,91]. As such, the inclusion of novel safe antimicrobial ingredients can offer a safer alternative with an improved safety profile.

4.2. Gel Ingredients

The stability of the gel is related to the pH of the compound, with a lower pH resulting in decreased mobility. Considering that ethanol and isopropanol have different polarities, and isopropanol exhibits a significantly lower polarity compared to ethanol, a greater amount of triethanolamine must be incorporated into isopropanol to produce a stable polymer with carbomer [92]. Studies have demonstrated that anionic thickeners (such as carbomer or acrylate) notably impair the antimicrobial persistence of ABHSGs on the skin. As an alternative, non-ionic polymer thickeners like hydroxypropyl cellulose may be considered to replace carbomer and improve antimicrobial persistence [93]. In addition, by optimizing the manufacturing process of the gel, it is possible to explore materials with good biocompatibility, fast volatilization, and even completely natural edible ingredients in HSGs, which not only contribute to the stability of the gel, but also provide enhanced biocompatibility, leading to faster volatilization. This exploration opens up the possibility of reducing reliance on synthetic ingredients and instead adopting entirely natural and edible ingredients. The trend towards using natural and edible ingredients in hand hygiene products is in line with consumers' growing preference for environmentally friendly and safe options.

5. Novel Antimicrobial Ingredients of HSGs

Due to various biocompatibility issues associated with antimicrobial components in traditional HSGs, there has been a notable shift towards the use of novel HSGs [88,90], as illustrated in Figure 3. Novel HSGs are embracing the use of natural and safe antimicrobial substances, such as essential oils, phenols, organic acids, and others, aiming to reduce or even completely replace traditional ingredients like ethanol or isopropanol. Additionally, NABHSGs represented by cationic surfactants have gained attention as an alternative [94]. The research community and consumers have also exhibited a growing interest in antimicrobial agents such as organic clay and nanoformulations. These agents not only provide antimicrobial effects but also offer certain anti-inflammatory and whitening properties, as shown in Figure 4. This dual functionality has contributed to their increasing popularity among researchers and consumers alike [95,96].

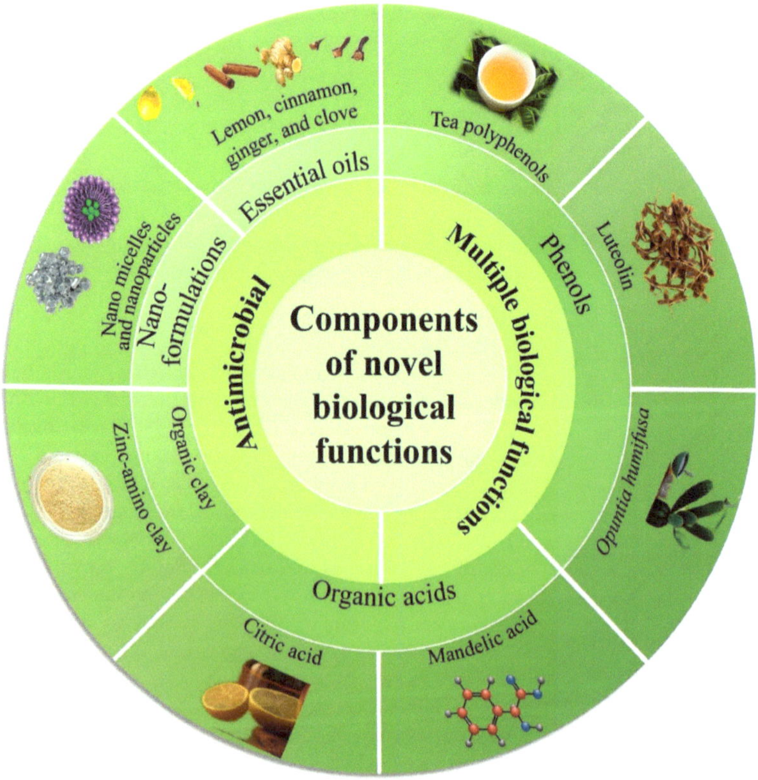

Figure 3. Novel components of hand sanitizer gels (HSGs) possess antimicrobial properties and exhibit multiple biological functions.

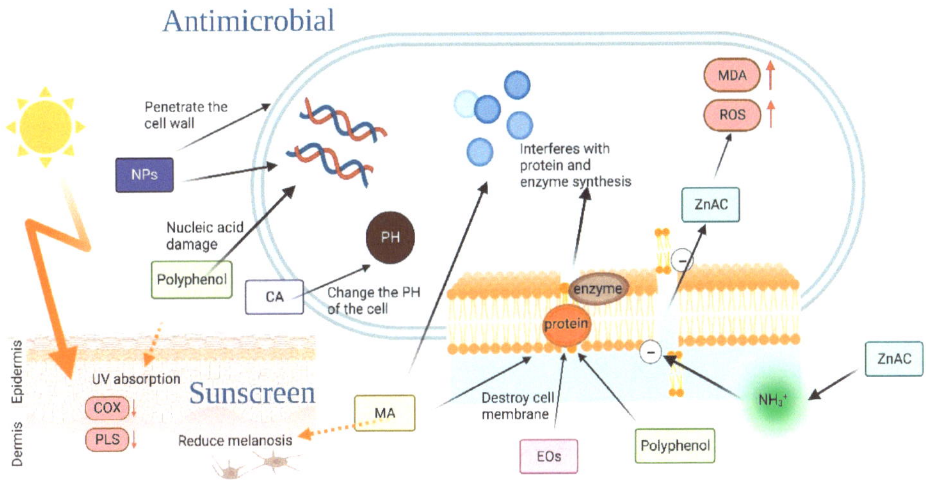

Figure 4. The antimicrobial mechanisms of essential oils, phenols, and organic acids, as well as the skin whitening and anti-inflammatory mechanisms of phenols and Mandelic acid (MA).

5.1. Essential Oils

Natural substances with antimicrobial activity have been extensively investigated and studied [97]. Among these, natural plant essential oils (EOs) have been found to affect the synthesis of bacterial lipid membranes, including certain proteins and enzymes on the membrane surface, due to their lipid-soluble nature. This makes the bacterial membranes more permeable and reduces the protein content of the bacterium, while increasing the hydrophobicity of the bacterial cell surface, affecting various cellular processes that ultimately lead to cell death [22]. Furthermore, EOs have a stronger impact on the cell wall of gram-positive bacteria than on the cell membrane [98]. By combining representative EOs, namely lemon, cinnamon, ginger, and clove, the novel NABHSGs can be developed with superior antimicrobial properties compared to 60% ethanol, as demonstrated by the synergistic action of the main components limonene, cinnamaldehyde, turmeric, and eugenol [99–101]. Additionally, M. Ismail et al. prepared a novel HSG with pepper essential oil, clove essential oil, and a small amount of ethanol. This novel HSG reduces the exposure risk of isopropanol and alcohol on the skin and exhibits good drug diffusion and antibacterial properties [102].

The utilization of natural substances, particularly essential oils, in the development of novel HSGs presents a promising approach to enhance their antimicrobial efficacy while minimizing potential risks associated with traditional antimicrobial components. By leveraging the synergistic effects and unique properties of these natural substances, researchers are able to explore novel avenues for the formulation of safer and more effective HSGs.

5.2. Phenols

Tea polyphenols have multiple biomedical functions, with active components like epigallocatechin gallate (EGCG) and epigallocatechin-3-gallate-palmitate (EC16) directly damaging the cell membrane, altering cell structure, and affecting the proteins and DNA of microorganisms to exhibit antimicrobial effects. In particular, EC16, the main component of tea polyphenol palmitate, has emerged as a novel safe and non-toxic HSG component [103,104]. Research by M. Xi et al. has found luteolin (LUT) to be a highly promising antimicrobial functional component of hand sanitizers. LUT increases cell membrane permeability to inhibit bacterial growth, causing significant and irreversible damage to bacterial structure and leading to K^+ extravasation. Additionally, LUT has strong antioxidant and free radical scavenging abilities, protecting cells from damage. As a novel HSG component, LUT not only exhibits antimicrobial effects, but also absorbs both ultraviolet A (UVA) and ultraviolet B (UVB) radiation, reducing their adverse effects on the skin. LUT can effectively mitigate the generation of reactive oxygen species (ROS) in the skin induced by ultraviolet (UV) radiation through mechanisms such as stabilizing free radicals, inhibiting pro-oxidant enzymes, and inducing antioxidant enzymes. Furthermore, LUT demonstrates effective inhibition of pro-inflammatory signaling pathways induced by UV radiation by targeting pro-inflammatory cytokines like IL-6, or COX-2, or NF-κB-dependent pathways [105,106].

Opuntia humifusa extract (OHE) is high in total polyphenols and flavonoids, giving it antioxidant and antimicrobial activity, as well as high moisture retention capabilities [107]. OHE stimulates the expression of epidermal hyaluronic acid synthases (HAS) and regulates the synthesis of hyaluronic acid (HA), thereby reducing the expression of the hyaluronidase transcript and contributing to skin hydration [25]. Additionally, OHE has whitening and anti-wrinkle effects. A study by Park et al. [108] demonstrated that OHE could regulate the ability of the skin to produce HA induced by UVB. OHE also inhibited the compensatory increase of HA, hyaluronic acid-binding protein (HABP), and CD44 protein expression under UV irradiation, all while reducing HAS percutaneous water loss and erythema. This suggests that OHE may protect the skin from the harmful effects of UVB radiation.

5.3. Organic Acids

Organic acids (OAs) act as broad-spectrum antimicrobial substances, inhibiting the growth of microorganisms through various mechanisms, and show synergistic antimicrobial effects with alcohol. Citric acid (CA), for instance, can alter the pH value of cells, thereby interfering with the normal metabolism of bacteria and exerting antimicrobial effects [109,110]. Using organic acids, alone, as antimicrobial ingredients in hand sanitizers is also significant. Mandelic acid (MA) has been identified as an antimicrobial component of NABHSGs, playing an antimicrobial role by affecting the bacterial cell membrane and inhibiting bacterial protein synthesis. Results of experiments by Pavlína Egner et al. suggest that MA has stronger antimicrobial properties and better stability than alcohol groups [111]. In addition to its antimicrobial properties, MA can also be employed as an active ingredient to reduce pigmentation and prevent skin erythema, sunburn, and skin cancer caused by ultraviolet rays. Lower concentrations of MA promote the regeneration of cells in different layers of the skin, reducing damage to the skin barrier and dehydration of the stratum corneum, thereby leading to maintenance of optimal pH values and a revamp of photoaging [112].

By harnessing the potential of these organic acids, researchers can develop novel HSGs that provide improved antimicrobial activity while promoting skin health and overall well-being.

5.4. Organic Clay

Zinc-amino clay (ZnAC) is a type of organic clay characterized by its strong antimicrobial properties and minimal toxic side effects. In an aqueous solution, ZnAC-HSG can interact with negatively charged bacterial lipid membranes through electrostatic interactions involving the quaternary ammonium (-NH_3^+) groups of organically modified clay. This interaction disrupts the permeability of the phospholipid bilayer and cell membrane, leading to an increased uptake of ZnAC by bacterial cells. Subsequently, intracellular ROS concentration rises, malondialdehyde (MDA) production is significantly increased, and ultimately, cell death occurs due to bacterial membrane disruption [25].

Moreover, the presence of organic clay in the sunscreen enhances the viscosity of its oil phase, resulting in the formation of a uniformly distributed film that improves the sunscreen's ability to absorb ultraviolet rays. Consequently, sunscreen products containing HSGs with organic clay components exhibit enhanced sun protection and whitening effects [113].

5.5. Nanoformulations

Nanoformulations have emerged as a promising approach for enhancing the stability and biocompatibility of antimicrobial agents, leading to their increasing adoption in novel NABHSGs. For instance: (I) Rhamnolipid nano-micelles exhibit antimicrobial activity comparable to NABHSGs, while maintaining non-cytotoxic properties. The inclusion of (biological) surfactants enhances the hydrophobicity of degrading microorganisms, thereby rendering the hydrophobic matrix more accessible to cells [114]. In comparison to ABHSGs, Rhamnolipid nano-micelles are safer and exhibit minimal impact on the skin [115,116]. Not only do Rhamnolipid nano-micelles possess antimicrobial properties, but they also hold potential as a therapeutic agent for COVID-19 treatment in future research [117]; (II) nanoparticles, measuring between 1 and 1000 nm in size, have been reported to exhibit antimicrobial activity against both viruses and bacteria [118]. Silver nanoparticles, in particular, are excellent materials in the biomedical field [119,120]. Synthesized from plant extracts, silver nanoparticles are not only safe in composition but also demonstrate potent antimicrobial effects [121]. Their high surface area-to-volume ratio enables them to penetrate bacterial cell walls, alter membrane structures, and even eliminate target cells. Silver nanoparticles exert their effects through the release of silver ions, increased cell membrane permeability, generation of reactive oxygen species, and interference with DNA replication [120]. For instance, moss and eucalyptus leaf extracts have been employed in the preparation of silver nanoparticles [122]. The eucalyptus leaf extract exhibited an inhibitory effect on bacterial biofilm formation and demonstrated more favorable antimicrobial effects [123,124]. Furthermore, chitosan serves as a coating on the surface of silver nanoparticles, acting as a stabilizer to prevent agglomeration and enhance the effectiveness of nanoparticles as antimicrobial agents. This property results in natural plant silver nanoparticles exhibiting superior stability and more potent antimicrobial effects [121]. The primary antimicrobial mechanism of zinc oxide nanoparticles (ZnO-NPs) involves the generation of ROS, which can target multiple sites, such as proteins, nucleic acids, lipids, and enzymes [125]. Compared to traditional ABHSGs, ZnO-NPs demonstrate enhanced antimicrobial activity and greater safety, along with high biocompatibility [118].

Although the systemic toxicity of nanoformulations as nanoscale materials has not been reported, some studies have shown that nanoformulations can easily interfere with biomolecules, cells, and organs [126]. Researchers are also concerned about the ability of nanoformulations to cross the blood–brain barrier via trans-synaptic transport, and accumulate in the brain [127]. Another issue with nanoformulations is that they pose a potential hazard to marine life if released into the environment, so future safety assessments for nanoformulations must be carried out to ensure they are safe for humans and the environment [128].

The utilization of nanoformulations in NABHSGs offers a promising avenue for developing hand hygiene products with improved antimicrobial efficacy, stability, and safety profiles. These advancements align with the growing demand for effective and safe hand hygiene solutions. Further research in this area can explore the optimization of nanoformulations, integration with natural active substances, and their potential applications in preventing the spread of infections and controlling disease transmission.

6. Future Research on HSGs

In the future, research in the field of novel HSGs should align with the increasing demands for safety and functionality among consumers [129]. The focus of such research may revolve around two main areas: Firstly, there is a need to screen natural components that possess low toxicity and demonstrate high antimicrobial performance. These components can be utilized as novel ingredients in HSGs [130]. Additionally, exploring the combination of multiple bioactive ingredients may further enhance the biological activities of HSGs. This approach allows for the development of multifunctional HSGs that not only sanitize hands effectively but also offer supplementary benefits [131]; secondly, optimizing the manufacturing processes of HSGs is crucial. This involves investigating gel materials that exhibit good biocompatibility, ensuring they are suitable for frequent use on the skin [132]. Rapid volatilization of the gel upon application can enhance the user experience and convenience. Moreover, exploring purely natural and edible components as gel materials holds the potential to increase the safety and acceptability of HSGs, addressing consumers' growing concerns about environmental and health impacts [133].

By addressing these research directions, we can continuously improve and innovate the performance of HSGs, thereby enhancing their practicality among consumers. Novel HSGs will be better equipped to meet the public's demand for safe, efficient, and eco-friendly hand cleaning solutions, contributing significantly to disease prevention and personal health maintenance [6]. These efforts will drive further advancements in the field of hand hygiene and provide us with more innovative and effective hand cleaning solutions.

7. Conclusions

Hand hygiene plays a key role in preventing and controlling infections, especially with the SARS-CoV-2 outbreak, which has seen a surge in demand for hand hygiene products. In this review, we provide a comprehensive overview of the various application scenarios, classifications, and challenges associated with HSGs. In addition, we highlight the emergence of novel ingredients with biological functions, which offer promising opportunities for advances in hand hygiene practice. By exploring multiple biological functions and ensuring ideal biosafety, future research could lead to the development of novel HSGs with excellent antimicrobial properties. Nevertheless, it is essential that we account for the potential risks of HSGs, such as the development of antimicrobial resistance and the impact of HSGs on the environment, while balancing their potential benefits. Better management and responsible hygiene practices should be promoted to ensure sustainable and effective hand hygiene solutions. In conclusion, this review provides valuable insights into hand hygiene and encourages further research and development of HSGs. As we continue to explore new possibilities and innovations in our human health goals, we remain committed to creating a safer, healthier, and more hygienic environment for all.

Author Contributions: Conceptualization, D.S., L.L. and Q.Y.; validation, M.C., W.W. and T.L.; formal analysis, D.S. and L.J.; writing—original draft preparation, Y.M. and J.Y.; writing—review and editing, Y.M., J.M. and H.Y.; funding acquisition, J.M. All authors have read and agreed to the published version of the manuscript.

Funding: This work was Supported by the Graduate Scientific Research Foundation of Wenzhou University (3162023003048, 3162023003049, 3162023003051 and 3162023004076).

Institutional Review Board Statement: Not applicable.

Informed Consent Statement: Not applicable.

Data Availability Statement: Not applicable.

Conflicts of Interest: The authors declare no conflict of interest.

Abbreviations

IHSs	Instant hand sanitizers
HSGs	Hand sanitizer gels
ABHSGs	Alcohol-based hand sanitizer gels
NABHSGs	Non-alcohol-based hand sanitizer gels
QACs	Quaternary ammonium compounds
BZK	Benzalkonium chloride
EOs	Essential oils
EGCG	Epigallocatechin gallate
EC16	Epigallocatechin-3-gallate-palmitate
LUT	Luteolin

UVA	Ultraviolet A
UVB	Ultraviolet B
UV	Ultraviolet
ROS	Reactive oxygen species
OHE	*Opuntia humifusa* extract
HAS	Hyaluronic acid synthases
HA	Hyaluronic acid
HABP	Hyaluronic acid-binding protein
OAs	Organic acids
CA	Citric acid
MA	Mandelic acid
ZnAC	Zinc-amino clay
MDA	Malondialdehyde
ZnO-NPs	Zinc oxide nanoparticles

References

1. Del Campo, R.; Martínez-García, L.; Sánchez-Díaz, A.M.; Baquero, F. Biology of Hand-to-Hand Bacterial Transmission. *Microbiol. Spectr.* **2019**, *7*, 1. [CrossRef] [PubMed]
2. Edmonds-Wilson, S.L.; Nurinova, N.I.; Zapka, C.A.; Fierer, N.; Wilson, M. Review of Human Hand Microbiome Research. *J. Dermatol. Sci.* **2015**, *80*, 3–12. [CrossRef] [PubMed]
3. Baquero, F.; Saralegui, C.; Marcos-Mencía, D.; Ballestero, L.; Vañó-Galván, S.; Moreno-Arrones, Ó.M.; Del Campo, R. Epidermis as a Platform for Bacterial Transmission. *Front. Immunol.* **2021**, *12*, 774018. [CrossRef] [PubMed]
4. Weber, T.P.; Stilianakis, N.I. Fomites, Hands, and the Transmission of Respiratory Viruses. *J. Occup. Environ. Hyg.* **2021**, *18*, 1–3. [CrossRef]
5. Lukashev, A.N.; Vakulenko, Y.A.; Turbabina, N.A.; Deviatkin, A.A.; Drexler, J.F. Molecular Epidemiology and Phylogenetics of Human Enteroviruses: Is There a Forest behind the Trees? *Rev. Med. Virol.* **2018**, *28*, e2002. [CrossRef]
6. Stadler, R.N.; Tschudin-Sutter, S. What Is New with Hand Hygiene? *Curr. Opin. Infect. Dis.* **2020**, *33*, 327–332. [CrossRef]
7. Howard, G.; Bartram, J.; Brocklehurst, C.; Colford, J.M.; Costa, F.; Cunliffe, D.; Dreibelbis, R.; Eisenberg, J.N.S.; Evans, B.; Girones, R.; et al. COVID-19: Urgent Actions, Critical Reflections and Future Relevance of 'WaSH': Lessons for the Current and Future Pandemics. *J. Water Health* **2020**, *18*, 613–630. [CrossRef]
8. Foddai, A.C.G.; Grant, I.R.; Dean, M. Efficacy of Instant Hand Sanitizers against Foodborne Pathogens Compared with Hand Washing with Soap and Water in Food Preparation Settings: A Systematic Review. *J. Food Prot.* **2016**, *79*, 1040–1054. [CrossRef]
9. Guidry, J.P.D.; O'Donnell, N.H.; Austin, L.L.; Coman, I.A.; Adams, J.; Perrin, P.B. Stay Socially Distant and Wash Your Hands: Using the Health Belief Model to Determine Intent for COVID-19 Preventive Behaviors at the Beginning of the Pandemic. *Health Educ. Behav.* **2021**, *48*, 424–433. [CrossRef]
10. Opatz, T.; Senn-Bilfinger, J.; Richert, C. Thoughts on What Chemists Can Contribute to Fighting SARS-CoV-2—A Short Note on Hand Sanitizers, Drug Candidates and Outreach. *Angew. Chem. Int. Ed. Engl.* **2020**, *59*, 9236–9240. [CrossRef]
11. Selam, M.N. Hand Sanitizers Marketed in the Streets of Addis Ababa, Ethiopia, in the Era of COVID-19: A Quality Concern. *Risk Manag. Healthc. Policy* **2020**, *13*, 2483–2487. [CrossRef] [PubMed]
12. Ory, J.; Zingg, W.; de Kraker, M.E.A.; Soule, H.; Pittet, D. Wiping Is Inferior to Rubbing: A Note of Caution for Hand Hygiene with Alcohol-Based Solutions. *Infect. Control Hosp. Epidemiol.* **2018**, *39*, 332–335. [CrossRef] [PubMed]
13. Rai, H.; Knighton, S.; Zabarsky, T.F.; Donskey, C.J. Comparison of Ethanol Hand Sanitizer versus Moist Towelette Packets for Mealtime Patient Hand Hygiene. *Am. J. Infect. Control* **2017**, *45*, 1033–1034. [CrossRef]
14. Jing, J.L.J.; Pei Yi, T.; Bose, R.J.C.; McCarthy, J.R.; Tharmalingam, N.; Madheswaran, T. Hand Sanitizers: A Review on Formulation Aspects, Adverse Effects, and Regulations. *Int. J. Environ. Res. Public Health* **2020**, *17*, 3326. [CrossRef]
15. Golin, A.P.; Choi, D.; Ghahary, A. Hand Sanitizers: A Review of Ingredients, Mechanisms of Action, Modes of Delivery, and Efficacy against Coronaviruses. *Am. J. Infect. Control* **2020**, *48*, 1062–1067. [CrossRef] [PubMed]
16. Saha, T.; Khadka, P.; Das, S.C. Alcohol-Based Hand Sanitizer—Composition, Proper Use and Precautions. *Germs* **2021**, *11*, 408–417. [CrossRef]
17. Abuga, K.; Nyamweya, N. Alcohol-Based Hand Sanitizers in COVID-19 Prevention: A Multidimensional Perspective. *Pharm. J. Pharm. Educ. Pract.* **2021**, *9*, 64. [CrossRef] [PubMed]
18. Selam, M.N.; Bayisa, R.; Ababu, A.; Abdella, M.; Diriba, E.; Wale, M.; Baye, A.M. Increased Production of Alcohol-Based Hand Rub Solution in Response to COVID-19 and Fire Hazard Potential: Preparedness of Public Hospitals in Addis Ababa, Ethiopia. *Risk Manag. Healthc. Policy* **2020**, *13*, 2507–2513. [CrossRef]
19. Edmonds-Wilson, S.; Campbell, E.; Fox, K.; Macinga, D. Comparison of 3 in Vivo Methods for Assessment of Alcohol-Based Hand Rubs. *Am. J. Infect. Control* **2015**, *43*, 506–509. [CrossRef]
20. Filipe, H.A.L.; Fiuza, S.M.; Henriques, C.A.; Antunes, F.E. Antiviral and Antibacterial Activity of Hand Sanitizer and Surface Disinfectant Formulations. *Int. J. Pharmaceut.* **2021**, *609*, 121139. [CrossRef]

21. Suchomel, M.; Fritsch, F.; Kampf, G. Bactericidal Efficacy of Two Modified WHO-Recommended Alcohol-Based Hand Rubs Using Two Types of Rub-in Techniques for 15 s. *J. Hosp. Infect.* **2021**, *111*, 47–49. [CrossRef]
22. Booq, R.Y.; Alshehri, A.A.; Almughem, F.A.; Zaidan, N.M.; Aburayan, W.S.; Bakr, A.A.; Kabli, S.H.; Alshaya, H.A.; Alsuabeyl, M.S.; Alyamani, E.J.; et al. Formulation and Evaluation of Alcohol-Free Hand Sanitizer Gels to Prevent the Spread of Infections during Pandemics. *Int. J. Environ. Res. Public Health* **2021**, *18*, 6252. [CrossRef] [PubMed]
23. Yüksel, Y.T.; Sonne, M.; Nørreslet, L.B.; Gundersen, G.; Fazli, M.M.; Agner, T. Skin Barrier Response to Active Chlorine Hand Disinfectant-An Experimental Study Comparing Skin Barrier Response to Active Chlorine Hand Disinfectant and Alcohol-Based Hand Rub on Healthy Skin and Eczematous Skin. *Ski. Res. Technol.* **2022**, *28*, 89–97. [CrossRef] [PubMed]
24. Advice for the Public on COVID-19—World Health Organization. Available online: https://www.who.int/emergencies/diseases/novel-coronavirus-2019/advice-for-public (accessed on 20 October 2022).
25. Hoang, H.T.; Van Tran, V.; Bui, V.K.H.; Kwon, O.-H.; Moon, J.-Y.; Lee, Y.-C. Novel Moisturized and Antimicrobial Hand Gel Based on Zinc-Aminoclay and *Opuntia humifusa* Extract. *Sci. Rep.* **2021**, *11*, 17821. [CrossRef] [PubMed]
26. Choi, K.; Sim, S.; Choi, J.; Park, C.; Uhm, Y.; Lim, E.; Kim, A.Y.; Yoo, S.J.; Lee, Y. Changes in Handwashing and Hygiene Product Usage Patterns in Korea before and after the Outbreak of COVID-19. *Environ. Sci. Eur.* **2021**, *33*, 79. [CrossRef] [PubMed]
27. Galvin, C.J.; Li, Y.-C.J.; Malwade, S.; Syed-Abdul, S. COVID-19 Preventive Measures Showing an Unintended Decline in Infectious Diseases in Taiwan. *Int. J. Infect. Dis. IJID Off. Publ. Int. Soc. Infect. Dis.* **2020**, *98*, 18–20. [CrossRef] [PubMed]
28. Rundle, C.W.; Presley, C.L.; Militello, M.; Barber, C.; Powell, D.L.; Jacob, S.E.; Atwater, A.R.; Watsky, K.L.; Yu, J.; Dunnick, C.A. Hand Hygiene during COVID-19: Recommendations from the American Contact Dermatitis Society. *J. Am. Acad. Dermatol.* **2020**, *83*, 1730–1737. [CrossRef]
29. Boyce, J.M.; Pittet, D.; Healthcare Infection Control Practices Advisory Committee; HICPAC/SHEA/APIC/IDSA Hand Hygiene Task Force Guideline for Hand Hygiene in Health-Care Settings. Recommendations of the Healthcare Infection Control Practices Advisory Committee and the HICPAC/SHEA/APIC/IDSA Hand Hygiene Task Force. Society for Healthcare Epidemiology of America/Association for Professionals in Infection Control/Infectious Diseases Society of America. *MMWR Recomm. Rep. Morb. Mortal. Wkly. Rep.* **2002**, *51*, 1–45, quiz CE1-4.
30. Guo, X.; Huang, S.; Sun, J.; Wang, F. Comparison of the Cytotoxicities and Wound Healing Effects of Hyaluronan, Carbomer, and Alginate on Skin Cells in Vitro. *Adv. Ski. Wound Care* **2015**, *28*, 410–414. [CrossRef]
31. Fresno Contreras, M.J.; Ramírez Diéguez, A.; Jiménez Soriano, M.M. Rheological Characterization of Hydroalcoholic Gels—15% Ethanol—Of Carbopol Ultrez 10. *Il Farm.* **2001**, *56*, 437–441. [CrossRef]
32. Berardi, A.; Perinelli, D.R.; Bisharat, L.; Sabbatini, B.; Bonacucina, G.; Tiboni, M.; Palmieri, G.F.; Cespi, M. Factors Affecting the Rheological Behaviour of Carbomer Dispersions in Hydroalcoholic Medium: Towards the Optimization of Hand Sanitiser Gel Formulations. *Int. J. Pharm.* **2022**, *616*, 121503. [CrossRef]
33. Islam, M.T.; Rodríguez-Hornedo, N.; Ciotti, S.; Ackermann, C. Rheological Characterization of Topical Carbomer Gels Neutralized to Different PH. *Pharm. Res.* **2004**, *21*, 1192–1199. [CrossRef]
34. Tamimi, A.H.; Carlino, S.; Edmonds, S.; Gerba, C.P. Impact of an Alcohol-Based Hand Sanitizer Intervention on the Spread of Viruses in Homes. *Food Environ. Virol.* **2014**, *6*, 140–144. [CrossRef] [PubMed]
35. Tamimi, A.H.; Maxwell, S.; Edmonds, S.L.; Gerba, C.P. Impact of the Use of an Alcohol-Based Hand Sanitizer in the Home on Reduction in Probability of Infection by Respiratory and Enteric Viruses. *Epidemiol. Infect.* **2015**, *143*, 3335–3341. [CrossRef] [PubMed]
36. Munoz-Figueroa, G.P.; Ojo, O. The Effectiveness of Alcohol-Based Gel for Hand Sanitising in Infection Control. *Br. J. Nurs. Mark Allen Publ.* **2018**, *27*, 382–388. [CrossRef] [PubMed]
37. Zhao, P.; Li, Y. Modeling and Experimental Validation of Microbial Transfer via Surface Touch. *Environ. Sci. Technol.* **2021**, *55*, 4148–4161. [CrossRef]
38. L'Huillier, A.G.; Tapparel, C.; Turin, L.; Boquete-Suter, P.; Thomas, Y.; Kaiser, L. Survival of Rhinoviruses on Human Fingers. *Clin. Microbiol. Infect. Off. Publ. Eur. Soc. Clin. Microbiol. Infect. Dis.* **2015**, *21*, 381–385. [CrossRef]
39. Edmonds, S.L.; Mann, J.; McCormack, R.R.; Macinga, D.R.; Fricker, C.M.; Arbogast, J.W.; Dolan, M.J. SaniTwice: A Novel Approach to Hand Hygiene for Reducing Bacterial Contamination on Hands When Soap and Water Are Unavailable. *J. Food Prot.* **2010**, *73*, 2296–2300. [CrossRef]
40. Lotfinejad, N.; Peters, A.; Tartari, E.; Fankhauser-Rodriguez, C.; Pires, D.; Pittet, D. Hand Hygiene in Health Care: 20 Years of Ongoing Advances and Perspectives. *Lancet Infect. Dis.* **2021**, *21*, e209–e221. [CrossRef]
41. Vermeil, T.; Peters, A.; Kilpatrick, C.; Pires, D.; Allegranzi, B.; Pittet, D. Hand Hygiene in Hospitals: Anatomy of a Revolution. *J. Hosp. Infect.* **2019**, *101*, 383–392. [CrossRef]
42. Finger, J.A.F.F.; Lima, E.M.F.; Coelho, K.S.; Behrens, J.H.; Landgraf, M.; Franco, B.D.G.M.; Pinto, U.M. Adherence to Food Hygiene and Personal Protection Recommendations for Prevention of COVID-19. *Trends Food Sci. Technol.* **2021**, *112*, 847–852. [CrossRef] [PubMed]
43. Sandora, T.J.; Taveras, E.M.; Shih, M.-C.; Resnick, E.A.; Lee, G.M.; Ross-Degnan, D.; Goldmann, D.A. A Randomized, Controlled Trial of a Multifaceted Intervention Including Alcohol-Based Hand Sanitizer and Hand-Hygiene Education to Reduce Illness Transmission in the Home. *Pediatrics* **2005**, *116*, 587–594. [CrossRef] [PubMed]
44. Montville, R.; Chen, Y.; Schaffner, D.W. Risk Assessment of Hand Washing Efficacy Using Literature and Experimental Data. *Int. J. Food Microbiol.* **2002**, *73*, 305–313. [CrossRef] [PubMed]

45. Feingold, K.R. Thematic Review Series: Skin Lipids. The Role of Epidermal Lipids in Cutaneous Permeability Barrier Homeostasis. *J. Lipid Res.* **2007**, *48*, 2531–2546. [CrossRef]
46. Singh, D.; Joshi, K.; Samuel, A.; Patra, J.; Mahindroo, N. Alcohol-Based Hand Sanitisers as First Line of Defence against SARS-CoV-2: A Review of Biology, Chemistry and Formulations. *Epidemiol. Infect.* **2020**, *148*, e229. [CrossRef]
47. Dear, K.; Grayson, L.; Nixon, R. Potential Methanol Toxicity and the Importance of Using a Standardised Alcohol-Based Hand Rub Formulation in the Era of COVID-19. *Antimicrob. Resist. Infect. Control.* **2020**, *9*, 129. [CrossRef] [PubMed]
48. Ho, H.J.; Poh, B.-F.; Choudhury, S.; Krishnan, P.; Ang, B.; Chow, A. Alcohol Handrubbing and Chlorhexidine Handwashing Are Equally Effective in Removing Methicillin-Resistant *Staphylococcus aureus* from Health Care Workers' Hands: A Randomized Controlled Trial. *Am. J. Infect. Control* **2015**, *43*, 1246–1248. [CrossRef] [PubMed]
49. Wolfe, M.K.; Gallandat, K.; Daniels, K.; Desmarais, A.M.; Scheinman, P.; Lantagne, D. Handwashing and Ebola Virus Disease Outbreaks: A Randomized Comparison of Soap, Hand Sanitizer, and 0.05% Chlorine Solutions on the Inactivation and Removal of Model Organisms Phi6 and *E. coli* from Hands and Persistence in Rinse Water. *PLoS ONE* **2017**, *12*, e0172734. [CrossRef]
50. Nyamadzawo, A.; Nishio, J.; Okada, S.; Nyamakura, R. Effect of Using Portable Alcohol-Based Handrub on Nurses' Hand Hygiene Compliance and Nasal Carriage of *Staphylococcus aureus* in a Low-Income Health Setting. *Am. J. Infect. Control* **2020**, *48*, 473–479. [CrossRef] [PubMed]
51. Andal, V.; Lakshmipathy, R.; Jose, D. Effect of Sanitizer on Obliteration of SARS-CoV2/COVID 19: A Mini Review. *Mater. Today Proc.* **2022**, *55*, 264–266. [CrossRef]
52. Ochwoto, M.; Muita, L.; Talaam, K.; Wanjala, C.; Ogeto, F.; Wachira, F.; Osman, S.; Kimotho, J.; Ndegwa, L. Anti-Bacterial Efficacy of Alcoholic Hand Rubs in the Kenyan Market, 2015. *Antimicrob. Resist. Infect. Control.* **2017**, *6*, 17. [CrossRef]
53. Kramer, A.; Arvand, M.; Christiansen, B.; Dancer, S.; Eggers, M.; Exner, M.; Müller, D.; Mutters, N.T.; Schwebke, I.; Pittet, D. Ethanol Is Indispensable for Virucidal Hand Antisepsis: Memorandum from the Alcohol-Based Hand Rub (ABHR) Task Force, WHO Collaborating Centre on Patient Safety, and the Commission for Hospital Hygiene and Infection Prevention (KRINKO), Robert Koch Institute, Berlin, Germany. *Antimicrob. Resist. Infect. Control.* **2022**, *11*, 93. [CrossRef] [PubMed]
54. Kratzel, A.; Todt, D.; V'kovski, P.; Steiner, S.; Gultom, M.; Thao, T.T.N.; Ebert, N.; Holwerda, M.; Steinmann, J.; Niemeyer, D.; et al. Inactivation of Severe Acute Respiratory Syndrome Coronavirus 2 by WHO-Recommended Hand Rub Formulations and Alcohols. *Emerg. Infect. Dis.* **2020**, *26*, 1592–1595. [CrossRef]
55. Kampf, G.; Hollingsworth, A. Comprehensive Bactericidal Activity of an Ethanol-Based Hand Gel in 15 Seconds. *Ann. Clin. Microbiol. Antimicrob.* **2008**, *7*, 2. [CrossRef] [PubMed]
56. Guilhermetti, M.; Marques Wiirzler, L.A.; Castanheira Facio, B.; da Silva Furlan, M.; Campo Meschial, W.; Bronharo Tognim, M.C.; Botelho Garcia, L.; Luiz Cardoso, C. Antimicrobial Efficacy of Alcohol-Based Hand Gels. *J. Hosp. Infect.* **2010**, *74*, 219–224. [CrossRef] [PubMed]
57. Gammon, J.; Hunt, J. COVID-19 and Hand Hygiene: The Vital Importance of Hand Drying. *Br. J. Nurs. Mark Allen Publ.* **2020**, *29*, 1003–1006. [CrossRef] [PubMed]
58. Voniatis, C.; Bánsághi, S.; Ferencz, A.; Haidegger, T. A Large-Scale Investigation of Alcohol-Based Handrub (ABHR) Volume: Hand Coverage Correlations Utilizing an Innovative Quantitative Evaluation System. *Antimicrob. Resist. Infect. Control.* **2021**, *10*, 49. [CrossRef] [PubMed]
59. Ren, S.; Ni, X.; Xu, H.; Den, J.; Wang, Y.; Sun, J.; Kong, Q.; Zha, J.; Jin, H.; Zhao, H.; et al. The Relationship between Hand Area and Hand Contamination. *Am. J. Infect. Control* **2011**, *39*, 66–68. [CrossRef]
60. Price, L.; Gozdzielewska, L.; Alejandre, J.C.; Jorgenson, A.; Stewart, E.; Pittet, D.; Reilly, J. Systematic Review on Factors Influencing the Effectiveness of Alcohol-Based Hand Rubbing in Healthcare. *Antimicrob. Resist. Infect. Control.* **2022**, *11*, 16. [CrossRef]
61. la Fleur, P.; Jones, S. *Non-Alcohol Based Hand Rubs: A Review of Clinical Effectiveness and Guidelines*; CADTH Rapid Response Reports; Canadian Agency for Drugs and Technologies in Health: Ottawa, ON, Canada, 2017.
62. Soltani, M.; Ravine, T.J.; Davis, J.H. Novel Boronium Salt Exhibits Substantial Antibacterial Activity When Compared to a Commercial Quaternary Ammonium Disinfectant. *Bioorg. Med. Chem. Lett.* **2021**, *36*, 127808. [CrossRef]
63. Sadakane, K.; Ichinose, T. Effect of the Hand Antiseptic Agents Benzalkonium Chloride, Povidone-Iodine, Ethanol, and Chlorhexidine Gluconate on Atopic Dermatitis in NC/Nga Mice. *Int. J. Med. Sci.* **2015**, *12*, 116–125. [CrossRef] [PubMed]
64. Dev Kumar, G.; Mishra, A.; Dunn, L.; Townsend, A.; Oguadinma, I.C.; Bright, K.R.; Gerba, C.P. Biocides and Novel Antimicrobial Agents for the Mitigation of Coronaviruses. *Front. Microbiol.* **2020**, *11*, 1351. [CrossRef] [PubMed]
65. Obłąk, E.; Piecuch, A.; Rewak-Soroczyńska, J.; Paluch, E. Activity of Gemini Quaternary Ammonium Salts against Microorganisms. *Appl. Microbiol. Biotechnol.* **2019**, *103*, 625–632. [CrossRef] [PubMed]
66. Herdt, B.L.; Black, E.P.; Zhou, S.S.; Wilde, C.J. Inactivation of SARS-CoV-2 by 2 Commercially Available Benzalkonium Chloride-Based Hand Sanitizers in Comparison with an 80% Ethanol-Based Hand Sanitizer. *Infect. Prev. Pract.* **2021**, *3*, 100191. [CrossRef]
67. Schrank, C.L.; Minbiole, K.P.C.; Wuest, W.M. Are Quaternary Ammonium Compounds, the Workhorse Disinfectants, Effective against Severe Acute Respiratory Syndrome-Coronavirus-2? *Acs Infect. Dis.* **2020**, *6*, 1553–1557. [CrossRef]
68. Ogilvie, B.H.; Solis-Leal, A.; Lopez, J.B.; Poole, B.D.; Robison, R.A.; Berges, B.K. SARS-CoV-2Alcohol-Free Hand Sanitizer and Other Quaternary Ammonium Disinfectants Quickly and Effectively Inactivate SARS-CoV-2. *J. Hosp. Infect.* **2021**, *108*, 142–145. [CrossRef]

69. Thaddeus, N.; Francis, E.; Jane, O.; Obumneme, A.; Okechukwu, E. Effects of Some Common Additives on the Antimicrobial Activities of Alcohol-Based Hand Sanitizers. *Asian Pac. J. Trop. Med.* **2018**, *11*, 222. [CrossRef]
70. Hakimi, A.A.; Armstrong, W.B. Hand Sanitizer in a Pandemic: Wrong Formulations in the Wrong Hands. *J. Emerg. Med.* **2020**, *59*, 668–672. [CrossRef]
71. Emami, A.; Javanmardi, F.; Keshavarzi, A.; Pirbonyeh, N. Hidden Threat Lurking behind the Alcohol Sanitizers in COVID-19 Outbreak. *Dermatol. Ther.* **2020**, *33*, e13627. [CrossRef]
72. Gloekler, L.E.; de Gandiaga, E.J.; Binczewski, N.R.; Steimel, K.G.; Massarsky, A.; Kozal, J.; Vincent, M.; Zisook, R.; LaGuardia, M.J.; Dotson, S.; et al. Evaluation of the Safety and Efficacy of Hand Sanitizer Products Marketed to Children Available during the COVID-19 Pandemic. *Int. J. Environ. Res. Public. Health* **2022**, *19*, 14424. [CrossRef]
73. Baye, A.M.; Ababu, A.; Bayisa, R.; Abdella, M.; Diriba, E.; Wale, M.; Selam, M.N. Perspectives of Compounding Pharmacists on Alcohol-Based Hand Sanitizer Production and Utilization for COVID-19 Prevention in Addis Ababa, Ethiopia: A Descriptive Phenomenology Study. *PLoS ONE* **2021**, *16*, e0250020. [CrossRef] [PubMed]
74. Chan, A.P.L.; Chan, T.Y.K. Methanol as an Unlisted Ingredient in Supposedly Alcohol-Based Hand Rub Can Pose Serious Health Risk. *Int. J. Environ. Res. Public. Health* **2018**, *15*, 1440. [CrossRef] [PubMed]
75. Matatiele, P.; Southon, B.; Dabula, B.; Marageni, T.; Poongavanum, P.; Kgarebe, B. Assessment of Quality of Alcohol-Based Hand Sanitizers Used in Johannesburg Area during the COVID-19 Pandemic. *Sci. Rep.* **2022**, *12*, 4231. [CrossRef] [PubMed]
76. Aschenbrenner, D.S. Is That Hand Sanitizer Safe? *Am. J. Nurs.* **2020**, *120*, 23. [CrossRef]
77. Basak, D.; Deb, S. Sensitivity of SARS-CoV-2 towards Alcohols: Potential for Alcohol-Related Toxicity in Humans. *Life* **2021**, *11*, 1334. [CrossRef]
78. Martin, G.C.; Le Roux, G.; Guindolet, D.; Boulanger, E.; Hasle, D.; Morin, E.; Vodovar, D.; Vignal, C.; Gabison, E.; Descatha, A.; et al. Pediatric Eye Injuries by Hydroalcoholic Gel in the Context of the Coronavirus Disease 2019 Pandemic. *JAMA Ophthalmol.* **2021**, *139*, 348–351. [CrossRef]
79. Akbas, E.; Korkmaz, I.; Palamar, M.; Barut Selver, O. Shifting Trends in Demographic Features of Chemical Eye Injuries during COVID-19 Pandemic. *Int. Ophthalmol.* **2022**, *42*, 2127–2132. [CrossRef]
80. Lopez, T.K.; Jones, K.; Roseberry-Lincoln, A.; Zidek, A.; MacKinnon, L.; Marro, L. Adult and Children's Use of Hand Sanitizer during a Pandemic—An Observational Study. *J. Expo. Sci. Environ. Epidemiol.* **2022**. [CrossRef]
81. Babić, Ž.; Turk, R.; Macan, J. Toxicological Aspects of Increased Use of Surface and Hand Disinfectants in Croatia during the COVID-19 Pandemic: A Preliminary Report. *Arh. Hig. Rada I Toksikol.* **2020**, *71*, 261–264. [CrossRef]
82. Gharpure, R.; Miller, G.F.; Hunter, C.M.; Schnall, A.H.; Kunz, J.; Garcia-Williams, A.G. Safe Use and Storage of Cleaners, Disinfectants, and Hand Sanitizers: Knowledge, Attitudes, and Practices among U.S. Adults during the COVID-19 Pandemic, May 2020. *Am. J. Trop. Med. Hyg.* **2021**, *104*, 496–501. [CrossRef]
83. Hsieh, S.; Sapkota, A.; Wood, R.; Bearer, C.; Kapoor, S. Neonatal Ethanol Exposure from Ethanol-Based Hand Sanitisers in Isolettes. *Arch. Dis. Child. Fetal Neonatal Ed.* **2018**, *103*, F55–F58. [CrossRef]
84. Ghosh, A.; Mahintamani, T.; Balhara, Y.P.S.; Roub, F.E.; Basu, D.; Bn, S.; Mattoo, S.K.; Mishra, E.; Sharma, B. Disulfiram Ethanol Reaction with Alcohol-Based Hand Sanitizer: An Exploratory Study. *Alcohol Alcohol. Oxf. Oxfs.* **2021**, *56*, 42–46. [CrossRef] [PubMed]
85. Lu, J.; Guo, J. Disinfection Spreads Antimicrobial Resistance. *Science* **2021**, *371*, 474. [CrossRef] [PubMed]
86. Daverey, A.; Dutta, K. COVID-19: Eco-Friendly Hand Hygiene for Human and Environmental Safety. *J. Environ. Chem. Eng.* **2021**, *9*, 104754. [CrossRef] [PubMed]
87. Assefa, D.; Melaku, T. Commercial Hand Sanitizers Use amid COVID-19 Pandemic: The Concerns of Antimicrobial Resistance. *Infect. Drug Resist.* **2021**, *14*, 2183–2185. [CrossRef]
88. Mahmood, A.; Eqan, M.; Pervez, S.; Alghamdi, H.A.; Tabinda, A.B.; Yasar, A.; Brindhadevi, K.; Pugazhendhi, A. COVID-19 and Frequent Use of Hand Sanitizers; Human Health and Environmental Hazards by Exposure Pathways. *Sci. Total Environ.* **2020**, *742*, 140561. [CrossRef]
89. Gaonkar, T.A.; Geraldo, I.; Caraos, L.; Modak, S.M. An Alcohol Hand Rub Containing a Synergistic Combination of an Emollient and Preservatives: Prolonged Activity against Transient Pathogens. *J. Hosp. Infect.* **2005**, *59*, 12–18. [CrossRef]
90. Loh, E.D.W.; Yew, Y.W. Hand Hygiene and Hand Eczema: A Systematic Review and Meta-Analysis. *Contact Dermat.* **2022**, *87*, 303–314. [CrossRef]
91. Dahmardehei, M.; Khadem Rezaiyan, M.; Safarnejad, F.; Ahmadabadi, A. An Unprecedented Increase in Burn Injuries Due to Alcohol-Based Hand Sanitizers during the COVID-19 Outbreak. *Med. J. Islam. Repub. Iran* **2021**, *35*, 107. [CrossRef]
92. Kolman, M.; Smith, C.; Chakrabarty, D.; Amin, S. Rheological Stability of Carbomer in Hydroalcoholic Gels: Influence of Alcohol Type. *Int. J. Cosmet. Sci.* **2021**, *43*, 748–763. [CrossRef]
93. Kaiser, N.; Klein, D.; Karanja, P.; Greten, Z.; Newman, J. Inactivation of Chlorhexidine Gluconate on Skin by Incompatible Alcohol Hand Sanitizing Gels. *Am. J. Infect. Control* **2009**, *37*, 569–573. [CrossRef] [PubMed]
94. Alghamdi, H.A. A Need to Combat COVID-19; Herbal Disinfection Techniques, Formulations and Preparations of Human Health Friendly Hand Sanitizers. *Saudi J. Biol. Sci.* **2021**, *28*, 3943–3947. [CrossRef] [PubMed]
95. Jain, V.M.; Karibasappa, G.N.; Dodamani, A.S.; Prashanth, V.K.; Mali, G.V. Comparative Assessment of Antimicrobial Efficacy of Different Hand Sanitizers: An in Vitro Study. *Dent. Res. J.* **2016**, *13*, 424–431. [CrossRef]

96. David, O.; Olatunji, F.; Alese, M.; Babalola, T.; Alese, O. Antimicrobial Activity, Safety and Acceptability of Formulated Ginger-Fortified Hand Sanitizer Gel. *Int. J. Trop. Dis. Health* **2017**, *22*, 1–11. [CrossRef]
97. Parham, S.; Kharazi, A.Z.; Bakhsheshi-Rad, H.R.; Nur, H.; Ismail, A.F.; Sharif, S.; RamaKrishna, S.; Berto, F. Antioxidant, Antimicrobial and Antiviral Properties of Herbal Materials. *Antioxidants* **2020**, *9*, 1309. [CrossRef] [PubMed]
98. Nazzaro, F.; Fratianni, F.; De Martino, L.; Coppola, R.; De Feo, V. Effect of Essential Oils on Pathogenic Bacteria. *Pharmaceuticals* **2013**, *6*, 1451–1474. [CrossRef]
99. Egner, P.; Pavlačková, J.; Sedlaříková, J.; Pleva, P.; Mokrejš, P.; Janalíková, M. Non-Alcohol Hand Sanitiser Gels with Mandelic Acid and Essential Oils. *Int. J. Mol. Sci.* **2023**, *24*, 3855. [CrossRef]
100. Chouhan, S.; Sharma, K.; Guleria, S. Antimicrobial Activity of Some Essential Oils-Present Status and Future Perspectives. *Medicines* **2017**, *4*, 58. [CrossRef]
101. Wang, X.; Shen, Y.; Thakur, K.; Han, J.; Zhang, J.-G.; Hu, F.; Wei, Z.-J. Antibacterial Activity and Mechanism of Ginger Essential Oil against Escherichia Coli and *Staphylococcus aureus*. *Molecules* **2020**, *25*, 3955. [CrossRef]
102. Ismail, M.; Penlap, V.; Mbacham, W.F.; Kamdem, S.L.S. Improvement of the Efficacy an WHO Alcohol-Based Hand Rub Formulation by Reducing Isopropanol Content and Associating Syzygium Aromarticum and Piper Nigrum Essential Oils. *J. Appl. Life Sci. Int.* **2019**, *21*, 1–11. [CrossRef]
103. Dickinson, D.; Tufo, M.D.; Liu, E.; Shao, X.; Hsu, S. Bactericidal and Fungicidal Activities of Novel ProtecTeaV Formulations—Alcohol-Based Hand Hygiene and Surface Disinfectant Prototypes Containing Epigallocatechin-3-Gallate-Palmitate (EC16). *Microbiol. Infect. Dis. Wilmington Del.* **2021**, *5*. [CrossRef] [PubMed]
104. Dickinson, D.; Marsh, B.; Shao, X.; Liu, E.; Sampath, L.; Yao, B.; Jiang, X.; Hsu, S. Virucidal Activities of Novel Hand Hygiene and Surface Disinfectant Formulations Containing EGCG-Palmitates (EC16). *Am. J. Infect. Control* **2022**, *50*, 1212–1219. [CrossRef]
105. Gendrisch, F.; Esser, P.R.; Schempp, C.M.; Wölfle, U. Luteolin as a Modulator of Skin Aging and Inflammation. *BioFactors Oxf. Engl.* **2021**, *47*, 170–180. [CrossRef]
106. Xi, M.; Hou, Y.; Wang, R.; Ji, M.; Cai, Y.; Ao, J.; Shen, H.; Li, M.; Wang, J.; Luo, A. Potential Application of Luteolin as an Active Antibacterial Composition in the Development of Hand Sanitizer Products. *Molecules* **2022**, *27*, 7342. [CrossRef] [PubMed]
107. Silva-Hughes, A.F.; Wedge, D.E.; Cantrell, C.L.; Carvalho, C.R.; Pan, Z.; Moraes, R.M.; Madoxx, V.L.; Rosa, L.H. Diversity and Antifungal Activity of the Endophytic Fungi Associated with the Native Medicinal Cactus *Opuntia humifusa* (Cactaceae) from the United States. *Microbiol. Res.* **2015**, *175*, 67–77. [CrossRef] [PubMed]
108. Park, K.; Choi, H.-S.; Hong, Y.H.; Jung, E.Y.; Suh, H.J. Cactus Cladodes (*Opuntia humifusa*) Extract Minimizes the Effects of UV Irradiation on Keratinocytes and Hairless Mice. *Pharm. Biol.* **2017**, *55*, 1032–1040. [CrossRef] [PubMed]
109. Burel, C.; Kala, A.; Purevdorj-Gage, L. Impact of PH on Citric Acid Antimicrobial Activity against Gram-Negative Bacteria. *Lett. Appl. Microbiol.* **2021**, *72*, 332–340. [CrossRef] [PubMed]
110. Ionidis, G.; Hübscher, J.; Jack, T.; Becker, B.; Bischoff, B.; Todt, D.; Hodasa, V.; Brill, F.H.H.; Steinmann, E.; Steinmann, J. Development and Virucidal Activity of a Novel Alcohol-Based Hand Disinfectant Supplemented with Urea and Citric Acid. *Bmc Infect. Dis.* **2016**, *16*, 77. [CrossRef] [PubMed]
111. Halla, N.; Fernandes, I.P.; Heleno, S.A.; Costa, P.; Boucherit-Otmani, Z.; Boucherit, K.; Rodrigues, A.E.; Ferreira, I.C.F.R.; Barreiro, M.F. Cosmetics Preservation: A Review on Present Strategies. *Molecules* **2018**, *23*, 1571. [CrossRef]
112. Bruce, S.; Roberts, W.; Teller, C.; Colvan, L. The Effects of a Daily Skincare Regimen on Maintaining the Benefits Obtained from Previous Chemical Resurfacing Treatments. *J. Drugs Dermatol.* **2016**, *15*, 1145–1150.
113. Yamaguchi, K.; Maeda, M.; Masaki, H.; Iwabuchi, T. Oil Thickening with Organoclay Enhances the Ultraviolet Absorption Ability of Sunscreen on a Skin-Mimicking Substrate. *J. Oleo Sci.* **2021**, *70*, 721–730. [CrossRef] [PubMed]
114. Banat, I.M.; Franzetti, A.; Gandolfi, I.; Bestetti, G.; Martinotti, M.G.; Fracchia, L.; Smyth, T.J.; Marchant, R. Microbial Biosurfactants Production, Applications and Future Potential. *Appl. Microbiol. Biotechnol.* **2010**, *87*, 427–444. [CrossRef] [PubMed]
115. Ali, A.M.; Hill, H.J.; Elkhouly, G.E.; Bakkar, M.R.; Raya, N.R.; Stamataki, Z.; Abo-Zeid, Y. Rhamnolipid Nano-Micelles Inhibit SARS-CoV-2 Infection and Have No Dermal or Eye Toxic Effects in Rabbits. *Antibiotics* **2022**, *11*, 1556. [CrossRef] [PubMed]
116. Abo-Zeid, Y.; Bakkar, M.R.; Elkhouly, G.E.; Raya, N.R.; Zaafar, D. Rhamnolipid Nano-Micelles versus Alcohol-Based Hand Sanitizer: A Comparative Study for Antibacterial Activity against Hospital-Acquired Infections and Toxicity Concerns. *Antibiotics* **2022**, *11*, 605. [CrossRef] [PubMed]
117. Bakkar, M.R.; Faraag, A.H.I.; Soliman, E.R.S.; Fouda, M.S.; Sarguos, A.M.M.; McLean, G.R.; Hebishy, A.M.S.; Elkhouly, G.E.; Raya, N.R.; Abo-zeid, Y. Rhamnolipids Nano-Micelles as a Potential Hand Sanitizer. *Antibiotics* **2021**, *10*, 751. [CrossRef] [PubMed]
118. Ismail, A.; Raya, N.R.; Orabi, A.; Ali, A.M.; Abo-zeid, Y. Investigating the Antibacterial Activity and Safety of Zinc Oxide Nanoparticles versus a Commercial Alcohol-Based Hand-Sanitizer: Can Zinc Oxide Nanoparticles Be Useful for Hand Sanitation? *Antibiotics* **2022**, *11*, 1606. [CrossRef]
119. Annamalai, J.; Nallamuthu, T. Green Synthesis of Silver Nanoparticles: Characterization and Determination of Antibacterial Potency. *Appl. Nanosci.* **2016**, *6*, 259–265. [CrossRef]
120. Yin, I.X.; Zhang, J.; Zhao, I.S.; Mei, M.L.; Li, Q.; Chu, C.H. The Antibacterial Mechanism of Silver Nanoparticles and Its Application in Dentistry. *Int. J. Nanomed.* **2020**, *15*, 2555–2562. [CrossRef]
121. Wulandari, I.O.; Pebriatin, B.E.; Valiana, V.; Hadisaputra, S.; Ananto, A.D.; Sabarudin, A. Green Synthesis of Silver Nanoparticles Coated by Water Soluble Chitosan and Its Potency as Non-Alcoholic Hand Sanitizer Formulation. *Materials* **2022**, *15*, 4641. [CrossRef]

122. Sugumar, S.; Ghosh, V.; Nirmala, M.J.; Mukherjee, A.; Chandrasekaran, N. Ultrasonic Emulsification of Eucalyptus Oil Nanoemulsion: Antibacterial Activity against *Staphylococcus aureus* and Wound Healing Activity in Wistar Rats. *Ultrason. Sonochem.* **2014**, *21*, 1044–1049. [CrossRef]
123. Ghosh, S.; Nandi, S.; Basu, T. Nano-Antibacterials Using Medicinal Plant Components: An Overview. *Front. Microbiol.* **2021**, *12*, 768739. [CrossRef] [PubMed]
124. Aleksic Sabo, V.; Knezevic, P. Antimicrobial Activity of Eucalyptus Camaldulensis Dehn. Plant Extracts and Essential Oils: A Review. *Ind. Crops Prod.* **2019**, *132*, 413–429. [CrossRef] [PubMed]
125. Sirelkhatim, A.; Mahmud, S.; Seeni, A.; Kaus, N.H.M.; Ann, L.C.; Bakhori, S.K.M.; Hasan, H.; Mohamad, D. Review on Zinc Oxide Nanoparticles: Antibacterial Activity and Toxicity Mechanism. *Nano-Micro Lett.* **2015**, *7*, 219–242. [CrossRef]
126. Palacios-Hernandez, T.; Diaz-Diestra, D.M.; Nguyen, A.K.; Skoog, S.A.; Vijaya Chikkaveeraiah, B.; Tang, X.; Wu, Y.; Petrochenko, P.E.; Sussman, E.M.; Goering, P.L. Cytotoxicity, Cellular Uptake and Apoptotic Responses in Human Coronary Artery Endothelial Cells Exposed to Ultrasmall Superparamagnetic Iron Oxide Nanoparticles. *J. Appl. Toxicol. JAT* **2020**, *40*, 918–930. [CrossRef]
127. Lebda, M.A.; Sadek, K.M.; Tohamy, H.G.; Abouzed, T.K.; Shukry, M.; Umezawa, M.; El-Sayed, Y.S. Potential Role of α-Lipoic Acid and Ginkgo Biloba against Silver Nanoparticles-Induced Neuronal Apoptosis and Blood-Brain Barrier Impairments in Rats. *Life Sci.* **2018**, *212*, 251–260. [CrossRef]
128. Liao, C.; Li, Y.; Tjong, S.C. Bactericidal and Cytotoxic Properties of Silver Nanoparticles. *Int. J. Mol. Sci.* **2019**, *20*, 449. [CrossRef]
129. Nisbar, N.D.; Jamal Khair, S.K.; Bujang, N.B.; Mohd Yusop, A.Y. Determination of Ethanol, Isopropyl Alcohol and Methanol in Alcohol-Based Hand Sanitiser to Ensure Product Quality, Safety and Efficacy. *Sci. Rep.* **2023**, *13*, 9478. [CrossRef]
130. Fallica, F.; Leonardi, C.; Toscano, V.; Santonocito, D.; Leonardi, P.; Puglia, C. Assessment of Alcohol-Based Hand Sanitizers for Long-Term Use, Formulated with Addition of Natural Ingredients in Comparison to WHO Formulation 1. *Pharmaceutics* **2021**, *13*, 571. [CrossRef]
131. Ramaiah, G.B.; Tegegne, A.; Melese, B. Developments in Nano-Materials and Analysing Its Role in Fighting COVID-19. *Mater. Today Proc.* **2021**, *47*, 4357–4363. [CrossRef]
132. Bisht, B.; Lohani, U.C.; Kumar, V.; Gururani, P.; Sinhmar, R. Edible Hydrocolloids as Sustainable Substitute for Non-Biodegradable Materials. *Crit. Rev. Food Sci. Nutr.* **2022**, *62*, 693–725. [CrossRef] [PubMed]
133. Ahmad Raus, R.; Wan Nawawi, W.M.F.; Nasaruddin, R.R. Alginate and Alginate Composites for Biomedical Applications. *Asian J. Pharm. Sci.* **2021**, *16*, 280–306. [CrossRef] [PubMed]

toxics MDPI

Article

Kinetics of Excretion of the Perfluoroalkyl Surfactant cC_6O_4 in Humans

Silvia Fustinoni [1,2,*], Rosa Mercadante [1], Giorgia Lainati [2], Simone Cafagna [2] and Dario Consonni [2]

1 Toxicology Lab, Department of Clinical Sciences and Community Health, University of Milan, 20122 Milano, Italy
2 Fondazione IRCCS Ca' Granda Ospedale Maggiore Policlinico, 20122 Milan, Italy
* Correspondence: silvia.fustinoni@unimi.it; Tel.: +39-02-50320158

Abstract: cC_6O_4 is a new-generation perfluoroalkyl surfactant used in the chemical industry for the synthesis of perfluoroalkyl polymers. It was introduced as a less biopersistent substitute of traditional perfluoroalkyl surfactants such as PFOA, but its kinetics in humans was never investigated. This work is aimed to investigate the kinetics of elimination of cC_6O_4 in exposed workers. Eighteen male individuals occupationally exposed to cC_6O_4 in the production of fluoropolymers volunteered for the study. Blood and urine samples were collected from the end of a work-shift for the following 5 days off work. Serum and urinary cC_6O_4 were measured by LC-MS/MS. Seventy-two samples with serum cC_6O_4 ranging from 0.38 to 11.29 μg/L were obtained; mean levels were 3.07, 2.82, 2.67 and 2.01 μg/L at times 0, 18, 42 and 114 h. Two hundred and fifty-four urine samples with cC_6O_4 ranging from 0.19 to 5.92 μg/L were obtained. A random-intercept multiple regression model was applied to serum data and a half-life of 184 (95% CI 162–213) h for a first-order kinetics elimination was calculated; a mean distribution volume of 80 mL/kg was also estimated. Pearson's correlation between ln-transformed serum and daily urine concentrations was good, with r ranging from 0.802 to 0.838. The amount of cC_6O_4 excreted daily in urine was about 20% of the amount present in serum. The study allowed calculating a half-life for cC_6O_4 in blood of about 8 days in humans, supporting its much shorter biopersistence in comparison with legacy PFAS. The good correlation between urine and serum cC_6O_4 suggests urine as a possible non-invasive matrix for biomonitoring. The amount of cC_6O_4 excreted daily in urine suggests urine as the sole elimination route.

Keywords: serum cC_6O_4; half-life; elimination route; biomonitoring; urine

Citation: Fustinoni, S.; Mercadante, R.; Lainati, G.; Cafagna, S.; Consonni, D. Kinetics of Excretion of the Perfluoroalkyl Surfactant cC_6O_4 in Humans. *Toxics* **2023**, *11*, 284. https://doi.org/10.3390/toxics11030284

Academic Editor: Carina Ladeira

Received: 23 February 2023
Revised: 14 March 2023
Accepted: 17 March 2023
Published: 20 March 2023

1. Introduction

cC_6O_4 (acetic acid, 2,2-difluoro-2-[[2,2,4,5-tetrafluoro-5-(trifluoromethoxy)-1,3-dioxolan-4-yl]oxy]-, ammonium salt (1:1) CAS 1190931-27-1) (see Figure 1) is a perfluoroalkyl surfactant used in aqueous emulsions for the synthesis of plastomer and elastomer fluoropolymers in the chemical plant of Solvay Specialty Polymers Italy S.p.A., located in Spinetta Marengo (AL), Italy.

This chemical was developed as a new polymerization surfactant to overcome the hazard issues associated with perfluorooctanoate (PFOA), mostly biopersistency. Indeed, toxicokinetics studies in rats provided evidence of a short half-life in serum (4 to 7 h), with the majority of the excretion via urine occurring usually within 24 h, and no evidence of biotransformation and bioaccumulation [1]. According to CLP (Regulation (EC) No. 1272/2008), cC_6O_4 was classified as harmful if swallowed (H302), able to cause skin irritation and serious eye damage (H315 and H318); moreover, it may cause damage to organs (liver) through prolonged or repeated exposure (oral) (H373). The substance does not accumulate nor exert toxic effects in the aquatic organisms; it is not a persistent bioaccumulative toxic (PBT) or a very persistent very bioaccumulative (vPvB) chemical on the basis of the Annex XIII REACH. cC_6O_4 has been produced since 2012 and it was introduced to replace

PFOA; today it is the candidate for the substitution of other polymerization surfactants in chemical processes and used as a bridging solution to a Non FluoroSurfactant technology.

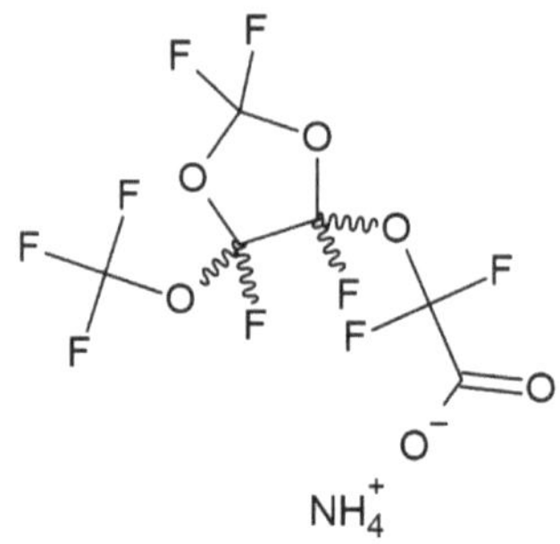

Figure 1. Molecular structure of cC_6O_4.

Human biological monitoring of exposure is a powerful tool to assess chemical exposure, integrating all exposure routes and sources, and taking into consideration characteristics of individuals [2]. Given the metabolic stability of perfluoroalkyl substances, biomonitoring is performed by measuring the unchanged chemicals in serum. Extensive literature is available on PFAS, with biomonitoring data in occupationally exposed individuals and in the general population, particularly regarding legacy PFAS, such as PFOA and perfluorooctane sulfonate (PFOS) [3]. Conversely, limited biological monitoring information has been reported on cC_6O_4. Particularly, data of workers of an Italian plant located in the Veneto region were retrieved in a technical report, with cC_6O_4 serum concentrations ranging from 0.5 to 932 µg/L [4]. More recently, a summary of 9 years of biomonitoring surveys in the chemical plant of Spinetta Marengo and in the research center of Bollate, both belonging to Solvay Specialty Polymers Italy SpA, Italy, was reported, with a large percentage of samples not quantifiable (below the analytical lower limit of quantification) and concentrations in the range of 0.1–56 µg/L [5].

The present work reports the results of a study on the kinetics of elimination of cC_6O_4 in occupationally exposed individuals. With this aim, blood and urine samples were collected immediately after a work-shift and during a period of 114 h while off work. The half-life of cC_6O_4 in serum and correlation between urine and serum concentrations were calculated, and a mass balance was performed by comparing the amount of cC_6O_4 excreted in urine and circulating in serum.

2. Materials and Methods

2.1. Field Study

The surveyed workers were employees of the chemical plant of Spinetta Marengo (AL), where cC_6O_4 is synthetized since 2012 and used as surfactant for the synthesis of fluorinated plastomers and elastomer polymers.

In this plant, the biological monitoring of cC_6O_4 started in 2013 and was performed annually (typically in February) in the frame of the occupational health and safety surveillance, according to the Italian law for the safety and health at workplaces [6], under the responsibility of occupational plant physicians; the workers signed an informed consent to agree to the survey [5].

The kinetics study was performed in collaboration with the occupational health physicians and industrial hygienists of the plant. From the historical data of the biological monitoring, it was observed that the highest exposure to cC_6O_4 was found in workers of the plastomers area. For this reason, the study was proposed to these workers. Eligible criteria were the following: being male; not affected by chronic diseases, particularly metabolic, hepatic and kidney diseases; and having a body mass index lower (BMI) than 30 kg/m^2. Among eligible male workers, all volunteered to participate in the study.

The sample collection lasted 114 h, during which 4 blood samples and a variable number of urine samples were collected.

For blood, the sampling protocol was as follows:

Day 1: A working day with a 6:00 to 14:00 work-shift; the first blood sample was collected at the end of the shift (time 0 h);
Day 2: A day off; at 8:00 the second blood sample was collected (time 18 h);
Day 3: A day off; at 8:00 the third blood sample was collected (time 42 h);
Day 4 and day 5: Two days off; no blood sample was collected;
Day 6: A working day with a 6:00 to 14:00 work-shift; before the start of the shift the fourth blood sample was collected (time 114 h).

About 5 mL of blood was drawn from the cubital vein in the infirmary of the site, by health personnel, using a tube with heparin as anticoagulant. Samples were centrifuged on site to separate serum, within 1 h after collection. Serum samples were frozen at −20 °C.

For urine, a sampling kit, including 25 numbered sterile recipients for urine (100 mL) and a void collection diary, was given to each participant with instructions for urine collection. The subject was asked to collect all urine of all voids starting from day 1 at 14:00 until day 3 at 8:00 (collection time 42 h). An additional sample was collected on day 6, before re-starting the shift, from most subjects (15 out of 18). Urine was self-collected and recorded in the form, indicating the void date, time and the recipient number/s. Initially, samples were kept at home, stored outside at ambient temperature, until the next morning, when they were given to the health personnel in the infirmary of the plant. Once there, urine samples were frozen at −20 °C.

A self-administered questionnaire was used to collect personal information such as height, weight, smoking habit and drug consumption.

The study was approved by the ethical committee of the University of Milan; each volunteer received the study protocol and signed the informed consent form.

The sample collection was performed from November 2021 to April 2022.

2.2. Analytical Measurements of Serum and Urinary cC_6O_4

The analytical work was performed following a previously published method with few modifications [7]. Briefly, each properly stored sample was thawed at room temperature. The sample was vortexed and an aliquot (20 µL of plasma or 40 µL of urine) was added with methanol (70 µL for plasma and 150 µL for urine) and 10 µL of the internal standard methanol working solution to obtain a 1:5 diluted sample containing 0.15 µg/L of internal standard (perfluoro-n-[1,2,3,4,6-13C5]hexanoic acid, chemical purity >98%, isotopic purity 99% Wellington Laboratories, Guelph, Canada). Plasma proteins were precipitated, and the clear supernatant was transferred to an autosampler vial; the same process was applied to urine samples. The injection volume was 10 µL. The chromatographic run was performed with an Agilent 1260 liquid chromatograph equipped with an Acquity HSS T3 C18 analytical column (2.1 × 100 mm, 1.8 µm, Waters) preceded by a SecurityGuard C18 pre-column (4 × 3 mm Phenomenex). A Hypersil GOLD delay column (3 × 50 mm, 3 µm, Thermo Fisher Scientific) was also installed between the pump and the injector. The separation was obtained with a binary pump linear gradient: mobile phases were aqueous 10 mM ammonium acetate buffer, pH 4.5 (eluent A), and acetonitrile (eluent B) pumped at a constant 0.2 mL/min flow rate. Detection was performed using a 5500 QTRAP Sciex triple quadrupole mass spectrometer equipped with an electrospray ionization source (ESI), operating in the negative ionization mode. Analytical standards were re-tuned and flow re-injected, allowing mass spectrometry parameter optimization; target compounds were detected by extracted ion chromatograms of monitored MRM transitions. The quantification was performed with calibration curves deriving from calibration standards that covered the cC_6O_4 linearity range concentrations (lower and upper limits of quantifications (LOQ): LLOQ 0.02 µg/L, ULOQ 1.25 µg/L). The calibration curves were built using serum or urine when appropriate to properly control for the specific matrix effect. The method was validated according to international guidelines for bioanalytical methods [8–10]. The quality requirements assessed in our previous work on serum were also achieved for urine matrix. Effect of dilution of serum and urine samples with methanol was assessed and deemed insignificant, enabling sample dilution when concentration was above the linearity range.

Urinary creatinine (crea) was determined using Jaffe's colorimetric method [11]. No criteria of acceptability based on urine dilution was applied.

2.3. Elimination Kinetics of Serum cC_6O_4 and Clearance

The one-compartment first-order kinetics model was applied to evaluate the elimination kinetics of serum cC_6O_4. The kinetic constant k (h^{-1}) was derived as the slope of the linear regression obtained plotting the natural logarithm of the concentration of serum cC_6O_4 at any time C (μg/L) on the y axis vs. the time elapsed since t_0, t (h) on the x axis, according to Equation (1).

$$\ln(C) = -k\,t + \ln(C_0) \tag{1}$$

The serum half-life of cC_6O_4 ($t_{1/2}$) was obtained applying Equation (2), obtained when the concentration was half its initial value, and therefore $C/C_0 = 2$.

$$t_{1/2} = \ln 2/k \tag{2}$$

Based on the 95% confidence interval of k, the 95% confidence interval of the $t_{1/2}$ (h) was also calculated.

The daily renal clearance (Cl) was calculated applying the following equation:

$$\mathrm{Cl}\ (\mathrm{mL/min}) = \frac{\textbf{amount } \mathbf{cC_6O_4} \textbf{ in urine} - \textbf{day 2 } (\mathbf{\mu g})}{\mathbf{cC_6O_4} \textbf{ serum concentration} - \textbf{day 2}} = \frac{\mathbf{Cu}\ (\mathbf{\mu g/L}) \times \mathbf{U}\ (\mathbf{L/day})}{\mathbf{Cs}\ (\mathbf{\mu g/L}) \times \mathbf{1440\ min/day}} \tag{3}$$

where 1440 is the daily minutes. Serum and urine concentration of cC_6O_4 (C_s and C_u, respectively), as well as the urine flow rate (U), were based on day 2, as in this day all parameters were available.

The apparent distribution volume (Vd) was calculated as

$$\mathrm{Vd}\ (\mathrm{mL}) = \frac{\mathbf{t1/2\ (min)} \times \mathbf{Cl\ (mL/min)}}{\mathbf{ln2}} \tag{4}$$

The daily renal clearance and the apparent distribution volume were also normalized for weight unit, according to the following equations:

$$\mathrm{Cl_w}\ (\mathrm{mL/min} \times \mathrm{kg}) = \frac{\mathbf{Cl\ (mL/min)}}{\mathbf{Body\ weight\ (kg)}} \tag{5}$$

and

$$\mathrm{Vd_w}\ (\mathrm{mL/kg}) = \frac{\mathbf{Vd\ (mL)}}{\mathbf{Body\ weight\ (kg)}} \tag{6}$$

2.4. Statistical Analysis

The concentrations of cC_6O_4 in serum and urine at different sampling times were described using the mean, standard deviation, minimum, 25th percentile, median, 75th percentile and maximum values. No sample below the LLOQ was found.

The urinary concentration was calculated for each collection day separately as the sum of the amounts of cC_6O_4 in the different voids divided by the total urine volume. Given that only for day 2 the urine collection was lasting for 24 h, this daily concentration was considered the most reliable and it was used for estimating the daily renal clearance (Cl), the distribution volume (Vd) and the daily excretion rate.

In order to ascertain whether the workers collected all urine voids, we estimated the expected daily excretion rate of urinary creatinine $\mathrm{Crea_{exp}}$, according to the formula [12]:

$$\mathrm{Crea_{exp}}\ (\mathrm{g/day}) = \mathbf{body\ weight(kg)} \times 24\ (\mathrm{g/(day \times kg)}) \tag{7}$$

From the % ratio between measured and expected daily amount of urinary creatinine, the percentage of incomplete urine collection was estimated (results reported in Table 1).

Table 1. Summary of characteristics of study subjects, available samples and urinary creatinine.

	Parameter (Unit)	Statistics	Data
Personal characteristics	Male subjects	n (%)	18 (100)
	Age, year	mean ± SD	44 ± 5
	Weight, kg	mean ± SD	79 ± 6
	Height, cm	mean ± SD	175 ± 6
	BMI, kg/m^2	mean ± SD	26.1 ± 2.3
	Current tobacco smoker	n (%)	4 (22)
	e-cig smoker	n (%)	1 (5)
	Use of drugs	n (%)	14 (78)
Available samples	Blood samples	n	72
	Blood sample/subject	n	4
	Urine samples	n	254
	Urine sample/subject (day 1–day 3)	n, mean ± SD	14 ± 4
	Lost urine samples	n	14
Urinary creatinine	Creat (g/day)	mean ± SD	1.465 ± 0.401
	Expected Crea (g/day)	mean ± SD	1.905 ± 0.144
	Crea ratio (%)	mean ± SD	77 ± 21
	Subjects with crea ratio < 60%	n (%)	4 (22)

This percentage is, however, affected by a large degree of variability, and only when it falls outside the range of 60–140% should urine collection be considered unreliable [13].

To calculate cC_6O_4 serum half-life we used random-intercept linear regression models applied to all pooled data to take into account intra-subject correlation. The potential effect modification of half-life by age and BMI (as continuous variables) were analyzed by introducing product (interaction) terms in the regression models and by calculating the corresponding Wald tests.

Statistical analyses were performed with Excel (Microsoft), SPSS (IBM) and Stata 17 (StataCorp. 2021).

3. Results

3.1. Study Subjects

In Table 1, the summary of personal and work characteristics of study workers are reported. Briefly, 18 male workers volunteered in the study. Their mean age was 44 years, and their BMI was 26.1 kg/m^2. Three subjects were current tobacco smokers, and one was an e-cig smoker. About 80% of them were assuming drugs (mostly for hypertension and hypercholesterolemia).

Workers were involved in the synthesis of plastomers, particularly of a melt-processable perfluoroalkoxy fluorocarbon resin (Hyflon®) and of poly-tetrafluoroethylene (PTFE) (Algoflon®).

Seventy-two samples of blood (four for subjects) and 254 urine voids were collected. The number of urine voids collected by each individual from 14:00 of day 1 to 8:00 of day 3 was 14 ± 4. Overall, 14 samples were missed during collection and five containers were delivered broken to the laboratory.

Through the calculation of expected daily excretion rate of creatine, the completeness of sample collection was estimated, returning a ratio of 77 ± 21%. This supports the missed samples (n = 14) declared by the volunteers and the delivered broken/empty containers (n = 5). Nevertheless, subjects with a percentage of urine collection below the normal range (60–140%) [13] were only four, with values ranging from 38 to 57%.

3.2. Biological Monitoring

In Table 2 the results of the concentration of cC_6O_4 in serum and urine samples, the last without and with creatinine correction, are reported for each sampling day.

3.2.1. cC_6O_4 Urinary Excretion

Nadler's equation was used to estimate the volume of blood in each study subject:

$$\text{Blood volume} = 0.3669 \times h^3 + 0.03219 \times bw + 0.6041 \quad (8)$$

where h and bw are subject's height (cm) and body weight (kg). Serum volume was assumed to be equal to the volume of plasma and estimated as 55% of the volume of blood. The volume of urine of day 2 was measured as the sum of the different voids in the 24 h. From the concentration of cC6O4 in serum and urine, and their volumes, the amount of cC6O4 in each matrix was estimated. The ratio between the amount of cC6O4 in urine and the amount in serum (day 1 and day 2) was evaluated as the percentage of cC_6O_4 excreted in urine. The mean value and SD were about 20 ± 10% for both days 1 and 2 (see Table 3).

Table 2. Concentration of cC_6O_4 in serum and urine for each sampling day.

ID Subject	cC_6O_4 Serum µg/L				cC_6O_4 Urine µg/L				cC_6O_4 Urine µg/g Crea			
	Day 1	Day 2	Day 3	Day 6	Day 1	Day 2	Day 3	Day 6	Day 1	Day 2	Day 3	Day 6
1	0.68	0.65	0.68	0.49	0.20	0.23	0.19		0.30	0.22	0.22	
2	0.77	0.64	0.61	0.38	0.71	0.50	0.49		0.48	0.44	0.43	
3	1.62	1.72	1.57	1.25	0.51	0.52	0.42		0.65	0.63	0.50	
4	0.75	0.77	0.70	0.58	0.61	0.46	0.38		0.32	0.31	0.26	
5	2.17	2.06	1.70	1.75	0.63	0.64	0.66		1.10	0.87	0.71	
6	2.89	2.57	2.37	1.88	1.76	1.32	1.47	1.95	1.11	0.85	0.37	0.46
7	2.52	2.91	2.30	2.08	1.60	1.54	1.76	0.68	0.56	0.57	0.55	0.23
8	2.42	2.40	2.03	1.45	1.45	1.62	1.71	1.43	1.25	1.05	1.02	0.78
9	2.26	2.24	2.32	1.84	0.45	0.57	0.70	0.59	0.60	0.46	0.23	0.33
10	1.75	1.79	1.55	1.03	1.06	1.50	0.64	0.39	0.87	0.88	1.55	0.44
11	11.29	8.01	9.57	7.46	4.20	4.82	4.02	3.48	3.73	3.14	1.89	1.63
12	1.80	1.69	1.55	1.30	0.64	0.87	0.60	0.50	0.63	0.66	0.43	0.35
13	3.97	3.65	3.07	1.86	3.89	4.63	6.08	3.76	5.12	3.59	2.59	1.18
14	1.42	1.21	1.09	0.78	0.70	0.66	0.68	0.58	0.59	0.63	0.32	0.44
15	3.83	3.74	3.50	3.17	0.78	0.95	1.09	0.65	0.49	0.64	0.93	0.47
16	3.86	3.60	3.22	2.03	2.40	1.12	1.45	2.09	1.60	1.56	1.49	0.97
18	4.31	3.88	3.66	2.28	5.92	4.90	2.48	3.12	2.11	1.78	1.49	1.27
19	7.00	7.15	6.62	4.50	2.45	2.38	1.71	2.94	2.19	1.30	1.48	1.65
Mean	3.07	2.82	2.67	2.01	1.66	1.62	1.47	1.70	1.32	1.09	0.91	0.78
SD	2.58	2.03	2.24	1.69	1.57	1.55	1.48	1.26	1.28	0.93	0.69	0.50
Min	0.68	0.64	0.61	0.38	0.20	0.23	0.19	0.39	0.30	0.22	0.22	0.23
p25	1.62	1.69	1.55	1.03	0.63	0.57	0.60	0.59	0.56	0.57	0.37	0.44
Median	2.34	2.32	2.17	1.80	0.92	1.04	0.90	1.43	0.76	0.76	0.63	0.47
p75	3.86	3.65	3.22	2.08	2.40	1.62	1.71	2.94	1.60	1.30	1.49	1.18
Max	11.29	8.01	9.57	7.46	5.92	4.90	6.08	3.76	5.12	3.59	2.59	1.65

Table 3. Amount of cC_6O_4 in plasma and urine and % of urinary excretion calculated considering the amount of cC_6O_4 in urine of day 2 vs. the amount in serum of day 1 or day 2.

	cC_6O_4 in Serum Day 1 µg	cC_6O_4 in Serum Day 2 µg	cC_6O_4 in Urine Day 2 µg	Urinary Excretion (vs. Serum Day 1) %	Urinary Excretion (vs. Serum Day 2) %
Mean	9.50	8.73	1.62	18.9	20.1
SD	7.95	6.36	1.42	10.1	11.0
Min	1.92	1.70	0.34	4.8	4.9
p25	5.51	5.38	0.74	12.5	12.8
Median	7.25	7.76	1.05	17.5	17.9
p75	10.92	10.42	1.85	24.1	25.7
Max	34.08	24.18	5.37	46.4	50.5

3.2.2. Pearson's Correlation and Linear Regression

The correlation analysis between the serum concentration of cC_6O_4 of day 1 or day 2 and the urine concentration of day 2, each transformed in the natural logarithm, was quite good, with Pearson's r correlation coefficients ranging from 0.803 to 0.838. The correlation coefficients were similar using urinary concentration both corrected and non-corrected for creatinine. The linear regressions corresponding to these correlations are illustrated in Figure 2. The correlation was very good also using the concentration of urinary cC_6O_4 in spot samples collected at the time of each blood drawing, with Pearson's r correlation coefficients ranging from 0.850 to 0.535 for corrected and non-corrected urinary concentration.

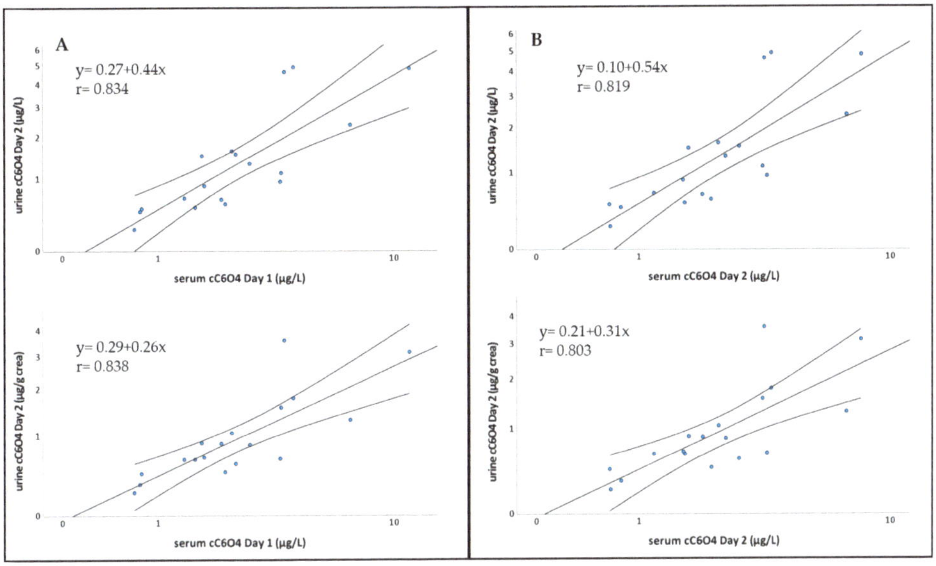

Figure 2. Correlation between the concentration of cC_6O_4 in day 2 urine not adjusted for creatinine(above) and adjusted for creatinine (below) in day 1 serum (**A**), or day 2 serum (**B**).

3.3. Kinetics of Elimination of cC_6O_4, Clearance and Distribution Volume

Considering day 1 (time 0 h) as the initial time t0, and the concentration of cC_6O_4 in serum of the day 1 sample as the initial concentration C0, the elimination kinetics of cC_6O_4 was investigated by applying a first-order model on natural logarithm (ln) transformed pooled data of all subjects. A mean half-life ($t_{1/2}$) of 184 (95% CI 162–213) h or 7.7 (95% CI 6.7–8.9) days was calculated. No effect modification by categories of age (p-interaction = 0.42) and BMI (p-interaction = 0.17) on the kinetics parameters was found. The elimination of cC_6O_4 from serum is reported in Figure 3. Data from each individual are reported separately with red dots connected by red lines, while the linear regression and 95% confidence interval are reported in bold black continuous and dotted lines.

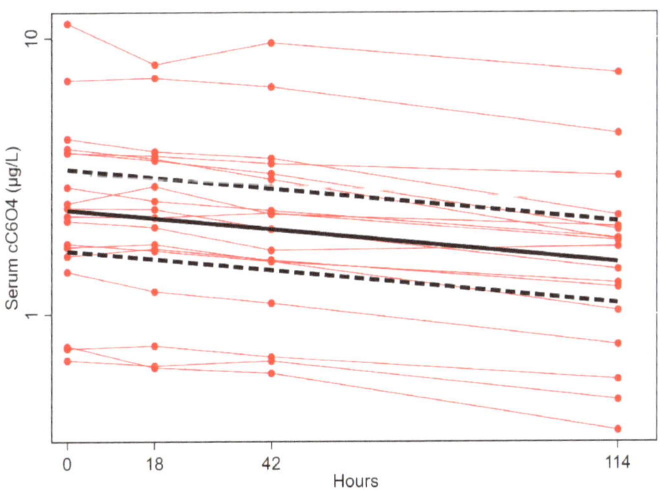

Figure 3. Elimination of cC_6O_4 from serum. Data from each individual are reported separately with red dots connected by red lines, while the linear regression and 95% confidence interval are reported in bold black continuous and dotted lines.

The renal clearance (Cl) and the apparent distribution volume (Vd) are reported in Table 4, also normalized for the individual's body weight.

Table 4. cC_6O_4 renal clearance (Cl) and distribution volume (Vd).

Statistics	Cl mL/min	Cl_w mL/min × kg	Serum Volume L	Vd, L	Vd_w L/kg
Mean	0.41	0.0052	3.1	6.6	0.084
SD	0.21	0.0026	0.4	3.4	0.042
Min	0.13	0.0015	2.6	2.1	0.023
p25	0.29	0.0037	2.9	4.6	0.06
Median	0.38	0.0051	3.0	6.1	0.081
p75	0.53	0.0064	3.1	8.6	0.103
Max	1.02	0.0128	4.0	16.3	0.204

4. Discussion

In the present work, the half-life of cC_6O_4 in occupationally exposed male individuals was estimated. The amount of cC_6O_4 excreted unchanged in urine was also investigated to evaluate the relevance of urine as the excretion route.

For the recruitment of volunteers, we evaluated the results of the annual biomonitoring campaigns performed from 2013 in the plant of Spinetta Marengo [5]. We noticed that a significant proportion of serum samples was below the lower limit of quantification (1 μg/L). Considering that, in the present study, we aimed to also investigate urinary cC_6O_4, expected to be in lower concentration compared to serum, we recruited our volunteers among plastomer workers as those with the highest exposures. Moreover, we adopted an analytical assay with much higher sensitivity, with an LLOQ of 0.02 μg/L for cC_6O_4 in both matrices [5]. The quantification of cC_6O_4 in 100% of samples testifies the goodness of the adopted strategy.

Comparing results of serum cC_6O_4 reported here with the previous work [5], we note a lowering of concentrations. In particular, the median and the maximum were 4 and 65 μg/L in Plastomer 2, and 4 and 20 μg/L in Plastomer 1 departments in 2021, while in the present study the median and maximum are to 2.34 and 11.3 μg/L. This difference might be explained by the increased sensitivity of the analytical assay that allowed to include all subjects in the statistical analysis, i.e., also those with cC_6O_4 in serum <1 μg/L (three individuals), which would have been excluded in 2021.

The concentrations of cC_6O_4 in serum and urine (corrected and non-corrected for creatinine) were highly correlated with Pearson's correlation coefficients $r > 0.80$ (Figure 2). This very good correlation suggests that biomonitoring of exposure to cC_6O_4 could be performed using urine instead of blood; this would be more convenient, as urine is non-invasive and can be collected autonomously, without the needs of health personnel. To further explore this possibility, we evaluated the correlation between cC_6O_4 in serum and in spot urine samples collected at the time of each blood drawing, obtaining a very good correlation when the concentration of cC_6O_4 was corrected for creatinine (Pearson's $r = 0.850$). This further supports the use of urine samples for the biological monitoring of the exposure to cC_6O_4.

The kinetics work was performed on volunteers who had blood sampled at fixed times and voids of urine collected over a defined period. This study design is convenient because it allows monitoring exposure in real-world conditions. At the same time, some drawbacks are present, and these include:

- Different doses experienced by study individuals, resulting in a large variability in the concentration of the chemical in the investigated specimens;
- Difficulties in adhering to the study protocol, particularly in collecting all voids, resulting in missing voids or incomplete urine collection. Moreover, this is often associated with an under-reporting of missing samples.

In the study of the kinetics of excretion of cC_6O_4 using the linear regression between ln-transformed concentrations of cC_6O_4 vs. time, different exposures result in different intercepts, but the slope of the linear regression is not affected. As the half-life is calculated from the kinetic constant k (see Equations (1) and (2)), which is derived as the slope of the linear regression, we can conclude that kinetics parameters are not influenced by different exposures. In fact, individual concentration lines were remarkably similar (parallel) across subjects (Figure 3). The second issue, associated with missing urine samples, was faced by calculating the excreted creatinine, as summarized in Table 1 [10]; this allowed us to estimate a rate of urine collection of 77% and to identify four subjects exceeding the low boundary of tolerance range [60–140%] [13]. Overall, we evaluated as "good" the compliance of our volunteers in collecting voids, but we acknowledge that the amount of cC_6O_4 excreted in urine is underestimated by about 23%. Considering this, the percentage of cC_6O_4 in serum daily excreted in urine would be closer to 25% instead of the 20% observed (Table 3); similarly, the renal clearance would be increased from 0.41 mL/min to 0.50 mL/min (Table 4).

cC_6O_4 was developed and introduced in the polymerization process by Solvay Specialty Polymers Italy S.p.A. to find alternatives with a better (eco)toxicological profile than persistent perfluoroalkyl chemicals such as PFOA. A kinetics study performed in rats showed a short half-life of 4–7 h [1]. The present study found a half-life in male humans of 184 (95% CI 162–213) h. This value is higher than that found in rats, as expected, as the metabolism in rodents is much faster than in humans.

This is the first study assessing the kinetics of elimination of cC_6O_4 in occupationally exposed individuals. Some studies on newly synthetized perfluoroalkyl chemicals, such as perfluorobutyrate (PFBA) [14] and GenX, were performed [15,16]. Table 5 provides an overview of the results of these studies together with a selection of studies investigating the kinetics of PFOA. The half-life of PFBA was evaluated in nine workers removed from the workplace for more than 7 days and was estimated to be 72 h (95% CI 43–101) [14]. The kinetics of GenX was investigated in 18 workers that took 3-4 days off: the half-life was 81 ± 55 h (mean ± SD) [15,16]. Noticeably, the half-lives of these compounds were all found to be within a few days (3 to 7.7 days), while a half-life of 3.16–3.5 years was reported for PFOA [4,5,17].

Table 5. Summary of studies investigating the half-life of some PFAS in occupationally exposed workers.

Chemical	Study	Study Subjects/ Company/Country	N Subjects	N Samples, N/Subject	Years of Study	C_0 Serum µg/L	$t_{1/2}$
PFOA	[5]	Fluorochemical workers after the ban of PFOA; Solvay, Italy	93 (15 F)	568, 2 to 9/subject	2013–2021	Median 750	GM 3.16 (95% CI 2.98–3.37) years
PFOA	[17]	Retired fluorochemical workers; 3M, US	26 (2 F)	4 to 8/subject	1998–2004	Mean 691	GM 3.5 (95% CI 3.0–4.1) years
PFOA	[4]	Retired fluorochemical workers; Miteni, Italy	35 M	81 2/subject	2000–2018	GM 1.489	Mean 3.35 (95% CI 2.89–3.99) years
Perfluorobutyrate (PFBA)	[14]	Fluorochemical workers removed from the workplace for >7 days; 3M, US	9 (2 F)	2/subject	2007	2–71	72 (95% CI 43–101) h
GenX	[15,16]	Fluorochemical workers off-work weekend (no exposure for 3–4 days); Chemours, NL	18	2/subject	Before 2021	Not reported	Mean ± SD 81 ± 55 h
cC_6O_4	Present study	Fluorochemical workers removed from the workplace for 5 days; Solvay, Italy	18 M	72 4/subject	2021–2022	Median 2.34	GM 184 (95% CI 162–213) h

The short half-life of cC_6O_4 and its urinary elimination are supported by a recent in vitro and in silico study evaluating the interactions of cC_6O_4 with renal basal and apical membrane transporters that may contribute to the reuptake from the tubule lumen. Results showed no significant interactions, suggesting that reabsorption from the proximal tubule is not likely to interfere with the urinary elimination of cC_6O_4 in humans [18].

Other kinetics parameters obtained in the present study, such as the distribution volume (Vd), calculated to be 6.6 L (mean value), can help in understanding the degree of distribution of cC_6O_4 in the body. Given that study subjects have a mean body weight of 79 kg, the volume of serum was estimated to be 3.1 L, the volume of the interstitial fluids to be 12.6 L (16% of body weight) and the total volume of body liquids to be 48 L (60% of body weight). Comparing these volumes with the Vd of cC_6O_4, it is possible to appreciate that this is more than twice the serum volume, suggesting that cC_6O_4 is not exclusively present in serum, but it is likely distributed also into other compartments. At the same time, given that the Vd is about 1/8 of the total volume of body liquids, we speculate that no accumulation in poorly perfused compartments and organs is present.

In conclusion, the present work evaluated the kinetics of excretion of cC_6O_4 in male workers. The half-life of 7.7 days, the daily urinary excretion of about 20% of the plasma dose, and the distribution volume supported the short biopersistence and the lack of bioaccumulation of cC_6O_4 in the human body. As this study does not includes females and younger or elder subjects, differences in the kinetics parameters could be present in individuals with different personal characteristics.

Author Contributions: Conceptualization, S.F.; methodology, S.F. and R.M.; formal analysis, D.C.; investigation, R.M., G.L. and S.C.; resources, S.F.; data curation, R.M. and G.L.; writing—original draft preparation, S.F., S.C. and R.M.; writing—review and editing, S.F. and D.C.; visualization, R.M. and D.C. All authors have read and agreed to the published version of the manuscript.

Funding: This research was funded by Solvay Specialty Polymers Italy S.p.A.

Institutional Review Board Statement: The study was conducted in accordance with the Declaration of Helsinki and approved by the Ethics Committee of the University of Milan (date of approval 16 February 2021).

Informed Consent Statement: Informed consent was obtained from all subjects involved in the study.

Data Availability Statement: The data presented in this study are available on request from the corresponding author. The data are not publicly available due to privacy reasons.

Acknowledgments: The authors would like to thank Ferruccio Casalino, Matteo Bandirali and the staff of the Occupational Health Service and infirmary of the plant for their valuable contribution to illustrate the study to workers and to organize the specimen handling and managing. The authors would also warmly like to thank the volunteers who participated in the study.

Conflicts of Interest: The authors declare no conflict of interest. The funders had no role in the analyses or interpretation of data; in the writing of the manuscript; or in the decision to publish the results.

References

1. ECHA, Registration Dossier. Acetic Acid, 2,2-difluoro-2-[[2,2,4,5-tetrafluoro-5-(trifluoromethoxy)-1,3-dioxolan-4-yl]oxy]-, Ammonium Salt (1:1)). According to CLP (Regulation (EC) No 1272/2008. Available online: https://echa.europa.eu/it/registration-dossier/-/registered-dossier/5712 (accessed on 10 October 2022).
2. Hopf, N.B.; Fustinoni, S. Biological monitoring of exposure to industrial chemicals. Part 2, Chemical Agents. In *Patty's Industrial Hygiene*; Cohrssen, B., Ed.; Wiley: Hoboken, NJ, USA, 2021. [CrossRef]
3. Jain, R.B.; Ducatman, A. Serum concentrations of selected perfluoroalkyl substances for US females compared to males as they age. *Sci Total Environ.* **2022**, *842*, 156891. [CrossRef] [PubMed]
4. Girardi, P.; Rosina, A.; Merler, E. Report of the Project: Bio-PFAS: La Concentrazione di Sostanze Perfluorurate nel Sangue dei Dipendenti ed Ex-Dipendenti Delle Ditte RIMAR e Miteni. Trissino. Vicenza. Italia [In Italian]. 2018. Available online: https://pfas.land/2019/03/02/02-marzo-2019-persistenza-dei-pfas-nel-sangue-dei-lavoratori-miteni/ (accessed on 15 December 2022).
5. Fustinoni, S.; Consonni, D. Historical Trend of Exposure to Perfluoroalkyl Surfactants PFOA, ADV, and cC6O4 and its Management in Two Perfluoroalkyl Polymers Plants, Italy. *Ann. Work Exp. Health* **2023**, wxac095. [CrossRef] [PubMed]
6. Decreto Legislativo 81 del 9 Aprile 2008. Available online: http://www.gazzettaufficiale.it/eli/id/2008/04/30/008G0104/sg (accessed on 7 July 2022).
7. Frigerio, G.; Cafagna, S.; Polledri, E.; Mercadante, R.; Fustinoni, S. Development and validation of an LC-MS/MS method for the quantitation of 30 legacy and emerging per- and polyfluoroalkyl substances (PFASs) in human plasma, including HFPO-DA, DONA, and cC6O4. *Anal. Bioanal. Chem.* **2022**, *414*, 1259–1278. [CrossRef] [PubMed]
8. U.S. Food and Drug Administration. Bioanalytical Method Validation Guidance for Industry. Available online: https://www.fda.gov/files/drugs/published/Bioanalytical-Method-Validation-Guidance-for-Industry.pdf (accessed on 8 January 2022).
9. EMA. ICH Guideline M10 on Bioanalytical Method Validation. 2022. Available online: https://www.ema.europa.eu/en/ich-m10-bioanalytical-method-validation-scientific-guideline#current-version-section (accessed on 8 January 2022).
10. D'Avolio, A.; Cantu, M.; Gervasoni, J.; Artusi, C.; Marinova, M.; Nonnato, A.; Cangemi, G.; Persichilli, S. Bioanalytical method validation of quantitative mass spectrometry based assay: Experimental protocols and regulations. *Biochim. Clin.* **2018**, *42*, 51–61. [CrossRef]
11. Kroll, M.H.; Chesler, R.; Hagengruber, C.; Blank, D.W.; Kestner, J.; Rawe, M. Automated determination of urinary creatinine without sample dilution: Theory and practice. *Clin. Chem.* **1986**, *32*, 446–452. [CrossRef] [PubMed]
12. Knuiman, J.T.; Hautvast, J.G.; van der Heijden, L.; Geboers, J.; Joossens, J.V.; Tornqvist, H.; Isaksson, B.; Pietinen, P.; Tuomilehto, J.; Flynn, A.; et al. A multi-centre study on within-person variability in the urinary excretion of sodium, potassium, calcium, magnesium and creatinine in 8 European centres. *Hum. Nutr. Clin. Nutr.* **1986**, *40*, 343–348. [PubMed]
13. Joossens, J.V.; Geboers, J. Monitoring Salt Intake of the Population: Methodological Considerations. In *Surveillance of the Dietary Habits of the Population with Regard to Cardiovascular Diseases, EURONUT Report 2*; De Backer, G.G., Pedoe, H.T., Ducimetiere, P., Eds.; Department of Human Nutrition, Agricultural University: Wageningen, The Netherlands, 1984; pp. 61–73.
14. Chang, S.C.; Das, K.; Ehresman, D.J.; Ellefson, M.E.; Gorman, G.S.; Hart, J.A.; Noker, P.E.; Tan, Y.M.; Lieder, P.H.; Lau, C.; et al. Comparative pharmacokinetics of perfluorobutyrate in rats. mice. monkeys. and humans and relevance to human exposure via drinking water. *Toxicol. Sci.* **2008**, *104*, 40–53. [CrossRef] [PubMed]

15. Clark, D.S. The Chemours Company. 2021, March 17. Letter to EPA, Office of Pollution Prevention and Toxics Regarding Propanoic Acid, 2,3,3,3-tetrafluoro-2-(1,1,2,2,3,3,3-heptafluoropropoxy)-CAS RN 13252-13-6 (also Known as HFPO-DA). Available online: https://heronet.epa.gov/heronet/index.cfm/reference/details/reference_id/8631852 (accessed on 6 June 2022).
16. U.S. Environmental Protection Agency. Human Health Toxicity Values for Hexafluoropropylene Oxide (HFPO) Dimer Acid and Its Ammonium Salt (CASRN 13252-13-6 and CASRN 62037-80-3), EPA Document Number: 822R-21-010. Available online: https://www.epa.gov/system/files/documents/2021-10/genx-chemicals-toxicity-assessment_tech-edited_oct-21-508.pdf (accessed on 2 October 2021).
17. Olsen, G.W.; Burris, J.M.; Ehresman, D.J.; Froehlich, J.W.; Seacat, A.M.; Butenhoff, J.L.; Zobel, L.R. Half-life of serum elimination of perfluorooctanesulfonate.perfluorohexanesulfonate. and perfluorooctanoate in retired fluorochemical production workers. *Environ. Health Perspect.* **2007**, *115*, 1298–1305. [CrossRef] [PubMed]
18. Bruno, S.; Bersani, M.; Astore, S.; Chiabotto, G.; Barge, A.; Binello, A.; Spyrakis, F. Lack of interaction of the fluorosurfactant C_6O_4 with human renal transporters: In vitro/in silico analysis. *Toxicology* **2022**, *476*, 153257. [CrossRef] [PubMed]

toxics

MDPI

Article

Using Urinary Biomarkers to Estimate the Benzene Exposure Levels in Individuals Exposed to Benzene

Shiwei Cui [1,2], Bo Pang [1], Huifang Yan [2], Bo Wu [3], Ming Li [4], Caihong Xing [2,*] and Juan Li [1,*]

1 Department of Hygienic Inspection, School of Public Health, Jilin University, Changchun 130021, China
2 National Institute for Occupational Health and Poison Control, Chinese Center for Disease Control and Prevention, Beijing 100050, China
3 Science and Technology Research Center of China Customs, Beijing 100026, China
4 Department of Occupational Health, Jinan Railway Disease Control and Prevention Center, Jinan 250001, China
* Correspondence: xingch@niohp.chinacdc.cn (C.X.); li_juan@jlu.edu.cn (J.L.)

Abstract: Urinary benzene metabolites trans, trans-muconic acid (*t, t*-MA), and S-phenyl mercapturic acid (S-PMA) are often used as biomarkers of internal exposure to benzene. However, there are few reports on using urinary benzene metabolites to estimate airborne benzene concentrations in individuals exposed to benzene. In this study, *t, t*-MA, and S-PMA were analyzed by UPLC-MS/MS, and a simple pharmacokinetic model was used to calculate the daily intake (DI) of benzene based on the levels of urinary *t, t*-MA, and S-PMA in occupational individuals. The back-calculated airborne benzene levels (BCABL) were obtained from the DI of benzene. Among the exposed subjects (n = 84), the median BCABL (3.67 mg/m^3) based on *t, t*-MA was very close to the median level of measured airborne benzene (3.27 mg/m^3, p = 0.171), and there was no effect of smoking or dietary habits on *t, t*-MA-based BCABL. In the control subjects (n = 49), the levels of measured airborne benzene were all below the quantitation limit (0.024 mg/m^3), and the BCABL (0.002–0.25 mg/m^3) calculated by S-PMA was close to this background level. Our study suggests that the *t, t*-MA-based BCABL can reflect the actual airborne benzene level in a range of 1.10–86.91 mg/m^3 and that the S-PMA-based BCABL is more reliable for non-professional benzene exposure.

Keywords: benzene exposure; biological monitoring; daily intake; S-PMA; *t, t*-MA

Citation: Cui, S.; Pang, B.; Yan, H.; Wu, B.; Li, M.; Xing, C.; Li, J. Using Urinary Biomarkers to Estimate the Benzene Exposure Levels in Individuals Exposed to Benzene. *Toxics* **2022**, *10*, 636. https://doi.org/10.3390/toxics10110636

Academic Editor: Carina Ladeira

Received: 11 October 2022
Accepted: 19 October 2022
Published: 23 October 2022

1. Introduction

Benzene is one of the simplest aromatic hydrocarbons. Because of its excellent thermodynamic stability and solubility, it has been widely used as a chemical synthesis intermediate and an industrial solvent in the industry. Nowadays, occupational exposure to benzene in China primarily occurs in the manufacturing industry, such as furniture manufacturing, printing, and the recording media reproduction industry [1]. Benzene is also found in automobile exhausts, flue gas, and cigarette smoke [2]. Chronic exposure to benzene can cause acute myeloid leukemia [3–5] and is associated with other blood-forming system diseases, such as aplastic anemia, myelodysplastic syndrome (MDS), and lymphocytic leukemias [6–8]. It is classified as a Group I carcinogen by the International Agency for Research on Cancer (IARC) [9].

In the past few decades, benzene exposure levels in the workplace have gradually decreased. The median concentration of benzene is below 3 mg/m^3 in more than 98% of benzene industries from six major benzene-exposed provinces in China [1]. The median concentration of benzene is 1.50 mg/m^3 (the conversion scale between ppm and mg/m^3 is 3.25 for benzene at 20 °C, the air concentration in this paper is displayed with mg/m^3 or μg/m^3) among petrochemical workers in Bulgarian [10] and is 32.6 μg/m^3 among refinery workers in Southern Italy [11]. However, benzene is harmful to human health even at a very low exposure level, and there is no evidence of a threshold. In a cross-sectional study of 390 workers, white blood cell and platelet counts were significantly decreased in workers exposed to <3.25 mg/m^3 of benzene [12]. Another study has shown that the levels of serum protein adducts increased in workers exposed to low concentrations of benzene (interquartile range: 6.96–43.6 mg/m^3), which indicated that low concentrations of benzene exposure might increase the risk of leukemia [13].

In order to effectively assess the health effects of benzene exposure, it is crucial to determine the actual absorbed dose of the benzene-exposed population. After benzene enters the human body, it is metabolized to benzene oxide by cytochrome P450 (CYP450). As products of the further metabolism of benzene oxide, trans, trans muconic acid (*t, t*-MA), and S-phenyl mercapturic acid (S-PMA) have been applied as internal exposure biomarkers to assess the risk of benzene exposure. Previous studies found that the urinary concentration of *t, t*-MA and S-PMA were consistently elevated when benzene exposure was above 0.65 mg/m^3 [14]. Additionally, both *t, t*-MA, and S-PMA positively correlated with airborne benzene among 0.2–396.5 mg/m^3 [15]. The individual daily intake (DI) of pollutants can be calculated using a simplified pharmacokinetic model based on the concentration of metabolites in urine. This model has been used to estimate the DI of phthalates [16,17] and benzene [18] in non-occupational exposed individuals. Therefore, the pharmacokinetic model successfully linked airborne benzene exposure to urinary benzene metabolite concentrations and, as a result, potentially contributed to the risk assessment of occupational benzene exposure.

This study aimed to use DI obtained by a simple pharmacokinetic model based on the levels of urinary *t, t*-MA, and S-PMA to back the calculated airborne benzene levels (BCABL) in individuals exposed to benzene. Under the factors of benzene exposure levels, smoking habits, and dietary habits, the correlation between airborne benzene levels and BCABL was analyzed.

2. Materials and Method

2.1. Study Population

The study population included 133 subjects. A total of 84 subjects, who were employed at a shoe factory in Shandong Province, China, formed part of the exposed group, and another 49 unexposed subjects from a food factory in the same geographic area were the control group. The unexposed subjects were frequency-matched with the exposed workers for age and gender. The workers involved in the study were informed of the study's aims, and each participant signed an informed consent form.

2.2. Personal Exposure

During the entire work shift of the subjects, benzene exposure was monitored using a personal diffusive sampler. Benzene was collected by active carbon particles (SKCINC®) filled in the sampler. Afterward, benzene was desorbed using carbon disulfide, then measured using gas chromatography with a flame ionization detector (GC-FID). The limit of the quantitation of the analysis procedure was 0.024 mg/m^3.

2.3. Urine Analysis

A sterile plastic bottle was used to collect the urine of each worker at the end of their work shift. Each collected urine sample was stored at −40 °C before being divided into two parts, one at 10 mL for the determination of urinary creatinine and the other at 20 mL for the determination of S-PMA and *t, t*-MA using Ultra Performance LC (Waters, Milford, MA, USA) tandem mass spectrometry (AB SCIEX QTRAP 5500, Framingham, MA, USA) and analytical methods with isotopic dilution.

As briefly described below, the method by Paci et al. [19] was modified for the determination of S-PMA. Firstly, a 1 mL urine sample was treated with 10 µL of a deuterated internal standard solution in methanol (S-PMA-d5, 0.5 mg/L) and 0.3 mL 9 M H_2SO_4 for the hydrolysis of the urinary precursor pre-SPMA for 10 min, and then 0.25 mL NaOH 50% in water was added to adjust the pH. Then, the urine samples were purified with solid phase extraction columns (C18 ODS, sorbent bed mass 500 mg, Agilent, Santa Clara, CA, USA), and the pretreatment method was optimized. The C18 ODS cartridge was balanced and pretreated with 3 mL of methanol and 3 mL 0.1% acetic acid. Then, the urine hydrolysate was loaded into the C18 ODS cartridge. The analytes of interest were eluted with 1 mL of methanol after washing the C18 ODS cartridge with 3 mL of 0.1% of acetic acid. The eluate was filtered on a 0.2 μm filter device and 10 μL was injected into the UPLC-MS/MS system, using a linear gradient of 20% methanol and 80% 0.1 mol/L acetic acids, with up to 90% methanol and an ACQUITY UPLC BEH C_{18} analytical column (2.1 × 50 mm, 1.7 μm, Waters, Milford, MA, USA). The negative ions and multiple reaction ion monitoring (MRM) modes were used. The ionic transitions monitored were as follows: m/z 238.1–m/z 109.1 for S-PMA and m/z 243.0–m/z 114.1 for the deuterated internal standard. If the content of S-PMA in the urine samples exceeded the linear range, the eluent would be diluted by methanol before analysis.

For *t*, *t*-MA a 1 mL portion of each sample was treated with 0.33 mL of phosphate buffer (0.59 g of KH_2PO_4 and 2.15 g Na_2HPO_4 in 500 mL water, pH = 7.4) and 50 μL of a deuterated internal standard solution in methanol (*t*, *t*-MA-d4, 4 mg/L). Then, the urine samples were purified with Bond Elut SAX cartridges (sorbent bed mass 500 mg, Agilent, Santa Clara, CA, USA). Bond Elut SAX cartridges were previously conditioned with 3 mL of methanol and 3 mL of 0.1% (v/v) acetic acid in water, and after loading the samples, the cartridges were washed with 3 mL of 0.1% (v/v) acetic acid in water and 3 mL of methanol. Finally the analytes of interest were eluted with 3 mL 10% (v/v) acetic acid in water. The eluate was filtered on a 0.2 μm filter device, and 10 μL was injected into the UPLC-MS/MS system using a linear gradient of 10% acetonitrile and 90% 0.1% acetic acid, with up to 90% acetonitrile [20]. An ACQUITY UPLC BEH Shield RP18 column (2.1 × 100 mm, 1.7 μm, Waters, USA) were used for the analysis of *t*, *t*-MA. The following m/z ion combinations (precursor—product) were monitored and the transitions (negative mode) were as follows: 141.0–97.0 for *t*, *t*-MA and m/z 145.0–100.0 for the deuterated internal standard. If the content of *t*, *t*-MA in the urine samples exceeded the linear range, the eluent would be diluted by 10% (v/v) acetic acid in water before analyzing.

The linear range was 0.17–50 μg/L for S-PMA and 3.3–1000 μg/L for *t*, *t*-MA. The precision and accuracy at the level of 5.0 μg/L were 2.1% and 99.8% for S-PMA, and at the level of 25.0 μg/L, it was 1.6% and 102%. For *t*, *t*-MA, precision and accuracy are 5.6% and 101% at the level of 50 μg/L, and at the level of 200 μg/L, it is 3.4% and 98.3%. The limits of quantitation (LOQs) were 0.17 μg/L and 3.3 μg/L for S-PMA and *t*, *t*-MA, respectively. A validated HPLC-UV method was used to determine the urinary creatinine for each subject. The urine was diluted appropriately, and the peak was detected at 254 nm [21].

2.4. Calculation of Benzene Daily Intake (DI)

Based on the urinary metabolites and urinary creatinine levels, the values of S-PMA and *t*, *t*-MA were converted to show the value of benzene daily intake (DI) by applying the following Equation (1), which has already been used to calculate the daily phthalate intake in the general population [22,23]:

$$\text{DI}\mu\text{g/day} = \frac{UE(\mu\text{g/g}) \times CE(\text{mg/kg/day}) \times M(\text{kg})}{F_{UE} \times 1000(\text{mg/g})} \times \frac{MW_d}{MW_m} \quad (1)$$

where UE is the urinary excretion of benzene metabolite S-PMA or *t*, *t*-MA in μg/g creatinine. CE is the creatinine excretion rate set to be 18 mg/kg/day for women and 23 mg/kg/day for men [16]. M is the weight of the subjects. F_{UE} is the fractional urinary excretion of benzene as S-PMA (0.11%) or *t*, *t*-MA (3.9%) in humans [24,25]. MW_d and MW_m are the molecular weights of benzene and metabolites. The molecular weights of benzene, S-PMA, and *t*, *t*-MA were 78.1, 239.3, and 142.1.

2.5. Back Calculated Airborne Benzene Levels

Based on the DI values of benzene, we back-calculated airborne benzene levels (BCABL) before applying the following Equation (2). An amount of 10 m^3 was the default human occupational volume of air inhaled in an 8 h work shift, and the inhalation absorption rate of benzene exposure was 50% [26,27]. Given that in a range of benzene exposure (median: 3.85 mg/m^3, range: 0.096–64 mg/m^3), the dose-specific levels of benzene metabolites (μmol/L/ppm benzene) declined with increasing benzene exposure, with a 3-fold reduction in *t, t*-MA, and no decreasing trend for S-PMA [14,28], we introduced a factor k (k was 3 for *t, t*-MA, and 1 for S-PMA) into the analysis of airborne benzene levels based on *t, t*-MA and S-PMA.

$$\text{BCABL}\left(\text{mg/m}^3\right) = \frac{\text{DI}}{10\text{m}^3 \times 1000 \times 50\%} \times k \tag{2}$$

2.6. Statistical Analysis

Statistical analysis was performed with the statistical package SPSS (Version 26.0). Non-detectable levels of airborne benzene were replaced by LOQ/2 (using the 49 control subjects). Airborne benzene and its two metabolites' values were transformed logarithmically to achieve an approximately normal distribution. A simple linear regression and Pearson's correlation were used to estimate the correlation coefficients. The Wilcoxon rank-sum test, Mann–Whitney U-test, and Chi-square test were applied to determine statistical differences between the groups. A p-value lower than 0.05 was considered to be statistically significant for all tests.

3. Results

3.1. Summary Statistics

As shown in Table 1, the characteristics of the exposed subjects and control subjects were similar in age, gender, weight, and working duration. The mean age for the benzene-exposed subjects was 34.2 (SD, 10.0) years, and for control subjects it was 32.5 (SD, 6.33) years; 46% of the exposed subjects and 51% of the controls were male. The mean working duration for the exposed subjects was 1.89 years (SD, 1.32). Subjects in the exposed group had higher rates of smoking compared with the unexposed group (23% vs. 8%; $p = 0.03$). Among the exposed subjects, the median measured airborne benzene level was 3.27 mg/m^3 (though it ranged from 1.10 to 86.2 mg/m^3). Among the control subjects, the levels of measured airborne benzene were all below the quantitation limit (0.024 mg/m^3).

Table 1. Descriptive parameters of control and exposed subjects.

	Controls (n = 49)	Exposure (n = 84)	p-Value
Age (years)	32.5 ± 6.33	34.2 ± 10.0	0.87 [a]
Weight (kg)	66.1 ± 9.56	64.8 ± 9.77	0.34 [a]
Working duration (years)	2.12 ± 1.11	1.89 ± 1.32	0.12 [a]
Gender [n (%)]			0.61 [b]
Male	25 (51)	39 (46)	
Female	24 (49)	45 (54)	
Current smokers [n (%)]			0.03 [b]
Yes	4 (8)	19 (23)	
No	45 (92)	65 (77)	
Measured airborne benzene (mg/m^3)			<0.001 [a]
Median	<0.024	3.27	
Range	<0.024	1.10–86.2	

[a] Wilcoxon rank-sum test between controls and exposed. [b] Chi-square test between controls and exposed.

The levels of benzene metabolites are summarized in Table 2. Among the exposed subjects, the median levels for S-PMA were 17.0 μg/g creatinine, and for *t, t*-MA, they were 371.4 μg/g creatinine, while, in the control subjects, the two values were 0.38 μg/g creatinine and 68.4 μg/g creatinine, respectively. No significant differences were observed between smokers and non-smokers for the benzene metabolite levels in the exposed subjects (p = 0.176 for S-PMA and p = 0.100 for *t*, *t*-MA). Among the control subjects, smokers had higher values of S-PMA (1.31 versus 0.36 μg/g creatinine, p = 0.001) and *t*, *t*-MA (117.2 versus 63.9 μg/g creatinine, p = 0.004) than non-smokers.

Table 2. Summary statistics for urinary metabolite excretion in control and exposed subjects.

Variable	Exposed			Control		
	Total (n = 84)	Smokers (n = 19)	Non-Smokers (n = 65)	Total (n = 49)	Smokers (n = 4)	Non-Smokers (n = 45)
S-PMA (μg/g creatinine)						
Median	17.0	15.3	20.2	0.38	1.31	0.36
Range	0.72–1908.7	1.26–150.8	0.72–1908.7	0.02–2.34	0.62–2.34	0.02–1.62
p-value	<0.001 [a]	0.176 [b]			0.001 [b]	
t, *t*-MA (μg/g creatinine)						
Median	371.4	264.8	404.0	68.4	117.2	63.9
Range	102.1–8331.2	102.1–2773.8	107.4–8331.2	26.1–252.4	41.1–186.6	26.1–252.4
p-value	<0.001 [a]	0.100 [b]			0.004 [b]	

A Mann–Whitney U-test *p*-value < 0.05 was considered as statistically significant. [a] exposed vs. controls. [b] Smokers vs. non-smokers, in exposed subjects or in controls.

As shown in the scatter plots for the measured airborne benzene levels versus S-PMA (Figure 1) and *t*, *t*-MA (Figure 2), measured airborne benzene concentrations were significantly correlated with S-PMA (r = 0.744, $p < 0.001$) and *t*, *t*-MA (r = 0.729, $p < 0.001$), respectively.

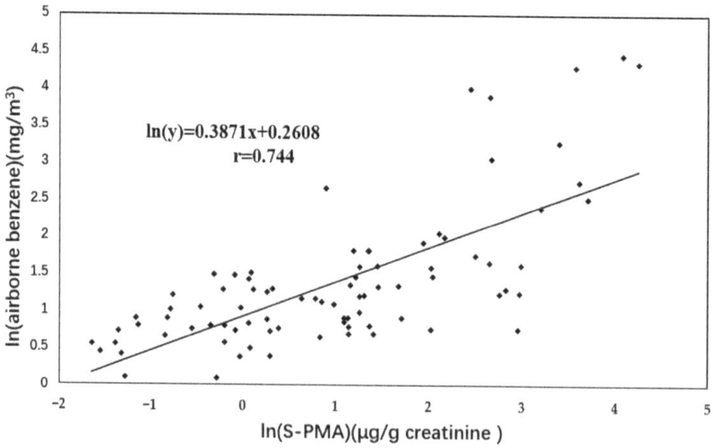

Figure 1. Correlation between natural logarithmic transformed benzene exposure levels (mg/m^3) and S-PMA values (μg/g creatinine).

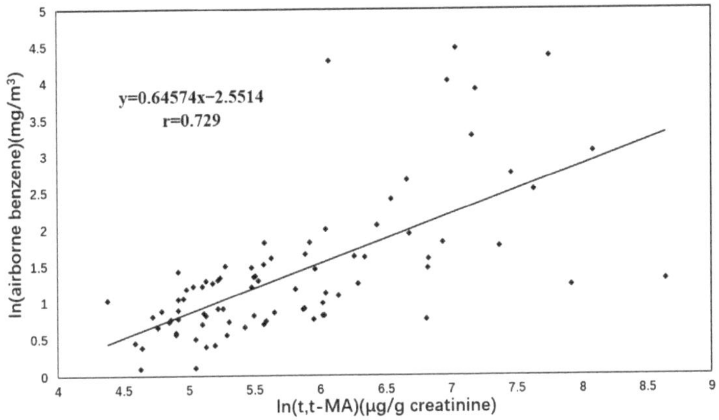

Figure 2. Correlation between natural logarithmic transformed benzene exposure levels (mg/m^3) and *t*, *t*-MA values (μg/g creatinine).

3.2. Comparison between Measured Airborne Benzene Levels and Back Calculated Airborne Benzene Levels

Table 3 shows the DI values for the benzene of the exposed and control groups. The median DI values among the exposed subjects were 6919.1 μg/day based on urinary S-PMA and 6123.8 μg/day based on urinary *t*, *t*-MA, while in the control subjects, the two values were 157.3 μg/day and 1330.6 μg/day, respectively.

Table 3. DI [a] values of benzene and BCABL [b] based on S-PMA and *t*, *t*-MA in control and exposed subjects.

Group (*n*)	Measured Airborne Benzene (mg/m^3)	SPMA-Based Calculation		*t*, *t*-MA-Based Calculation	
		DI (μg/day)	BCABL (mg/m^3)	DI (μg/day)	BCABL (mg/m^3)
Controls (49)	<0.024	157.3 (9.62–1259.8)	0.03 (0.002–0.25)	1330.6 (405.1–4536.5)	0.80 (0.24–2.72)
Exposed (84)	3.27 (1.10–86.2)	6919.1 (211.0–530,199.6)	1.39 [c] (0.04–106.0)	6123.8 (1447.0–131,036.1)	3.67 [c] (0.87–78.6)

[a] DI—daily intake. [b] BCABL—back calculated airborne benzene levels. [c] *p*-value between airborne benzene and BCABL (based on S-PMA or *t*, *t*-MA): $p < 0.01$ for S-PMA and $p = 0.171$ for *t*, *t*-MA.

BCABL was calculated based on DI using Formula (2). Among the exposed subjects, the median BCABL based on urinary *t*, *t*-MA was 3.67 mg/m^3 (range: from 0.87 to 78.6 mg/m^3), which was very close to the median of the measured airborne benzene level of 3.27 mg/m^3 (range: from 1.10 to 86.2 mg/m^3) and there was no significant difference between the BCABL and the measured airborne benzene level ($p = 0.171$). The measured airborne benzene concentrations were significantly correlated with BCABL based on *t*, *t*-MA (r = 0.729, $p < 0.001$). However, the median BCABL based on urinary S-PMA was 1.39 mg/m^3 (range: from 0.04 to 106.0 mg/m^3), which was significantly lower than the measured median airborne benzene level of 3.27 mg/m^3 ($p < 0.01$). In the controls, the levels of measured airborne benzene were all below the limit of detection (0.024 mg/m^3) and median BCABLs, which were 0.03 mg/m^3 based on S-PMA and 0.80 mg/m^3 based on *t*, *t*-MA.

The above results indicate that the benzene exposure level was able to be accurately extrapolated by urinary *t*, *t*-MA. However, the urinary *t*, *t*-MA concentration might be affected by smoking and dietary habits [29]. In order to further clarify the accuracy of BCABL calculated by *t*, *t*-MA, we further investigated the effects of smoking habits and dietary habits on BCABL results.

3.3. Comparison of BCABL Based on t, t-MA with Respect to Smoking in EXPOSED Subjects

As shown in Table 4, the median levels of BCABL in smokers (3.20 mg/m^3) and non-smokers (3.85 mg/m^3) are close to the median measured airborne benzene level (2.49 mg/m^3 in smokers and 3.36 mg/m^3 in non-smokers, respectively). Therefore, there are no significant differences between BCABL and the measured airborne benzene levels (all $p > 0.05$). Considering that smoking may exhibit more impact on *t, t*-MA at low levels of benzene exposure, the subjects were categorized according to the median measured airborne benzene level of 3.27 mg/m^3. For the subjects exposed to ≤3.27 mg/m^3, the median level of BCABL was 2.58 mg/m^3 in smokers compared to 2.34 mg/m^3 in non-smokers, which is close to the median measured airborne benzene level (2.04 mg/m^3 in smokers, $p > 0.05$; 2.24 mg/m^3 in non-smokers, $p > 0.05$, respectively). For the subjects exposed to >3.27 mg/m^3, there are no significant differences between BCABL and the measured airborne benzene level in smokers and non-smokers (all $p > 0.05$). The results indicate that the BCABL based on *t, t*-MA in benzene-exposed subjects was not affected by smoking habits.

Table 4. Summary of airborne benzene and BCABL based on *t, t*-MA in smokers and non-smokers.

Subgroup (*n*)	Airborne Benzene (mg/m^3)	BCABL (mg/m^3)	*p*-Value
Smokers (19)	2.49 (1.47–15.6)	3.20 (1.40–34.5)	0.091 [a]
≤3.27 mg/m^3 (12)	2.04 (1.47–3.08)	2.58 (1.39–7.58)	0.200 [a]
>3.27 mg/m^3 (7)	3.85 (3.63–15.6)	5.03 (2.96–34.5)	0.710 [a]
Non-Smokers (65)	3.36 (1.10–86.2)	3.85 (0.87–78.6)	0.483 [a]
≤3.27 mg/m^3 (30)	2.24 (1.10–3.22)	2.34 (0.87–19.8)	0.525 [a]
>3.27 mg/m^3 (35)	5.03 (3.32–86.2)	6.52 (1.01–78.6)	0.883 [a]

[a] *p*-value between airborne benzene and BCABL (based on *t, t*-MA).

3.4. Comparison of BCABL Based on t, t-MA with Respect to Dietary Habit in Exposed Subjects

Sorbic acid, a wildly used food preservative, can be metabolized to *t, t*-MA [30]. So, we further analyzed the effect of dietary habits on BCABL. The subjects were categorized according to the median measured airborne benzene level of 3.27 mg/m^3. As shown in Table 5, there was no significant difference between the BCABL and measured airborne benzene level (all $p > 0.05$) between the subjects who ate preserved food and the subjects who did not eat preserved food. In the group of less than 3.27 mg/m^3, the median BCABLs was 2.62 mg/m^3 for the subjects who ate preserved food and 2.51 mg/m^3 for the subjects who did not eat preserved food, which was close to the median measured airborne benzene level of 2.24 mg/m^3 and 2.13 mg/ m^3, respectively. Similar conclusions were found in the group of more than 3.27 mg/m^3. The median levels of BCABL in subjects who ate preserved food and subjects who did not eat preserved food (4.08 mg/m^3 and 8.62 mg/m^3, respectively) were close to the median measured airborne benzene level (4.46 mg/m^3 and 4.99 mg/m^3, respectively). There was no significant difference between the BCABL and the measured airborne benzene level (all $p > 0.05$) in subjects who ate preserved food compared to subjects who did not eat preserved food, no matter if the airborne benzene exposure level was below or above 3.27 mg/m^3.

Table 5. Summary of airborne benzene and BCABL based on *t, t*-MA considering dietary habit.

Subgroup (*n*)	Airborne Benzene (mg/m^3)	BCABL (mg/m^3)	*p*-Value
Eat preserved food (30)	3.57 (1.73–73.3)	3.39 (0.98–78.6)	0.836 [a]
≤3.27 mg/m^3 (11)	2.24 (1.73–2.98)	2.62 (0.98–11.1)	0.401 [a]
>3.27 mg/m^3 (19)	4.46 (3.32–73.3)	4.08 (1.01–78.6)	0.246 [a]
Not eat preserved food (54)	2.84 (1.10–86.2)	3.87 (0.87–57.0)	0.073 [a]
≤3.27 mg/m^3 (31)	2.13 (1.10–3.22)	2.51 (0.87–19.8)	0.120 [a]
>3.27 mg/m^3 (23)	4.99 (3.43–86.2)	8.62 (2.68–57.0)	0.089 [a]

[a] *p*-value between airborne benzene and BCABL (based on *t, t*-MA).

4. Discussion

CYP450 enzymes play an essential role in benzene metabolism. Different metabolic pathways of benzene compete for CYP450. The production of *t, t*-MA requires the secondary oxidation of CYP450 enzymes, while S-PMA is formed by the conjugation of benzene oxide with glutathione, which does not require the secondary oxidation of CYP450 enzymes [31]. For metabolites that require secondary oxidation by CYP450 enzymes, their production rate may decrease with increasing benzene exposure due to the saturation of benzene metabolism [32]. In support of this is the evidence that the dose-specific levels (μmol/L/ppm benzene) of *t, t*-MA decrease about 3-fold when benzene exposure levels increase in the range of 0.096–64 mg/m^3, with no decreasing trend for S-PMA [28]. Interestingly, considering the effect of the 3-fold-reduced metabolic rate on *t, t*-MA, we found that the *t, t*-MA-based BCBAL (median:3.67 mg/m^3) was very close to the measured airborne benzene level (median: 3.27 mg/m^3, $p = 0.171$) of subjects exposed to benzene in the range of 1.10–86.2 mg/m^3. Furthermore, the stratified analysis did not find the effect of smoking and dietary habit on *t, t*-MA concentration. Our findings suggest that airborne benzene levels can be accurately estimated by BCABL based on *t, t*-MA. As far as we know, this method has been used for the first time to estimate the benzene exposure level for occupational benzene-exposed populations.

Upon the findings for the S-PMA-based BCBAL examined here, we noted that the S-PMA-based BCBAL in control subjects (range: 2–250 μg/m^3) was close to the ambient levels of benzene from 2001 to 2016 (0.19–53.90 μg/m^3) in China [33], and similar to the benzene exposure concentration in the general environment of the United States (0–47.3 μg/m^3) [34]. However, the S-PMA-based BCBAL in the benzene-exposed subjects (median: 1.39 mg/m^3) was far from the measured airborne benzene level (median: 3.27 mg/m^3). One possible interpretation for the above results is that the amount of S-PMA in urine is related to the saturable enzymatic metabolic pathway too. However, the results of previous studies were inconsistent. Melikian et al. [15] found that urinary S-PMA displayed a downward trend of dose-related production. The dose-specific levels (μmol/L/ppm benzene) of S-PMA decreased about 2.7-fold with the increase of benzene exposure levels from a range of 1.1 to 73.5 mg/m^3. The report of Lung-Cheng Lin et al. [35] showed the same phenomenon of S-PMA production. However, the report of Kim et al. [14] did not show a decreasing trend for S-PMA, with benzene exposure levels increasing in the range of 0.096–64 mg/m^3. Our results also show that the maximum S-PMA result for the non-smokers of the exposed subjects was very high (1908.7 μg/g creatinine). This may be because the worker was exposed to high concentrations of benzene in the workplace. Another possible reason is that the glutathione S-transferase (GST) genetic polymorphism influenced the excretion levels of S-PMA. A previous study noted that the levels of S-PMA in the subjects with the GSTT1-positive genotype were higher than those with GSTT1-deficient genotypes [35]. In this study, S PMA is suitable to extrapolate low benzene exposure levels (below 250 μg/m^3), and the BCABL calculated by S-PMA in the control subjects can be used as a background reference for benzene occupational exposure. However, S-PMA and the BCABL calculated by S-PMA produced a lack of assessment for the benzene exposure of the exposed subjects.

A study has indicated that as a biomarker of benzene exposure, *t, t*-MA might not be able to reliably evaluate the benzene level at very low exposure concentrations (below 1.6 mg/m^3) [36]. Consistently, we found that the measured airborne benzene concentration of the control group was below the quantitation limit (0.024 mg/m^3), but BCBAL calculated by *t, t*-MA (median: 0.80 mg/m^3) was much higher than the quantitation limit. This may be due to the effect of non-benzene sources on *t, t*-MA at low levels of benzene. The background concentration of *t, t*-MA is not only derived from the metabolism of benzene but is also affected by dietary habits and smoking habits. In addition, *t, t*-MA is a metabolite of sorbic acid that can be present in various foodstuffs. The percentage of *t, t*-MA in non-smokers, which was attributed to dietary sorbic acid, was about 10–50% [37]. This study found that the levels of *t, t*-MA in smokers of the control group were significantly higher than those in the non-smoker's group. Similar results were reported in previous studies [11,38]. Consistently, in the control group of our study, smokers had higher values of t, t-MA (117.2 versus 63.9 μg/g creatinine, p = 0.004) than non-smokers. Therefore, BCABL, when calculated based on t, t-MA, may not accurately reflect benzene exposure for the non-occupational benzene-exposed populations.

5. Conclusions

Our study confirmed that the BCABL determined from t, t-MA-based DI reflected the actual airborne benzene levels that ranged from 1.10 mg/m^3 to 86.2 mg/m^3, and the BCABL from the S-PMA-based DI was more reliable for non-professional benzene exposure.

Author Contributions: Conceptualization, S.C., C.X. and J.L.; Data curation, S.C., B.P.; Formal analysis, H.Y. and B.P.; Investigation, S.C., B.W. and M.L.; Methodology, C.X. and J.L.; Project administration, H.Y.; Resources, B.W. and M.L.; Writing—original draft, S.C.; Writing—review and editing, B.P., C.X. and J.L. All authors have read and agreed to the published version of the manuscript.

Funding: This work was supported by the National Natural Science Foundation of China (No. 82070116).

Institutional Review Board Statement: The study was approved by the medical ethical review committee of the National Institute for Occupational Health and Poison Control, Chinese Center for Disease Control and Prevention (2018, Approval NO. NIOHP201812).

Informed Consent Statement: Informed consent was obtained from all subjects involved in the study.

Data Availability Statement: The data presented in this study are available on request from the corresponding author. The data are not publicly available due to the privacy of subjects.

Acknowledgments: We would like to thank colleagues at the Chinese Center for Disease Control and Prevention and colleagues at Jilin University.

Conflicts of Interest: The authors declare no conflict of interest.

References

1. Wang, X.; Zhou, J.; Han, L.; Cheng, X.; Shao, H.; Jia, Q.; Xu, P.; Liu, J.; Ren, J.; Li, J.; et al. The Distribution and Concentration Monitoring of Benzene Industries—Six PLADs, China, 2020. *China CDC Wkly.* **2021**, *3*, 897–900. [CrossRef] [PubMed]
2. Wang, J.; Guo, X.; Chen, Y.; Zhang, W.; Ren, J.; Gao, A. Association between benzene exposure, serum levels of cytokines and hematological measures in Chinese workers: A cross-sectional study. *Ecotoxicol. Environ. Saf.* **2021**, *207*, 111562. [CrossRef] [PubMed]
3. Khalade, A.; Jaakkola, M.S.; Pukkala, E.; Jaakkola, J.J. Exposure to benzene at work and the risk of leukemia: A systematic review and meta-analysis. *Environ. Health A Glob. Access Sci. Source* **2010**, *9*, 31. [CrossRef] [PubMed]
4. Smith, M.T.; Zhang, L.; McHale, C.M.; Skibola, C.F.; Rappaport, S.M. Benzene, the exposome and future investigations of leukemia etiology. *Chem. Biol. Interact.* **2011**, *192*, 155–159. [CrossRef]
5. Shallis, R.M.; Weiss, J.; Deziel, N.C.; Gore, S.D. A clandestine culprit with critical consequences: Benzene and acute myeloid leukemia. *Blood Rev.* **2021**, *47*, 100736. [CrossRef]
6. Smith, M.T. Advances in Understanding Benzene Health Effects and Susceptibility. *Annu. Rev. Public Health* **2010**, *31*, 133–148. [CrossRef]
7. Schnatter, A.R.; Glass, D.; Tang, G.; Irons, R.D.; Rushton, L. Myelodysplastic Syndrome and Benzene Exposure Among Petroleum Workers: An International Pooled Analysis. *JNCI J. Natl. Cancer Inst.* **2012**, *104*, 1724–1737. [CrossRef]
8. Li, W.; Schnatter, A.R. Benzene risk assessment: Does new evidence on myelodysplastic syndrome justify a new approach? *Crit. Rev. Toxicol.* **2018**, *48*, 417–432. [CrossRef]
9. International Agency for Research on Cancer. Monograph of carcinogenic risks in human. In *Overall Evaluations of Carcinogenicity: An Updating of Iarc Monographs*; IARC: Lyon, France, 1987; Volume 42, pp. 120–122.
10. Seow, W.J.; Pesatori, A.; Dimont, E.; Farmer, P.B.; Albetti, B.; Ettinger, A.S.; Bollati, V.; Bolognesi, C.; Roggieri, P.; Panev, T.I.; et al. Urinary Benzene Biomarkers and DNA Methylation in Bulgarian Petrochemical Workers: Study Findings and Comparison of Linear and Beta Regression Models. *PLoS ONE* **2012**, *7*, e50471. [CrossRef]
11. Carrieri, M.; Spatari, G.; Tranfo, G.; Sapienza, D.; Scapellato, M.L.; Bartolucci, G.B.; Manno, M. Biological monitoring of low level exposure to benzene in an oil refinery: Effect of modulating factors. *Toxicol. Lett.* **2018**, *298*, 70–75. [CrossRef]
12. Lan, Q.; Zhang, L.; Li, G.; Vermeulen, R.; Weinberg, R.S.; Dosemeci, M.; Rappaport, S.M.; Shen, M.; Alter, B.P.; Wu, Y.; et al. Hematotoxicity in Workers Exposed to Low Levels of Benzene. *Science* **2004**, *306*, 1774–1776. [CrossRef] [PubMed]

13. Rappaport, S.M.; Waidyanatha, S.; Yeowell-O'Connell, K.; Rothman, N.; Smith, M.T.; Zhang, L.; Qu, Q.; Shore, R.; Li, G.; Yin, S. Protein adducts as biomarkers of human benzene metabolism. *Chem. Biol. Interact.* **2005**, *153*, 103–109. [CrossRef] [PubMed]
14. Kim, S.; Vermeulen, R.; Waidyanatha, S.; Johnson, B.A.; Lan, Q.; Rothman, N.; Smith, M.T.; Zhang, L.; Li, G.; Shen, M.; et al. Using urinary biomarkers to elucidate dose-related patterns of human benzene metabolism. *Carcinogenesis* **2006**, *27*, 772–781. [CrossRef] [PubMed]
15. Melikian, A.A.; Qu, Q.; Shore, R.; Li, G.; Li, H.; Jin, X.; Cohen, B.; Chen, L.; Li, Y.; Yin, S.; et al. Personal exposure to different levels of benzene and its relationships to the urinary metabolites S-phenylmercapturic acid and trans,trans-muconic acid. *J. Chromatogr. B Analyt. Technol. Biomed. Life Sci.* **2002**, *778*, 211–221. [CrossRef]
16. Koch, H.M.; Drexler, H.; Angerer, J. An estimation of the daily intake of di(2-ethylhexyl)phthalate (DEHP) and other phthalates in the general population. *Int. J. Hyg. Environ. Health* **2003**, *206*, 77–83. [CrossRef]
17. Dewalque, L.; Charlier, C.; Pirard, C. Estimated daily intake and cumulative risk assessment of phthalate diesters in a Belgian general population. *Toxicol. Lett.* **2014**, *231*, 161–168. [CrossRef]
18. Schettgen, T.; Ochsmann, E.B.; Alt, A.; Kraus, T.W. A biomarker approach to estimate the daily intake of benzene in non-smoking and smoking individuals in Germany. *J. Expo. Sci. Environ. Epidemiol.* **2010**, *20*, 427–433. [CrossRef]
19. Paci, E.; Pigini, D.; Cialdella, A.M.; Faranda, P.; Tranfo, G. Determination of free and total *S*-phenylmercapturic acid by HPLC/MS/MS in the biological monitoring of benzene exposure. *Biomarkers* **2007**, *12*, 111–122. [CrossRef]
20. Tranfo, G.; Paci, E.; Sisto, R.; Pigini, D. Validation of an HPLC/MS/MS method with isotopic dilution for quantitative determination of trans,trans-muconic acid in urine samples of workers exposed to low benzene concentrations. *J. Chromatogr. B Analyt. Technol. Biomed. Life Sci.* **2008**, *867*, 26–31. [CrossRef]
21. Wijemanne, N.; Soysa, P.; Wijesundara, S.; Perera, H. Development and Validation of a Simple High Performance Liquid Chromatography/UV Method for Simultaneous Determination of Urinary Uric Acid, Hypoxanthine, and Creatinine in Human Urine. *Int. J. Anal. Chem.* **2018**, *2018*, 1647923. [CrossRef]
22. David, R.M. Exposure to phthalate esters. *Environ. Health Perspect.* **2000**, *108*, A440. [CrossRef] [PubMed]
23. Koch, H.M.; Becker, K.; Wittassek, M.; Seiwert, M.; Angerer, J.; Kolossa-Gehring, M. Di-n-butylphthalate and butylbenzylphthalateurinary metabolite levels and estimated daily intakes: Pilot study for the German Environmental Survey on children. *J. Expo. Sci. Environ. Epidemiol.* **2007**, *17*, 378–387. [CrossRef] [PubMed]
24. Boogaard, P.J.; van Sittert, N.J. Suitability of S-phenyl mercapturic acid and trans-trans-muconic acid as biomarkers for exposure to low concentrations of benzene. *Environ. Health Perspect.* **1996**, *104*, 1151–1157. [CrossRef]
25. Boogaard, P.J.; Van Sittert, N.J. Biological monitoring of exposure to benzene: A comparison between S-phenylmercapturic acid, trans,trans-muconic acid, and phenol. *Occup. Environ. Med.* **1995**, *52*, 611–620. [CrossRef]
26. Sabourin, P.J.; Chen, B.; Lucier, G.; Birnbaum, L.S.; Fisher, E.; Henderson, R.F. Effect of dose on the absorption and excretion of [14C]benzene administered orally or by inhalation in rats and mice. *Toxicol. Appl. Pharmacol.* **1987**, *87*, 325–336. [CrossRef]
27. Inoue, O.; Seiji, K.; Nakatsuka, H.; Watanabe, T.; Yin, S.N.; Li, G.L.; Cai, S.X.; Jin, C.; Ikeda, M. Urinary t,t-muconic acid as an indicator of exposure to benzene. *Br. J. Ind. Med.* **1989**, *46*, 122–127. [CrossRef] [PubMed]
28. Kim, S.; Vermeulen, R.; Waidyanatha, S.; Johnson, B.A.; Lan, Q.; Smith, M.T.; Zhang, L.; Li, G.; Shen, M.; Yin, S.; et al. Modeling Human Metabolism of Benzene Following Occupational and Environmental Exposures. *Cancer Epidemiol. Biomark. Prev.* **2006**, *15*, 2246–2252. [CrossRef] [PubMed]
29. Weaver, V.M.; Buckley, T.; Groopman, J.D. Lack of specificity of trans,trans-muconic acid as a benzene biomarker after ingestion of sorbic acid-preserved foods. *Cancer Epidemiol. Biomark. Prev.* **2000**, *9*, 749–755.
30. Yu, R.; Weisel, C.P. Measurement of the urinary benzene metabolite trans,trans-muconic acid from benzene exposure in humans. *J. Toxicol. Environ. Health Part A* **1996**, *48*, 453–477. [CrossRef]
31. Schettgen, T.; Musiol, A.; Alt, A.; Kraus, T. Fast determination of urinary S-phenylmercapturic acid (S-PMA) and S-benzylmercapturic acid (S-BMA) by column switching liquid chromatography–tandem mass spectrometry. *J. Chromatogr. B Analyt. Technol. Biomed. Life Sci.* **2008**, *863*, 283–292. [CrossRef]
32. Waidyanatha, S.; Rothman, N.; Fustinoni, S.; Smith, M.T.; Hayes, R.B.; Bechtold, W.; Dosemeci, M.; Guilan, L.; Yin, S.; Rappaport, S.M. Urinary benzene as a biomarker of exposure among occupationally exposed and unexposed subjects. *Carcinogenesis* **2001**, *22*, 279–286. [CrossRef]
33. Ji, Y.; Gao, F.; Wu, Z.; Li, L.; Li, D.; Zhang, H.; Zhang, Y.; Gao, J.; Bai, Y.; Li, H. A review of atmospheric benzene homologues in China: Characterization, health risk assessment, source identification and countermeasures. *J. Environ. Sci.* **2020**, *95*, 225–239. [CrossRef] [PubMed]
34. Teras, L.R.; Diver, W.R.; Deubler, E.L.; Krewski, D.; Flowers, C.R.; Switchenko, J.M.; Gapstur, S.M. Residential ambient benzene exposure in the United States and subsequent risk of hematologic malignancies. *Int. J. Cancer* **2019**, *145*, 2647–2660. [CrossRef] [PubMed]
35. Lin, L.-C.; Chen, W.-J.; Chiung, Y.-M.; Shih, T.-S.; Liao, P.-C. Association between *GST* Genetic Polymorphism and Dose-Related Production of Urinary Benzene Metabolite Markers, *trans, trans*-Muconic Acid and S-Phenylmercapturic Acid. *Cancer Epidemiol. Biomark. Prev.* **2008**, *17*, 1460–1469. [CrossRef]
36. Jalai, A.; Ramezani, Z.; Ebrahim, K. Urinary Trans, Trans -Muconic Acid is Not a Reliable Biomarker for Low-level Environmental and Occupational Benzene Exposures. *Saf. Health Work* **2017**, *8*, 220–225. [CrossRef] [PubMed]

37. Ruppert, T.; Scherer, G.; Tricker, A.R.; Adlkofer, F. trans,trans -Muconic acid as a biomarker of non-occupational environmental exposure to benzene. *Int. Arch. Occup. Environ. Health* **1997**, *69*, 247–251. [CrossRef]
38. Fustinoni, S.; Consonni, D.; Campo, L.; Buratti, M.; Colombi, A.; Pesatori, A.C.; Bonzini, M.; Bertazzi, P.A.; Foà, V.; Garte, S.; et al. Monitoring Low Benzene Exposure: Comparative Evaluation of Urinary Biomarkers, Influence of Cigarette Smoking, and Genetic Polymorphisms. *Cancer Epidemiol. Biomark. Prev.* **2005**, *14*, 2237–2244. [CrossRef]

toxics

MDPI

Article

The Hematological and Biochemical Effects from Pesticide Exposure on Thai Vegetable Farmers

Siriphan Bunsri [1], Nutnichawan Muenchamnan [1], Warangkana Naksen [1,*] and Parichat Ong-Artborirak [1,2,*]

[1] Faculty of Public Health, Chiang Mai University, Chiang Mai 50200, Thailand; siriphan_bunsri@hotmail.com (S.B.); nutnichawan@gmail.com (N.M.)

[2] Department of Research and Medical Innovation, Faculty of Medicine Vajira Hospital, Navamindradhiraj University, Bangkok 10300, Thailand

* Correspondence: warangkana.n@cmu.ac.th (W.N.); ong.artborirak@gmail.com (P.O.-A.); Tel.: +66-5-394-2507 (W.N.); +66-2-244-3125 (P.O.-A.)

Abstract: Pesticide-related health concerns are a global public health issue. Few studies in Thailand have explored the hematological and biochemical effects of occupational pesticide exposure. The goal of this study was to investigate the effects of pesticides on the hematology, hepatic, and renal function of Thai vegetable farmers. A cross-sectional study was carried out in Chiang Mai, northern Thailand. A total of 124 apparently healthy vegetable farmers were interviewed about their lifetime exposure to agricultural pesticides. Blood samples were collected via venipuncture to be tested for complete blood count (CBC), liver function, and kidney function. Approximately 46% of the farmers were pesticide users who reported a history of pesticide use for their crops, while 54% were non-pesticide users. In the male farmers, the activities of aspartate aminotransferase (AST), mean corpuscular hemoglobin (MCH), and mean corpuscular hemoglobin concentration (MCHC) were significantly higher in the pesticide users compared to the non-pesticide users, while the estimated glomerular filtration rate (eGFR), hematocrit (HCT), and red blood cells (RBC) were significantly lower ($p < 0.05$). In the females, the pesticide users had significantly higher levels of alanine aminotransferase (ALT), alkaline phosphatase (ALP), and MCHC than the non-pesticide users ($p < 0.05$). Pesticide use among Thai vegetable farmers may cause hematological alterations and increase the risk of hepatic and renal dysfunction. Some hematological and biochemical parameters may be used for monitoring to protect them from the adverse health effects of occupational exposure to pesticides.

Keywords: pesticide; herbicide; insecticide; farmers; blood; liver; kidney

Citation: Bunsri, S.; Muenchamnan, N.; Naksen, W.; Ong-Artborirak, P. The Hematological and Biochemical Effects from Pesticide Exposure on Thai Vegetable Farmers. *Toxics* **2023**, *11*, 707. https://doi.org/10.3390/toxics11080707

Academic Editor: Carina Ladeira

Received: 24 July 2023
Revised: 11 August 2023
Accepted: 14 August 2023
Published: 17 August 2023

1. Introduction

Pesticides are still widely used in agriculture around the world. Thailand is an agricultural country that imports a large amount of chemicals, mostly glyphosate for herbicides and abamectin for insecticides [1]. Illegal and excessive pesticide use, as well as poor exposure prevention behaviors, such as using personal protective equipment (PPE) during pesticide application, have been noted among Thai farmers, particularly those who grow vegetables [2,3]. The use of pesticides can have both acute and chronic negative effects on human health, with a direct impact on farmers, and can also cause ecosystem damage, mainly affecting animals [4–6]. It is estimated that about 44% of farmers are poisoned by pesticides annually [7]. Exposure to pesticides may cause oxidative stress and DNA damage [8,9]. Many studies have also found an increased risk of cancers such as non-Hodgkin lymphoma, leukemia, brain, prostate, kidney, and liver [10–14].

Exposure to different pesticides may impact organ functions due to their cytotoxic effects, and multiple biomarkers may be used to monitor the early adverse effects of pesticides [15,16]. Numerous studies have revealed that individuals who are exposed to pesticides experience significant hemotoxic effects [16–22]. A study found significant hematological changes in Thai rice farmers [23]. However, little is known about the hematological effects of pesticide exposure in Thailand.

In addition, the effects of pesticides may include liver and kidney dysfunction, as measured by blood biochemical parameters. Many studies have found changes in liver function indicators, such as serum aspartate aminotransferase (AST), alanine aminotransferase (ALT), and alkaline phosphatase (ALP), associated with occupational pesticide exposure [15,16,20,22,24,25]. Alterations in these enzyme activities among Thai pesticide-using farmers are unclear due to previous studies yielding insignificant results [23,26,27].

According to literature reviews, pesticide exposure is a risk factor for the development of nephrotoxic effects, which include diminished renal function and chronic kidney disease (CKD) [28–31]. Exposure to certain pesticides may also increase the risk of end-stage renal disease [32]. Several studies have observed associations between exposure to herbicides and insecticides, including carbamates and organophosphates, and a lower estimated glomerular filtration rate (eGFR) [33–36]. Kidney function alterations have been reported in Thai farmers who use organophosphates and carbamates, depending on their use of PPE [37].

As evidence from research in Thailand is limited, this study was conducted to investigate the effects of pesticides on the hematological and biochemical parameters in vegetable farmers. The findings may be useful for the surveillance and prevention of pesticide-related adverse health effects among Thai farmers.

2. Materials and Methods

This cross-sectional study was conducted on farmers from the Mae On District in the Chiang Mai Province, northern Thailand (Figure 1). Both conventional and organic farmers are located in this agricultural region. The sample size for the study was calculated using the G*Power program for two mean independent comparisons, with a medium effect size of 0.5, power of 80%, and alpha of 5%. We increased the sample size to 25%, totaling 128 farmers. The inclusion criteria included being a male or female vegetable farmer between the ages of 18 and 65, having worked for at least a year, being free of any underlying medical conditions (such as hypertension, diabetes, cirrhosis, or kidney failure), not being pregnant, and being able to communicate in Thai. Convenience sampling was used to recruit study participants who were eligible through public relations by village health volunteers. Due to a sample loss during the data collection, we obtained completed data for 124 farmers at the end. This study protocol was approved by the Committee of Research Ethics, Faculty of Public Health, Chiang Mai University (Document No. ET004/2021). Prior to the data collection, all the study participants provided written informed consent.

The data were collected in 2021 between February and March. A face-to-face interview with the vegetable farmers was conducted by trained research assistants using a questionnaire. The questionnaire contained demographic variables (e.g., gender, age, body mass index, marital status, education, monthly family income, and distance from home to nearest farm), work-related information (e.g., work experience, farm size, and types of vegetables cultivated), health behaviors (e.g., smoking, drinking alcohol, caffeine consumption, supplement use, exercise, food consumption, and daily water consumption), and a question about their lifetime history of mixing or applying agricultural pesticides (ever-/never-use), which showed a good reliability of self-reported information on pesticide use [38]. The farmers were divided into two groups: pesticide users who reported having used pesticides on their crops and non-pesticide users who had never used pesticides. The pesticide users were also asked for additional information on the duration of their pesticide use, pesticide application, pesticide types and trade names, use of personal protective equipment (PPE), keeping and cleaning of agricultural equipment and pesticide containers, and pesticide container disposal.

Figure 1. Location of study area in Mae On District in Chiang Mai, Thailand. Adapted from https://commons.m.wikimedia.org/wiki/File:Thailand_Chiang_Mai_locator_map.svg, (accessed on 31 July 2023).

Blood sampling was performed by nurses via venipuncture in the morning to evaluate the hematological and biological parameters. The blood samples were then transferred to the laboratory of Bangkok RIA Lab Co., Ltd. (Medical Technology Clinic) Chiang Mai Branch located in Chiang Mai, Thailand, which is certified by the Bureau of Laboratory Quality Standards, Ministry of Public Health. The hematological parameters from the complete blood count (CBC) tests included white blood cells (WBC), hemoglobin (HGB), hematocrit (HCT), mean corpuscular volume (MCV), mean corpuscular hemoglobin (MCH), mean corpuscular hemoglobin concentration (MCHC), red blood cells (RBC), and platelets (PLT). The liver function test measured the serum AST, ALT, and ALP. To assess the kidney function, the eGFR was calculated from the creatinine using the CKD-EPI Equation for Whites [39].

The SPSS statistical program (SPSS Inc., Chicago, IL, USA) was used to analyze the data. The results of the study were described using frequency (n), percentage (%), arithmetic mean, standard deviation (SD), and median. The Chi-square test was used to compare the categorical variables, including the demographic characteristics, work-related information, and health behaviors between two farmer groups: pesticide users and non-pesticide users. The independent *t*-test and Mann–Whitney U test were used to compare differences in the continuous outcomes, such as the hematological and biological parameters for normal and non-normal data distributions, respectively, between the two farmer groups. A *p*-value of 0.05 was considered to be statistically significant. In this study, the *p*-values were not adjusted for multiple comparisons to avoid rejecting the null hypothesis, as this may lead to fewer errors of interpretation when the data are actual observations on nature, whereas an adjustment is required in confirmatory studies [40,41].

3. Results

Of the 124 vegetable farmers, 57 (46.0%) were pesticide users who reported a history of using pesticides for their crops, while 67 (54.0%) were non-pesticide users. The personal characteristics and work-related information of the two groups of farmers are shown in Table 1. It was found that the pesticide and non-pesticide users had significantly different sex proportions ($p < 0.001$). There were no differences in terms of proportions of age, body mass index, marital status, education, monthly family income, farm proximity, work experience, farm size, growing babycorn, and other vegetable cultivation between the two groups of farmers. In terms of health, a significant difference in the proportion of smokers was observed between the farmers who used pesticides and those who did not ($p < 0.001$), while there was no difference between the two groups in terms of the following health behaviors: alcohol consumption (no/sometimes, always), caffeine drinking (no/sometimes, always), supplement use (no, yes), exercise (no, yes), eating salty foods (very salty, moderately salty, less salty), plant food consumption (occasionally, more than once a week), meat consumption (rarely, every week, every day), and daily water consumption (<5 glasses, 5–7 glasses, 8–10 glasses, 11–13 glasses, >13 glasses). Because of the differences in sex proportions between the two groups, smoking was also observed to be different. As a result, subanalyses of the blood outcomes based on sex were performed.

Table 1. Demographic and work characteristics of pesticide and non-pesticide users among vegetable farmers (n = 124).

Factor	Pesticide Users (n = 57)	Non-Users (n = 67)	p-Value
Sex			
Male	35 (61.4%)	12 (17.9%)	<0.001
Female	22 (38.6%)	55 (82.1%)	
Age			
≤35 years	5 (8.8%)	13 (19.4%)	
36–55 years	34 (59.6%)	40 (59.7%)	0.153
>55 years	18 (31.6%)	14 (20.9%)	
Body mass index (BMI)			
Underweight (<18.5 kg/m^2)	1 (1.8%)	2 (3.0%)	
Normal (18.5–22.9 kg/m^2)	21 (36.8%)	25 (37.3%)	0.261
Overweight (23.0–29.9 kg/m^2)	21 (36.8%)	15 (22.4%)	
Obese (≥30 kg/m^2)	14 (24.6%)	25 (37.3%)	
Marital status			
Single/Divorced/Widowed	6 (10.5%)	15 (22.4%)	0.079
Married	51 (89.5%)	52 (77.6%)	
Education			
Primary school	35 (61.4%)	32 (47.8%)	0.129
Secondary school or higher	22 (38.6%)	35 (52.2%)	
Monthly family income			
<10,000 Baht (<285 USD)	25 (43.9%)	28 (41.8%)	0.816
≥10,000 Baht (≥285 USD)	32 (56.1%)	39 (58.2%)	
Farm proximity			
<300 m.	23 (40.4%)	23 (34.3%)	0.489
>300 m.	34 (59.6%)	44 (65.7%)	
Work experience			
≤5 years	11 (19.3%)	23 (34.3%)	
6–10 years	9 (15.8%)	14 (20.9%)	0.079
11–20 years	13 (22.8%)	15 (22.4%)	
≥21 years	24 (42.1%)	15 (22.4%)	

Table 1. *Cont.*

Factor	Pesticide Users (n = 57)	Non-Users (n = 67)	p-Value
Farm size			
<8000 m^2	27 (47.4%)	40 (59.7%)	0.170
≥8000 m^2	30 (52.6%)	27 (40.3%)	
Growing babycorn			
No	23 (40.4%)	35 (52.2%)	0.186
Yes	34 (59.6%)	32 (47.8%)	
Other vegetable cultivation			
No	22 (38.6%)	31 (46.3%)	0.389
Yes	35 (61.4%)	36 (53.7%)	
Growing rice			
No	40 (70.2%)	53 (79.1%)	0.252
Yes	17 (29.8%)	14 (20.9%)	

The pesticide users were 61.4% male (n = 35) and 38.6% female (n = 22). The average duration of pesticide use was 12.2 ± 10.1 years (14.1 ± 10.8 years for males and 9.1 ± 8.3 years for females). There were 89.5% mixers (91.4% for males and 86.4% for females) and 89.5% sprayers (94.3% for males and 81.8% for females). The history of the individual pesticides used by the farmers is shown in Table 2. Paraquat (24.6%) and glyphosate (19.3%) were found to be the herbicides most commonly used by the farmers. Acetamiprid (26.3%) and carbaryl (21.0%) were the most commonly reported insecticides. Approximately 47.4% of the farmers always used all types of PPE when applying pesticides, including hats (93.0%), chemical respirators (49.1%), gloves (91.3%), long-sleeved shirts and long pants (100%), and boots (96.5%). Most kept their agricultural equipment and pesticides in a space separate from their home (54.4%), cleaned their agricultural equipment and containers on the farm (68.4%), and disposed of the pesticide packaging in a landfill (64.9%).

Table 2. History of individual pesticide use among vegetable farmers.

Chemical Class	Common Name	WHO Class	Pesticide Users (n = 57)
Herbicide			30 (52.6%)
Bipyridylium	Paraquat	II	14 (24.6%)
Glycine	Glyphosate	III	11 (19.3%)
Phenoxy	2,4-D	II	3 (5.3%)
Chloroacetamide/Anilide	Alachlor	II	1 (1.8%)
Triazines	Atrazine	III	1 (1.8%)
Insecticide			48 (84.2%)
Organophosphate	Chlorpyrifos	II	1 (1.8%)
	Methamidophos	Ib	1 (1.8%)
	Dichlorvos	Ib	1 (1.8%)
	Profenofos	II	1 (1.8%)
Carbamate	Carbaryl	II	12 (21.0%)
	Methomyl	Ib	8 (14.0%)
	Carbofuran	Ib	3 (5.3%)
	Carbosulfan	II	2 (3.5%)
Neonicotinoid	Acetamiprid	II	15 (26.3%)
	Imidacloprid	II	8 (14.0%)
	Dinotefuran	III	1 (1.8%)
Avermectin	Abamectin	Ib	7 (12.3%)
	Emamectin	II	4 (7.0%)
Pyrethroid	Cypermethrin	II	4 (7.0%)
Phenylpyrazole	Fipronil	II	5 (8.8%)
Spinosyn	Spinetoram	U	4 (7.0%)

Note: Ib = Highly hazardous; II = Moderately hazardous; III = Slightly hazardous; and U = Unlikely to present acute hazard in normal use.

More than 90% of the vegetable farmers had CBC, ALP, AST, ALT, and eGFR values within the normal range. Comparisons of the hematological and biochemical parameters between the pesticide users and non-users are shown in Tables 3 and 4. In the males, the mean HCT and RBC were significantly lower in the pesticide users compared to the non-users ($p < 0.05$). There were differences in the medians of the following outcomes between the pesticide users and non-users: MCH (28.7 vs. 27.55 pg), MCHC (32.6 vs. 31.3 g/dL), eGFR (102.55 vs. 112.58 mL/min/1.73 m^2), and AST (27 vs. 19 units/L). In the females, the pesticide users had a significantly higher mean ALP than the non-users ($p < 0.05$). The pesticide users and non-users also had different medians for MCHC (31.2 vs. 30.9 g/dL) and ALT (19 vs. 15 units/L).

Table 3. Hematological parameters between the two groups of vegetable farmers.

Outcome		Pesticide Users (n = 57)	Non-Users (n = 67)	p-Value
		Mean (SD)	Mean (SD)	
WBC (mm^3)	All	6979 (1596)	6671 (1669)	0.299 [a]
	Male	7021 (1612)	6818 (1323)	0.696 [a]
	Female	6912 (1605)	6639 (1744)	0.528 [a]
HGB (g/dL)	All	13.76 (1.56)	12.92 (1.71)	0.005 [a]
	Male	14.44 (1.20)	14.73 (1.42)	0.497 [a]
	Female	12.69 (1.48)	12.52 (1.51)	0.666 [a]
HCT (%)	All	42.84 (4.37)	41.64 (5.10)	0.166 [a]
	Male	44.51 (3.45)	47.08 (4.17)	0.040 [a]
	Female	40.18 (4.44)	40.46 (4.50)	0.810 [a]
MCV (fL)	All	84.91 (10.91)	81.22 (12.44)	0.149 [b]
	Male	87.76 (9.33)	83.08 (13.07)	0.294 [b]
	Female	80.36 (11.88)	80.81 (12.39)	0.588 [b]
MCH (pg)	All	27.31 (4.05)	25.24 (4.07)	0.005 [b]
	Male	28.56 (3.60)	26.01 (4.20)	0.047 [b]
	Female	25.31 (4.00)	25.08 (4.06)	0.822 [b]
MCHC (g/dL)	All	32.11 (1.29)	31.05 (0.67)	<0.001 [b]
	Male	32.48 (1.34)	31.29 (0.48)	0.017 [b]
	Female	31.53 (0.96)	31.00 (0.69)	0.014 [b]
RBC ($10^6/mm^3$)	All	5.10 (0.62)	5.18 (0.70)	0.502 [a]
	Male	5.12 (0.66)	5.74 (0.60)	0.006 [a]
	Female	5.08 (0.57)	5.06 (0.67)	0.942 [a]
PLT (mm^3)	All	249,947 (58,755)	267,567 (58,711)	0.098 [a]
	Male	234,971 (49,762)	240,417 (30,485)	0.724 [a]
	Female	273,773 (65,024)	273,491 (61,839)	0.986 [a]

Abbreviations: WBC, white blood cell; HGB, hemoglobin; HCT, hematocrit; MCV, mean corpuscular volume; MCH, mean corpuscular hemoglobin; MCHC, mean corpuscular hemoglobin concentration; RBC, red blood cell; and PLT, platelets. [a] Independent t-test; [b] Mann–Whitney U test.

Table 4. Liver and kidney function tests between the two groups of vegetable farmers.

Outcome		Pesticide Users (n = 57)	Non-Users (n = 67)	p-Value
		Mean (SD)	Mean (SD)	
ALP (U/L)	All	73.84 (24.64)	65.30 (20.11)	0.036 [a]
	Male	71.63 (23.72)	71.83 (14.48)	0.978 [a]
	Female	77.36 (26.22)	63.87 (20.98)	0.020 [a]
AST (U/L)	All	29.65 (24.86)	18.85 (8.58)	<0.001 [b]
	Male	35.91 (29.87)	24.08 (13.88)	0.033 [b]
	Female	19.68 (5.87)	17.71 (6.58)	0.115 [b]

Table 4. *Cont.*

Outcome		Pesticide Users (n = 57)	Non-Users (n = 67)	p-Value
ALT (U/L)	All	29.28 (24.06)	19.57 (14.30)	<0.001 [b]
	Male	35.83 (28.31)	30.17 (20.35)	0.366 [b]
	Female	18.86 (7.85)	17.25 (11.63)	0.029 [b]
eGFR (mL/min/1.73 m^2)	All	102.75 (13.59)	106.54 (11.56)	0.169 [b]
	Male	99.66 (14.60)	110.00 (11.07)	0.020 [b]
	Female	107.66 (10.31)	105.79 (11.63)	0.569 [b]

Abbreviations: ALP, alkaline phosphatase; AST, aspartate aminotransferase; ALT, alanine aminotransferase; and eGFR, estimated glomerular filtration rate. [a] Independent t-test; [b] Mann–Whitney U test.

4. Discussion

This study demonstrated the hematological and biochemical effects from pesticide exposure on Thai vegetable farmers. In terms of the hematological parameters, the pesticide users had lower HCT and RBC values, higher MCH and MCHC values in the male farmers, and only lower MCHC values in the female farmers. This may have been due to pesticide exposure altering iron homeostasis, which affects the red blood cells [42]. Long-term exposure could disrupt the balance between free radical production and antioxidant defenses, resulting in DNA damage [20]. This significant difference is consistent with previous findings for pesticide-exposed people, such as male pesticide sprayers and agricultural workers, including greenhouse workers [16–22]. In Thailand, a study found that rice farmers had significantly lower MCH and MCHC values [23], whereas other studies have found no difference in the CBC parameters for pesticide-using farmers [26,27]. These inconsistent results may be due to differences in the pesticide type, frequency, and duration in each study, as well as differences in individual immune responses [31]. Our findings indicated a higher risk of blood disorders such as anemia for vegetable farmers who are repeatedly exposed to pesticides, particularly men.

In our study, the pesticide users had significantly higher levels of serum enzymes that indicate hepatic damage, including AST for males and ALT and ALP for females. Our results are consistent with the findings of earlier studies, which have found elevated levels of these enzyme activities in farmers and agricultural workers, including pesticide sprayers [15,16,20,24,25,43]. High ALT and AST levels may indicate lysis of the liver cells and leakage of enzymes into the blood, causing cytotoxic effects [25]. Elevated ALP levels may be caused by an increased plasma membrane permeability or cellular necrosis [16,44]. Pesticide exposure, such as to organophosphate, may disrupt the liver metabolism and mitochondrial metabolic pathways caused by oxidative damage [45]. According to Khan et al., (2008) [15], pesticide residues such as methomyl, imidacloprid, cypermethrin, and methamidophos were positively correlated with ALT or AST in pesticide-exposed tobacco farmers. In line with these findings, the vegetable growers in our study were also exposed to these pesticides. Even though previous studies on Thai orchid farmers and rice farmers who used pesticides observed no difference in liver function tests such as AST, ALT, and ALP [23,26,27], our findings indicated the potential effects of pesticide exposure on hepatotoxicity among vegetable farmers.

In terms of renal function, eGFR was found to be lower in the male farmers who used pesticides. This significant difference was in agreement with previous findings for pesticide-exposed individuals, in which associations with lower eGFR were observed for organophosphates and carbamates, as well as herbicides such as atrazine and glyphosate [33–36]. Organophosphate exposure may alter the glomeruli, as well as the proximal and distal tubules [46,47]. Neonicotinoids have been identified as a possible risk factor for renal tubular dysfunction [48]. Renal tubule injury, resulting in inflammation and fibrosis, may also lead to the progression of CKD [49]. Previous research has found that lifetime exposure to pesticides such as glyphosate, paraquat, and organophosphates in paddy fields was associated with CKD in farmers [28]. Among Thai farmers, exposure to organophosphates and carbamates, as well as PPE use, were significantly associated with an increased risk of CKD [37]. This indicates that the farmers investigated in our study may be at a higher risk of renal injury as a result of not always using PPE, in particular chemical respirators. Another study found that glyphosate use and the intensity of exposure were related to serum creatinine, but not to eGFR [50]. Some pesticides used by farmers in this study, such as alachlor, atrazine, glyphosate, paraquat, and permethrin, have been linked to CKD [51]. According to our findings, long-term pesticide exposure may cause nephrotoxic changes and increase the risk of CKD in vegetable farmers.

In addition, these findings emphasize the importance of strict pesticide legislations and preventive measures for Thai farmers. In comparison to other countries, the pesticide regulation in Thailand is characterized by the lack of a consolidated, uniform system designed specifically for pesticide management, which has weakened the enforcement of the current regulations, resulting in pesticide misuse/overuse [2]. Regarding the study's limitations, the cross-sectional design may not be able to draw a causal relationship between pesticide exposure and changes in the hematological and biochemical parameters. A larger sample size is required to increase the power of the analysis, as the data were subanalyzed based on gender. The study did not look into specific pesticides and the intensity of exposure. A pesticide residue analysis to assess internal exposure, such as urinary metabolites, is recommended for future research. Other factors, such as work clothing and heat stress, may also be considered. However, our findings add to the evidence on the effects of pesticide exposure on hematological and biochemical alterations among Thai farmers.

5. Conclusions

Our findings emphasized that occupational exposure to pesticides may alter hematological and biochemical parameters, reflecting hemotoxic, nephrotoxic, and hepatotoxic effects among Thai vegetable farmers who mixed or sprayed pesticides. These parameters might be useful in tracking the early negative effects of pesticides on Thai farmers' health. An intensive intervention to provide knowledge about the potential risks of occupational pesticide exposure and promote the appropriate use of PPE, particularly chemical respirators, should be considered to minimize pesticide exposure and prevent health effects in Thai farmers. Moreover, pesticide-related problems must be recognized and prioritized by the government in order to develop effective policies and strategies, as well as to continue extremely strict actions for strengthening the weak legislation and regulations, ultimately leading to an actual reduction in these problems.

Author Contributions: Conceptualization, W.N. and P.O.-A.; methodology, W.N. and P.O.-A.; software, S.B.; validation, W.N. and P.O.-A.; formal analysis, S.B. and P.O.-A.; investigation, W.N. and P.O.-A.; resources, S.B. and N.M.; data curation, S.B. and P.O.-A.; writing—original draft preparation, S.B. and N.M.; writing—review and editing, W.N. and P.O.-A.; visualization, S.B. and P.O.-A.; supervision, W.N. and P.O.-A.; project administration, P.O.-A.; funding acquisition, N.M. and P.O.-A. All authors have read and agreed to the published version of the manuscript.

Funding: This research was supported by CMU Junior Research Fellowship Program, Chiang Mai University.

Institutional Review Board Statement: The study was conducted in accordance with the Declaration of Helsinki, and approved by the Committee of Research Ethics, Faculty of Public Health, Chiang Mai University (Document No. ET004/2021).

Informed Consent Statement: Informed consent was obtained from all subjects involved in the study.

Data Availability Statement: The data are available upon request from the corresponding author.

Acknowledgments: The authors would like to thank all of the farmers for their participation, as well as the public health officers and village health volunteers for their assistance in this study.

Conflicts of Interest: The authors declare no conflict of interest.

References

1. Statistics of Hazardous Chemicals Imported into Thailand. Available online: https://www.doa.go.th/ard/?page_id=386 (accessed on 20 May 2022).
2. Panuwet, P.; Siriwong, W.; Prapamontol, T.; Ryan, P.B.; Fiedler, N.; Robson, M.G.; Barr, D.B. Agricultural Pesticide Management in Thailand: Situation and Population Health Risk. *Environ. Sci. Policy* **2012**, *17*, 72–81. [CrossRef] [PubMed]
3. Kongtip, P.; Nankongnab, N.; Mahaboonpeeti, R.; Bootsikeaw, S.; Batsungnoen, K.; Hanchenlaksh, C.; Tipayamongkholgul, M.; Woskie, S. Differences among Thai agricultural workers' health, working conditions, and pesticide use by farm type. *Ann. Work. Expo. Health* **2018**, *62*, 167–181. [CrossRef] [PubMed]
4. Köhler, H.R.; Triebskorn, R. Wildlife ecotoxicology of pesticides: Can we track effects to the population level and beyond? *Science* **2013**, *341*, 759–765. [CrossRef] [PubMed]
5. Tang, F.H.M.; Lenzen, M.; McBratney, A.; Maggi, F. Risk of pesticide pollution at the global scale. *Nat. Geosci.* **2021**, *14*, 206–210. [CrossRef]
6. Ziliotto, M.; Kulmann-Leal, B.; Roitman, A.; Bogo Chies, J.A.; Ellwanger, J.H. Pesticide Pollution in the Brazilian Pampa: Detrimental Impacts on Ecosystems and Human Health in a Neglected Biome. *Pollutants* **2023**, *3*, 280–292. [CrossRef]
7. Boedeker, W.; Watts, M.; Clausing, P.; Marquez, E. The global distribution of acute unintentional pesticide poisoning: Estimations based on a systematic review. *BMC Public Health* **2020**, *20*, 1875. [CrossRef]
8. Kisby, G.E.; Muniz, J.F.; Scherer, J.; Lasarev, M.R.; Koshy, M.; Kow, Y.W.; McCauley, L. Oxidative stress and DNA damage in agricultural workers. *J. Agromed.* **2009**, *14*, 206–214. [CrossRef]
9. Ledda, C.; Cannizzaro, E.; Cinà, D.; Filetti, V.; Vitale, E.; Paravizzini, G.; Di Naso, C.; Iavicoli, I.; Rapisarda, V. Oxidative stress and DNA damage in agricultural workers after exposure to pesticides. *J. Occup. Med. Toxicol.* **2021**, *16*, 1. [CrossRef]
10. Alavanja, M.C.; Hoppin, J.A.; Kamel, F. Health effects of chronic pesticide exposure: Cancer and neurotoxicity. *Annu. Rev. Public Health* **2004**, *25*, 155–197. [CrossRef]
11. Bassil, K.L.; Vakil, C.; Sanborn, M.; Cole, D.C.; Kaur, J.S.; Kerr, K.J. Cancer health effects of pesticides: Systematic review. *Can. Fam. Physician* **2007**, *53*, 1704–1711.
12. Xie, B.; Hu, Y.; Liang, Z.; Liu, B.; Zheng, X.; Xie, L. Association between pesticide exposure and risk of kidney cancer: A meta-analysis. *Onco Targets Ther.* **2016**, *28*, 3893–3900. [CrossRef] [PubMed]
13. VoPham, T.; Bertrand, K.A.; Hart, J.E.; Laden, F.; Brooks, M.M.; Yuan, J.M.; Talbott, E.O.; Ruddell, D.; Chang, C.H.; Weissfeld, J.L. Pesticide exposure and liver cancer: A review. *Cancer Causes Control* **2017**, *28*, 177–190. [CrossRef] [PubMed]
14. Andreotti, G.; Beane Freeman, L.E.; Shearer, J.J.; Lerro, C.C.; Koutros, S.; Parks, C.G.; Blair, A.; Lynch, C.F.; Lubin, J.H.; Sandler, D.P.; et al. Occupational pesticide use and risk of renal cell carcinoma in the Agricultural Health Study. *Environ. Health Perspect.* **2020**, *128*, 67011. [CrossRef] [PubMed]
15. Khan, D.A.; Bhatti, M.M.; Khan, F.A.; Naqvi, S.T.; Karam, A. Adverse effects of pesticides residues on biochemical markers in Pakistani tobacco farmers. *Int. J. Clin. Exp. Med.* **2008**, *1*, 274–282.
16. Araoud, M.; Neffeti, F.; Douki, W.; Hfaiedh, H.B.; Akrout, M.; Hassine, M.; Najjar, M.F.; Kenani, A. Adverse effects of pesticides on biochemical and haematological parameters in Tunisian agricultural workers. *J. Expo. Sci. Environ. Epidemiol.* **2012**, *22*, 243–247. [CrossRef]
17. Patil, J.A.; Patil, A.J.; Govindwar, S.P. Biochemical effects of various pesticides on sprayers of grape gardens. *Indian J. Clin. Biochem.* **2003**, *18*, 16–22. [CrossRef]
18. Rastogi, S.K.; Singh, V.K.; Kesavachandran, C.; Jyoti Siddiqui, M.K.; Mathur, N.; Bharti, R.S. Monitoring of plasma butyrylcholinesterase activity and hematological parameters in pesticide sprayers. *Indian J. Occup. Environ. Med.* **2008**, *12*, 29–32. [CrossRef]
19. Fareed, M.; Pathak, M.K.; Bihari, V.; Kamal, R.; Srivastava, A.K.; Kesavachandran, C.N. Adverse respiratory health and hematological alterations among agricultural workers occupationally exposed to organophosphate pesticides: A cross-sectional study in North India. *PLoS ONE* **2013**, *8*, e69755. [CrossRef]
20. Ahmadi, N.; Mandegary, A.; Jamshidzadeh, A.; Mohammadi-Sardoo, M.; Mohammadi-Sardo, M.; Salari, E.; Pourgholi, L. Hematological abnormality, oxidative stress, and genotoxicity induction in the greenhouse pesticide sprayers; investigating the role of NQO1 gene polymorphism. *Toxics* **2018**, *6*, 13. [CrossRef]
21. Leili, M.; Ghafiuri-Khosroshahi, A.; Poorolajal, J.; Samiee, F.; Smadi, M.T.; Bahrami, A. Pesticide residues levels as hematological biomarkers-a case study, blood serum of greenhouse workers in the city of Hamadan, Iran. *Environ. Sci. Pollut. Res. Int.* **2022**, *29*, 38450–38463. [CrossRef]
22. Khan, M.; Nazir, I.; Nazir, S.; Wadood, M.; Irfan, G.; Nasir, S.S. Impact of chronic exposure of organophosphorus pesticide on hematological and biochemical parameters of agriculture workers: A cross sectional study. *J. Pharm. Negat. Results* **2023**, *14*, 3059–3067.
23. Suwannahong, K.; Sridon, A.; Pongstaporn, W.; Thongmuang, P.; Sudjaroen, Y. Biochemical and Hematological Status in Rice farmers with Chronic Pesticide Exposure, Suphan Buri, Thailand. *Indian J. Public Health Res. Dev.* **2020**, *11*, 1285–1290.
24. Patil, J.A.; Patil, A.J.; Sontakke, A.V.; Govindwar, S.P. Occupational pesticides exposure of sprayers of grape gardens in western Maharashtra (India): Effects on liver and kidney function. *J. Basic. Clin. Physiol. Pharmacol.* **2009**, *20*, 335–355. [CrossRef] [PubMed]

25. Manfo, F.P.T.; Mboe, S.A.; Nantia, E.A.; Ngoula, F.; Telefo, P.B.; Moundipa, P.F.; Cho-Ngwa, F. Evaluation of the effects of agro pesticides use on liver and kidney function in farmers from Buea, Cameroon. *J. Toxicol.* **2020**, *2020*, 2305764. [CrossRef] [PubMed]
26. Aroonvilairat, S.; Kespichayawattana, W.; Sornprachum, T.; Chaisuriya, P.; Siwadune, T.; Ratanabanangkoon, K. Effect of pesticide exposure on immunological, hematological and biochemical parameters in Thai orchid farmers—A cross-sectional study. *Int. J. Environ. Res. Public Health* **2015**, *12*, 5846–5861. [CrossRef]
27. Sudjaroen, Y.; Suwannahong, K. Comparison of biochemical, hematological parameters and pesticide expose-related symptoms among organic and non-organic farmers, Singburi, Thailand. *Asian J. Pharm.* **2017**, *11*, 235–241.
28. Jayasumana, C.; Paranagama, P.; Agampodi, S.; Wijewardane, C.; Gunatilake, S.; Siribaddana, S. Drinking well water and occupational exposure to Herbicides is associated with chronic kidney disease, in Padavi-Sripura, Sri Lanka. *Environ. Health* **2015**, *14*, 6. [CrossRef]
29. Valcke, M.; Levasseur, M.E.; Soares da Silva, A.; Wesseling, C. Pesticide exposures and chronic kidney disease of unknown etiology: An epidemiologic review. *Environ. Health* **2017**, *16*, 49. [CrossRef]
30. Prudente, I.R.G.; Cruz, C.L.; Nascimento, L.C.; Kaiser, C.C.; Guimarães, A.G. Evidence of risks of renal function reduction due to occupational exposure to agrochemicals: A systematic review. *Environ. Toxicol. Pharmacol.* **2018**, *63*, 21–28. [CrossRef]
31. Wan, E.T.; Darssan, D.; Karatela, S.; Reid, S.A.; Osborne, N.J. Association of pesticides and kidney function among adults in the US population 2001–2010. *Int. J. Environ. Res. Public Health* **2021**, *18*, 10249. [CrossRef]
32. Lebov, J.F.; Engel, L.S.; Richardson, D.; Hogan, S.L.; Hoppin, J.A.; Sandler, D.P. Pesticide use and risk of end-stage renal disease among licensed pesticide applicators in the Agricultural Health Study. *Occup. Environ. Med.* **2016**, *73*, 3–12. [CrossRef]
33. García-Trabanino, R.; Jarquín, E.; Wesseling, C.; Johnson, R.J.; González-Quiroz, M.; Weiss, I.; Glaser, J.; José Vindell, J.; Stockfelt, L.; Roncal, C.; et al. Heat stress, dehydration, and kidney function in sugarcane cutters in El Salvador--A cross-shift study of workers at risk of Mesoamerican nephropathy. *Environ. Res.* **2015**, *142*, 746–755. [CrossRef] [PubMed]
34. Abdul, K.S.M.; De Silva, P.M.C.S.; Ekanayake, E.M.D.V.; Thakshila, W.A.K.G.; Gunarathna, S.D.; Gunasekara, T.D.K.S.C.; Jayasinghe, S.S.; Asanthi, H.B.; Chandana, E.P.S.; Chaminda, G.G.T.; et al. Occupational paraquat and glyphosate exposure may decline renal functions among rural farming communities in Sri Lanka. *Int. J. Environ. Res. Public Health* **2021**, *18*, 3278. [CrossRef] [PubMed]
35. Prudente, I.R.G.; Souza, B.R.S.; Nascimento, L.C.; Gonçalves, V.S.D.S.; Silva, D.S.D.; Rabelo, T.K.; Alves, O.; Junior Santos, A.C.; de Rezende Neto, J.M.; Tavares, D.D.S.; et al. Nephrotoxic effects caused by occupational exposure to agrochemicals in a region of Northeastern Brazil: A cross-sectional study. *Environ. Toxicol. Chem.* **2021**, *40*, 1132–1138. [CrossRef] [PubMed]
36. Shearer, J.J.; Sandler, D.P.; Andreotti, G.; Murata, K.; Shrestha, S.; Parks, C.G.; Liu, D.; Alavanja, M.C.; Landgren, O.; Beane Freeman, L.E.; et al. Pesticide use and kidney function among farmers in the Biomarkers of Exposure and Effect in Agriculture study. *Environ. Res.* **2021**, *199*, 111276. [CrossRef] [PubMed]
37. Sombatsawat, E.; Barr, D.B.; Panuwet, P.; Robson, M.G.; Siriwong, W. Pesticide-induced changes in cholinesterase activity and chronic kidney disease of unknown etiology among farmers in Nakhon Ratchasima, Thailand. *Hum. Ecol. Risk Assess.* **2021**, *27*, 2038–2050. [CrossRef]
38. Blair, A.; Tarone, R.; Sandler, D.; Lynch, C.F.; Rowland, A.; Wintersteen, W.; Steen, W.C.; Samanic, C.; Dosemeci, M.; Alavanja, M.C. Reliability of reporting on life-style and agricultural factors by a sample of participants in the Agricultural Health Study from Iowa. *Epidemiology* **2002**, *13*, 94–99. [CrossRef]
39. Levey, A.S.; Stevens, L.A.; Schmid, C.H.; Zhang, Y.L.; Castro, A.F.; Feldman, H.I.; Kusek, J.W.; Eggers, P.; Van Lente, F.; Greene, T.; et al. CKD-EPI (Chronic Kidney Disease Epidemiology Collaboration). A new equation to estimate glomerular filtration rate. *Ann. Intern. Med.* **2009**, *150*, 604–612. [CrossRef]
40. Rothman, K.J. No adjustments are needed for multiple comparisons. *Epidemiology* **1990**, *1*, 43–46. [CrossRef]
41. Bender, R.; Lange, S. Adjusting for multiple testing—When and how? *J. Clin. Epidemiol.* **2001**, *54*, 343–349. [CrossRef]
42. Schreinemachers, D.M.; Ghio, A.J. Effects of environmental pollutants on cellular iron homeostasis and ultimate links to human disease. *Environ. Health Insights* **2016**, *10*, 35–43. [CrossRef] [PubMed]
43. Wafa, T.; Nadia, K.; Amel, N.; Ikbal, C.; Insaf, T.; Asma, K.; Hedi, M.A.; Mohamed, H. Oxidative stress, hematological and biochemical alterations in farmers exposed to pesticides. *J. Environ. Sci. Health B* **2013**, *48*, 1058–1069. [CrossRef] [PubMed]
44. Celik, I.; Yilmaz, Z.; Turkoglu, V. Hematotoxic and hepatotoxic effects of dichlorvos at sublethal dosages in rats. *Environ. Toxicol.* **2009**, *24*, 128–132. [CrossRef] [PubMed]
45. Karami-Mohajeri, S.; Ahmadipour, A.; Rahimi, H.R.; Abdollahi, M. Adverse effects of organophosphorus pesticides on the liver: A brief summary of four decades of research. *Arh. Za Hig. Rada Toksikol.* **2017**, *68*, 261–275. [CrossRef] [PubMed]
46. Desai, S.N.; Desai, P.V. Changes in renal clearance and renal tubular function in albino mice under the influence of Dichlorvos. *Pestic. Biochem. Physiol.* **2008**, *91*, 160–169. [CrossRef]
47. Sobolev, V.E.; Sokolova, M.O.; Jenkins, R.O.; Goncharov, N.V. Molecular mechanisms of acute organophosphate nephrotoxicity. *Int. J. Mol. Sci.* **2022**, *23*, 8855. [CrossRef]
48. Taira, K.; Kawakami, T.; Weragoda, S.K.; Herath, H.M.A.S.; Ikenaka, Y.; Fujioka, K.; Hemachandra, M.; Pallewatta, N.; Aoyama, Y.; Ishizuka, M.; et al. Urinary concentrations of neonicotinoid insecticides were related to renal tubular dysfunction and neuropsychological complaints in Dry-zone of Sri Lanka. *Sci. Rep.* **2021**, *11*, 22484. [CrossRef]
49. Liu, B.C.; Tang, T.T.; Lv, L.L.; Lan, H.Y. Renal tubule injury: A driving force toward chronic kidney disease. *Kidney Int.* **2018**, *93*, 568–579. [CrossRef]

50. Mueangkhiao, P.; Siviroj, P.; Sapbamrer, R.; Khacha-Ananda, S.; Lungkaphin, A.; Seesen, M.; Jaikwang, P.; Wunnapuk, K. Biological variation in kidney injury and kidney function biomarkers among farmers in Lamphun province, Thailand. *Environ. Sci. Pollut. Res. Int.* **2020**, *27*, 12386–12394. [CrossRef]
51. Kshirsagar, A.V.; Zeitler, E.M.; Weaver, A.; Franceschini, N.; Engel, L.S. Environmental Exposures and Kidney Disease. *Kidney360* **2022**, *3*, 2174–2182. [CrossRef]

toxics

MDPI

Article

Associations between Urinary Phthalate Metabolites with BDNF and Behavioral Function among European Children from Five HBM4EU Aligned Studies

Elena Salamanca-Fernández [1,2,†], Lydia Espín-Moreno [3,†], Alicia Olivas-Martínez [2], Ainhoa Pérez-Cantero [1], José L. Martín-Rodríguez [2,4,*], Rafael M. Poyatos [5], Fabio Barbone [6], Valentina Rosolen [7], Marika Mariuz [6], Luca Ronfani [6], L'ubica Palkovičová Murínová [8], Lucia Fábelová [8], Tamás Szigeti [9], Réka Kakucs [9], Amrit K. Sakhi [10], Line S. Haug [10], Birgitte Lindeman [10], Janja Snoj Tratnik [11], Tina Kosjek [11], Griet Jacobs [12], Stefan Voorspoels [12], Helena Jurdáková [13], Renáta Górová [13], Ida Petrovičová [14], Branislav Kolena [14], Marta Esteban [15], Susana Pedraza-Díaz [15], Marike Kolossa-Gehring [16], Sylvie Remy [17], Eva Govarts [17], Greet Schoeters [18], Mariana F. Fernández [1,2,3,‡] and Vicente Mustieles [1,2,3,4,*,‡]

1 Biomedical Research Center (CIBM), Department of Radiology and Physical Medicine, University of Granada, 18012 Granada, Spain; esalamanca@ugr.es (E.S.-F.); apercan@ugr.es (A.P.-C.); marieta@ugr.es (M.F.F.)
2 Instituto de Investigación Biosanitaria (ibs.GRANADA), 18012 Granada, Spain; aolivas@ugr.es
3 CIBER de Epidemiología y Salud Pública (CIBERESP), 28034 Madrid, Spain; lydia.espin@ciberesp.es
4 Servicio de Radiodiagnóstico, Hospital Universitario Clínico San Cecilio, 18012 Granada, Spain
5 Unidad de Gestión Clínica de Laboratorios, Hospital Universitario Clínico San Cecilio, 18012 Granada, Spain
6 Department of Medicine, Surgery and Health Sciences, University of Trieste, Strada di Fiume 447, 34149 Trieste, Italy
7 Central Directorate for Health, Social Policies and Disability, Friuli Venezia Giulia Region, Via Cassa di Risparmio 10, 34121 Trieste, Italy
8 Department of Environmental Medicine, Faculty of Public Health, Slovak Medical University, 831 04 Bratislava, Slovakia
9 Center for Public Health and Pharmacy, Albert Flórián út 2-6, 1097 Budapest, Hungary
10 Norwegian Institute of Public Health, P.O. Box 222, Skøyen, N-0213 Oslo, Norway
11 Jožef Stefan Institute, 1000 Ljubljana, Slovenia
12 VITO GOAL, Flemish Institute for Technological Research (VITO), Boeretang 200, 2400 Mol, Belgium
13 Department of Analytical Chemistry, Faculty of Natural Sciences, Comenius University, Ilkovičova 6, Mlynská Dolina, 84215 Bratislava, Slovakia
14 Department of Zoology and Anthropology, Faculty of Natural Sciences and Informatics, Constantine the Philosopher University in Nitra, Nabrezie mladeze 91, 94974 Nitra, Slovakia
15 National Centre for Environmental Health, Instituto de Salud Carlos III, 28034 Madrid, Spain
16 German Environment Agency (UBA), Corrensplatz 1, 14195 Berlin, Germany
17 VITO Health, Flemish Institute for Technological Research (VITO), 2400 Mol, Belgium
18 Department of Biomedical Sciences, University of Antwerp, 2610 Antwerp, Belgium
* Correspondence: joseluismartin.rx@hotmail.com (J.L.M.-R.); vmustieles@ugr.es (V.M.)
† These authors contributed equally to this work as co-first authors.
‡ These authors contributed equally to this work as co-last authors.

Citation: Salamanca-Fernández, E.; Espín-Moreno, L.; Olivas-Martínez, A.; Pérez-Cantero, A.; Martín-Rodríguez, J.L.; Poyatos, R.M.; Barbone, F.; Rosolen, V.; Mariuz, M.; Ronfani, L.; et al. Associations between Urinary Phthalate Metabolites with BDNF and Behavioral Function among European Children from Five HBM4EU Aligned Studies. *Toxics* **2024**, *12*, 642. https://doi.org/10.3390/toxics12090642

Academic Editor: Megan Mahoney

Received: 30 July 2024
Revised: 20 August 2024
Accepted: 30 August 2024
Published: 31 August 2024

Abstract: Based on toxicological evidence, children's exposure to phthalates may contribute to altered neurodevelopment and abnormal regulation of brain-derived neurotrophic factor (BDNF). We analyzed data from five aligned studies of the Human Biomonitoring for Europe (HBM4EU) project. Ten phthalate metabolites and protein BDNF levels were measured in the urine samples of 1148 children aged 6–12 years from Italy (NACII-IT cohort), Slovakia (PCB-SK cohort), Hungary (InAirQ-HU cohort) and Norway (NEBII-NO). Serum BDNF was also available in 124 Slovenian children (CRP-SLO cohort). Children's total, externalizing and internalizing behavioral problems were assessed using the Child Behavior Checklist at 7 years of age (only available in the NACII-IT cohort). Adjusted linear and negative binomial regression models were fitted, together with weighted quantile sum (WQS) regression models to assess phthalate mixture associations. Results showed that, in boys but not girls of the NACII-IT cohort, each natural-log-unit increase in mono-n-butyl phthalate (MnBP) and Mono(2-ethyl-5-oxohexyl) phthalate (MEOHP) was cross-sectionally associated with higher externalizing problems [incidence rate ratio (IRR): 1.20; 95% CI: 1.02, 1.42 and 1.26; 95% CI: 1.03,

1.55, respectively]. A suggestive mixture association with externalizing problems was also observed per each tertile mixture increase in the whole population (WQS—IRR = 1.15; 95% CI: 0.97, 1.36) and boys (IRR = 1.20; 95% CI: 0.96, 1.49). In NACII-IT, PCB-SK, InAirQ-HU and NEBII-NO cohorts together, urinary phthalate metabolites were strongly associated with higher urinary BDNF levels, with WQS regression confirming a mixture association in the whole population (percent change (PC) = 25.9%; 95% CI: 17.6, 34.7), in girls (PC = 18.6%; 95% CI: 7.92, 30.5) and mainly among boys (PC = 36.0%; 95% CI: 24.3, 48.9). Among CRP-SLO boys, each natural-log-unit increase in ∑DINCH concentration was associated with lower serum BDNF levels (PC: −8.8%; 95% CI: −16.7, −0.3). In the NACII-IT cohort, each natural-log-unit increase in urinary BDNF levels predicted worse internalizing scores among all children (IRR: 1.15; 95% CI: 1.00, 1.32). Results suggest that (1) children's exposure to di-n-butyl phthalate (DnBP) and di(2-ethylhexyl) phthalate (DEHP) metabolites is associated with more externalizing problems in boys, (2) higher exposure to DINCH may associate with lower systemic BDNF levels in boys, (3) higher phthalate exposure is associated with higher urinary BDNF concentrations (although caution is needed since the possibility of a "urine concentration bias" that could also explain these associations in noncausal terms was identified) and (4) higher urinary BDNF concentrations may predict internalizing problems. Given this is the first study to examine the relationship between phthalate metabolite exposure and BDNF biomarkers, future studies are needed to validate the observed associations.

Keywords: phthalate; behavior; BDNF; effect biomarker; exposure biomarker; HBM4EU; PARC

1. Introduction

Phthalates are a family of industrial chemicals produced at high volumes and used as plasticizers in food packaging materials, medical equipment, toys, furniture and as solubilizers in cosmetics [1]. High-molecular-weight phthalates, such as di(2-ethylhexyl) phthalate (DEHP), are mainly used as plasticizers to impart flexibility in polyvinyl chloride (PVC) materials [2], while low-molecular-weight phthalates, such as di-ethyl phthalate (DEP), are often used as solvents in personal care products and in adhesives, varnishes and coatings [3]. In recent years, other phthalates including diisononyl phthalate (DiNP), and the non-phthalate plasticizer di-iso-nonyl-cyclohexane-1,2-dicarboxylate (DINCH®) are replacing DEHP and other phthalate compounds in several applications [4–7]. Human phthalate exposure occurs through multiple pathways [8], and although these compounds are rapidly metabolized and excreted in urine, exposure is chronic on a daily basis. Thus, more than 90% of the European population shows detectable urinary phthalate metabolite concentrations [9,10], and recent human biomonitoring studies also show increasing temporal trends for some phthalate replacements [9,11–13].

Some phthalates and their metabolites are known endocrine-disrupting chemicals (EDCs) [14–17], that is, chemicals or mixtures of chemicals able to interfere with different aspects of hormone action [18]. Additionally, some phthalates are known or suspected neurotoxicants [19]. Thus, exposure to both individual phthalate compounds and their mixtures has been shown to alter brain function and behavior in rodent models [20], often in a sex-specific manner [21]. In humans, exposure to specific phthalate metabolites has been associated with altered behavior, although more evidence is needed to confirm this relationship [22,23].

Phthalates and their metabolites can act through diverse modes of action, including anti-androgenic, estrogenic, thyroid-disrupting and/or oxidative stress pathways [24–26]. Using an adverse outcome pathway (AOP) network, we showed that the regulation of brain-derived neurotrophic factor (BDNF) is a downstream key central event of these mechanisms, having a key role in neurodevelopment [27]. BDNF is strongly involved in neuronal plasticity mechanisms, promoting neuronal growth, synaptic development and neuronal survival [28]. Peripheral BDNF biomarkers measured in plasma or serum have been shown to correlate with central BDNF levels in rodents [29]. Bisphenol A, certain heavy metals and nonpersistent pesticides have been associated with altered BDNF regulation in adolescent boys from the Spanish INMA-Granada cohort [28,30,31], and BDNF was shown to mediate the longitudinal association between BPA and altered behavior [28]. There is also preliminary evidence that altered BDNF regulation may also be an intermediate step between phthalate and neurodevelopmental impairments in rodents [32,33], although data are still lacking in epidemiological studies.

The European Human Biomonitoring Initiative [34] had among its objectives the assessment of exposure to priority environmental chemicals among Europeans in a harmonized manner. In practice, this was achieved in the so-called HBM4EU aligned studies [35–37], among other European populations. HBM4EU included phthalates as priority substances in order to further review their regulation in the EU [38]. In addition to exposure biomarkers, HBM4EU implemented effect biomarkers in relation to some of the health effects of concern. Indeed, BDNF was selected in relation to neurodevelopment [27,39,40]. This work is framed as a continuation of the research lines initiated in HBM4EU, now included within the objectives of the European project Partnership for the Assessment of Risks from Chemicals (PARC) [41]. The aim of this study was to investigate (1) whether children's exposure to phthalates, both individually and as a mixture, is associated with alterations in their behavior, (2) if children's exposure to phthalates is associated with BDNF levels and (3) whether BDNF levels can predict children's behavioral problems.

2. Materials and Methods

2.1. Study Population

The present work used individual data concerning the assessment of child behavior, urinary and serum BDNF levels and urinary phthalate concentrations in five European cohorts/countries participating in the HBM4EU aligned studies. Studies included the Italian Northern Adriatic Cohort II (NACII-IT) study [42,43], the Slovak PCB-SK cohort [44], the Hungarian Integrated Indoor Air Quality Management (InAirQ-HU) cohort [45,46], the Norwegian Environmental Biobank Part II (NEBII-NO) cohort [47] and the Exposure of Children and Adolescents to Selected Chemicals Through Their Habitat Environment in Slovenia (CRP-SLO) cohort [48,49]. A detailed description of the sampling methods for the inclusion of participants and data harmonization in the HBM4EU aligned studies can be found elsewhere [35,36]. In brief, the study participants were 6–8-years-old children enrolled in the NACII-IT cohort, 8–11 years in the InAirQ-HU cohort, 10–12 years in the PCB-SK cohort, 7–11 years in the NEBII-NO cohort and 7–10 years in the CRP-SLO cohort. Information on the urinary concentrations of phthalate metabolites and urinary BDNF levels was available in four cohorts (NACII-IT, PCB-SK, InAirQ-HU and NEBII-NO), one cohort had data on urinary phthalate metabolites and serum BDNF levels (CRP-SLO) and NACII-IT also had data for child behavior. The selection of the participants of each cohort was performed following a step-wise selection procedure previously described [35]. Informed consent was obtained from the caregivers of all children involved, in accordance with the Declaration of Helsinki. The research protocols of the cohorts were approved by their respective ethics committees.

2.2. Exposure Biomarker Assessment

Concurrently with the behavioral evaluation, metabolites of the following parent phthalate compounds were measured in spot urine samples, previously collected in propylene tubes and stored between −20 and −80 °C: di-isobutyl phthalate (DiBP), di-n-butyl phthalate (DnBP), benzyl-butyl phthalate (BBzP), di-ethyl phthalate (DEP), di-(2-ethylhexyl) phthalate (DEHP) and di(isononyl)cyclohexane-1,2-dicarboxylate (DINCH)/HEXAMOLL® metabolites. The following biomarkers of exposure were measured in the children's urine: mono-isobutyl phthalate (MiBP) for DiBP, mono-n-butyl phthalate (MnBP) for DnBP, mono-benzyl phthalate (MBzP) for BBzP, mono-ethyl phthalate (MEP) for DEP, mono(2-ethyl-5-hydroxyhexyl) phthalate (MEHHP), mono(2-ethyl-5-oxohexyl) phthalate (MEOHP) and mono(2-ethyl-5-carboxypentyl) phthalate (MECPP) for DEHP. Mono(2-

ethylhexyl) phthalate (MEHP) was not measured in the NACII-IT cohort, and therefore in this work, we focused on the other three DEHP metabolites (MEHHP, MEOHP and MECPP). Cyclohexane-1,2-dicarboxylic acid monohydroxy isononyl ester (MHINCH) and cyclohexane-1,2-dicarboxylic acid monocarboxyisooctyl ester (MCOCH) metabolites were measured to cover exposure to the non-phthalate replacement DINCH (HEXAMOLL®). The NEBII-NO cohort had data on MHINCH but not on MCOCH.

The quality and inter-comparability of results from the laboratories that measured the urinary phthalate and DINCH metabolites were assured through the HBM4EU quality control (QA/QC) program [50,51]. The participating laboratories in the HBM4EU analyzed standard test materials using their own method, in all cases using liquid chromatography-tandem mass spectrometry (LC-MS/MS). Laboratories reported the concentrations in ng/mL through a web tool and provided the details of their methods (e.g., regarding deconjugation, extraction, instrumental analysis, use of internal standards, method of quantification and identification parameters). Extended information of the phthalates and DINCH exercises can be found elsewhere [51]. All samples were analyzed in the laboratories that participated and obtained successful results in the QA/QC HBM4EU program [50,51]. Details for each specific cohort have been previously described [52]. In order to improve comparability, urinary creatinine was measured to standardize phthalate metabolites by the voiding volume (urine dilution) of the participants.

2.3. Effect Biomarker Assessment: Urinary and Serum BDNF Levels

Urinary BDNF protein levels were measured in four out of the five included cohorts: NACII-IT, PCB-SK, InAirQ-HU and NEBII-NO, at the Hospital Universitario Clinico-San Cecilio (HUSC) of Granada (Spain), using the commercial RayBio® ELISA kit (Raybiotech, Norcross, GA, USA). This ELISA kit was the one that proved to offer the best performance among the several kits tested within a previous HBM4EU validation program for effect biomarkers [53]. Briefly, urine samples were thawed, vortexed and pretreated following the protocol and procedure described in Olivas-Martínez et al. [53]. Each urine sample was assessed in duplicate in different batches, and the mean value was calculated. The intra- and inter-assay coefficients of variability were <5% and <15%, respectively.

Serum total BDNF protein levels were measured in the CRP-SLO cohort using the commercial Quantikine® enzyme-linked immunosorbent assay (ELISA) kit (R&D Systems, Minneapolis, MN, USA). Serum samples were defrosted, vortexed, separated in two aliquots of 10 μL, diluted 100 times and tested according to the manufacturer's instructions at the Biomedical Research Center (CIBM) of the University of Granada (Spain). Each sample was tested in duplicate in different plates, and the mean of these two values was calculated in order to reduce measurement variation. The intra- and inter-assay coefficients of variability were <5% and <15%, respectively.

2.4. Behavioral Assessment in the NACII-IT Cohort

In the NACII-IT cohort, and concurrently with the collection of urine samples, children's behavioral function was assessed using the parent-reported Child Behavior Checklist (CBCL/6-18), a validated questionnaire that evaluates parental perception about the behavior of their children and/or adolescents during the previous six months [54]. The CBCL includes 118 items rated on a three-point Likert scale (0 = "Not True", 1 = "Somewhat or Sometimes True" or 2 = "Very/Often True") which are grouped into eight syndrome scales (anxious/depressed, withdrawn/depressed, somatic complaints, social problems, thought problems, attention problems, rule-breaking behavior and aggressive behavior). These scales are also grouped into two empirically derived composite scales: (i) the internalizing domain as a measure of emotional problems (sum of scores on the anxious/depressed, withdrawn/depressed and somatic complaints scales) and (ii) the externalizing domain as a measure of behavioral problems (sum of scores on the rule-breaking behavior and aggressive behavior scales). Three other scales are considered mixed-syndrome scales that do not belong to either domain: social, thought and attention problems. The total problems composite scale finally quantifies general impairment and corresponds to the sum of scores from the eight syndrome scales, together with a group of 17 "Other problems" items that do not belong to any specific syndrome scale. Our analysis was focused on continuous raw scores for the total [range 0–87], externalizing [range 0–27] and internalizing [range 0–34] composite scales, with higher scores meaning higher behavioral problems.

2.5. Covariates

As previously described [55], based on information obtained from the questionnaires applied in the different HBM4EU aligned studies [35,36], covariates were harmonized. The level of education of the mothers was based on the International Standard Classification of Education (ISCED) developed by the United Nations Educational, Scientific and Cultural Organization (UNESCO) [56]. Mothers with less than secondary education were included in the lower-education level (ISCED 0–2), mothers with secondary but not postsecondary education represented the medium level (ISCED 3–4) and those with college and/or university degrees and higher education were included in the high-education level (ISCED $\geq$ 5). At urine sample collection, the children's height and weight were measured by the healthcare staff in each cohort (with the exception of the NEBII-NO cohort in which self-reported measures were provided). BMI (kg/m^2) was calculated as the ratio between weight (kg) and height squared (m^2). In order to improve comparability among studies, age- and sex-specific BMI z-scores were calculated using the 2007 World Health Organization (WHO) growth reference standards [57]. Child sex, age and season of urine collection were obtained from questionnaires. Maternal intelligence quotient (IQ) was assessed using Raven's Progressive Matrices Test and was considered in models examining CBCL scores, only available in the NACII-IT cohort.

2.6. Statistical Analysis

Study participant characteristics were described using measures of central tendency and dispersion for numerical variables and frequencies for categorical variables.

2.6.1. Phthalate Exposure Biomarkers

All selected phthalate and DINCH metabolites were detected in most urine samples (>93%). The few concentrations below the limit of detection (LOD) were singly and randomly imputed considering the distribution of quantified values [58]. For phthalate metabolites belonging to a same parent compound, molar sums (µmol/L) were also computed (i.e., ∑DEHP and ∑DINCH). The ∑DINCH measure was estimated without considering the NEBII-NO cohort which only had data on the MHINCH but not the MCOCH metabolite. Urinary phthalate metabolite concentrations were divided or standardized by urinary creatinine levels and expressed as micrograms of chemical per gram of creatinine. Creatinine-standardized urinary phthalate metabolites were natural log transformed to reduce the skewness of distributions and the influence of extreme values. Correlation among phthalate metabolites was examined using Spearman's rank correlation coefficients.

2.6.2. Urinary and Serum BDNF Protein Levels

Urinary BDNF protein levels were standardized by urinary creatinine levels and expressed as micrograms of BDNF per gram of creatinine. Both serum BDNF protein levels and creatinine-standardized urinary BDNF levels were natural log transformed to reduce the skewness of distributions and the influence of extreme values.

2.6.3. Covariates and Confounding Factors

Potential confounders of phthalate–BDNF behavior associations were identified a priori based on causal scientific knowledge [59]. All models were adjusted for the same set of covariates: maternal education [low (0–2), middle (3–4) or high (>=5) ISCED classification], child sex, age at urine collection or CBCL assessment, child body mass index z-score (BMI z-score), season of urine collection (summer/autumn vs. winter/spring, since specific

CBCL scores and levels of some phthalates differed according to season) and creatinine (log transformed). The phthalate–CBCL and BDNF–CBCL models were additionally adjusted for maternal IQ (but only in the NACII-IT cohort, which had data on CBCL behavior scores concurrently with phthalate and BDNF measurements). Phthalate associations with urinary BDNF levels were examined in four cohorts (NACII-IT, PCB-SK, InAirQ-HU and NEBII-NO), including "cohort" as an additional covariate in the statistical models. The slight number of missing values for some covariates (in all cases <4%) was singly imputed with the variable median or mode value.

2.6.4. Cross-Sectional Associations between Phthalates, BDNF Concentrations and Children's Behavior

As a first step, we only analyzed data from the NACII-IT cohort, since it had data on phthalate metabolites, urinary BDNF concentrations and CBCL scores: (1) phthalate–CBCL associations were examined using adjusted negative binomial regression (NBR) [60–62]. Since phthalate concentrations were natural log transformed prior to analysis, the incidence rate ratios (IRRs) and 95% confidence intervals (95% CIs) corresponded to the multiplicative change in the probability of CBCL scores for each natural log-unit increase in exposure biomarkers; (2) the phthalate–BDNF associations were examined using adjusted linear regression models. Since both phthalate metabolites and BDNF were natural log transformed, regression estimates were expressed as the percent change (PC) for each natural-log-unit increase in standardized urinary phthalate biomarkers using the following formula: $([e^{(\beta)} - 1] \times 100)$; (3) BDNF–CBCL associations were also examined using NBR models.

As a second step, the phthalate–BDNF associations were investigated together in the four cohorts that had data on urinary BDNF levels, using linear regression models and expressing results as the PC for each natural-log-unit increase in standardized urinary phthalate biomarkers.

A phthalate–sex interaction term was added to the models to identify potential sex-specific associations, since previous studies have shown effect modification by this variable [21,23]. Nevertheless, considering the modest sample size, especially for behavior analyses only available in the NACII-IT cohort, an interaction term *p*-value below 0.10 was considered as potential evidence of effect modification.

2.6.5. Phthalate's Mixture Models

The overall association between childhood exposure to mixtures of phthalates and DINCH metabolites on urinary BDNF levels and behavioral CBCL scores was estimated using adjusted linear and negative binomial weighted quantile sum (WQS) regressions for BDNF and CBCL scores, respectively. WQS regression was selected since it allowed to model mixture effects using negative binomial regression. Consistent with single-pollutant models, WQS regressions were adjusted for the same set of covariates, and we fitted models for the whole population, as well as in boys and girls separately. Phthalate and DINCH metabolite concentrations were modeled in tertiles, and thus the results represent the change in estimates per tertile increase in the phthalate mixture.

WQS is a quantile-based approach that combines multiple independent variables additively into a weighted index, with weights estimating the contribution of each component to the mixture. WQS regression assumes that all exposures included in the index are associated with the outcome in the same direction [63,64]. Therefore, after observing no mixture effects in the negative direction, only WQS indexes in the positive direction were considered, in line with our initial hypothesis expecting deleterious effects of phthalate exposure on child behavior. WQS performs a random partition of the data into training and validation datasets, used to estimate the relative weights of each chemical/metabolite and test the statistical significance of the WQS index, respectively. Such splitting reduces statistical power and may lead to unstable results. To counteract this issue, we implemented a repeated holdout validation [65], consisting in repeating "n" times the random partitioning and WQS analyses and averaging the "n" estimates. We repeated WQS regressions 100 times; the study population was split into training (40%) and validation (60%) sets; and weights were calculated based on 100 bootstrap samples. The scatterplots of the WQS index vs. the outcome (adjusted for covariate model residuals) were checked to examine the shape of the mixture associations, which in this analysis did not show relevant departures from linearity.

2.6.6. Sensitivity Analysis

Phthalate associations with urinary BDNF levels were studied separately in each of the four HBM4EU aligned studies to assess robustness.

Since different methods to account for urinary dilution could provide different results, we tested and compared five different approaches to adjust for urinary creatinine levels in models of urinary phthalates and urinary BDNF levels in the four HBM4EU aligned studies combined. The approaches tested include (a) raw phthalates and BDNF levels controlling for creatinine as a covariate, (b) phthalates and BDNF levels standardized by creatinine, (c) phthalate and BDNF levels standardized by creatinine in addition to adjusting for creatinine as a covariate (main model), (d) covariate-adjusted standardization (CAS) of both phthalates and BDNF levels, following the method described by O'Brien et al. [66], and (e) CAS standardization controlling for creatinine as a covariate.

2.6.7. Interpretation and Software

Statistical significance was set as p-value < 0.05. A p-value between 0.05 and 0.10 was considered as possibly suggestive of an association. However, results were interpreted not solely depending on statistical significance but also considering patterns of associations and the previous evidence available to contextualize the observed findings [67]. Regression analyses were carried out with Stata version 14.2 (Stata Corp., College Station, TX, USA), while WQS models were implemented using the "gWQS" package (https://cran.r-project.org/web/packages/gWQS/gWQS.pdf, accessed on 28 August 2024) in RStudio version 4.0.3 (RStudio Team, 2020).

3. Results

3.1. Study Population Characteristics

A total of 1148 children were included in the current analysis (Table 1). Half (50%) of the children were boys, and the mean age (standard deviation—SD) and BMI z-score at the time of urine collection was 9.23 (1.7) years and 0.51 (1.17) kg/m^2, respectively. Almost half (45%) of the mothers had a high-education level, with PCB-SK mothers having the lowest percentage (9% highly educated). The mean (SD) total CBCL raw score in NACII-IT was 23 (15), ranging from 0 to 87 points. In the four cohorts combined, urinary BDNF median concentrations (percentiles P25, P75) were 2.31 (1.52, 3.46) μg/g creatinine. In the SLO-CRP cohort, median (P25, P75) serum BDNF levels were 45.1 (38.9, 52.6) ng/mL. The highest median (P25, P75; μg/g creatinine) concentrations of the investigated phthalate and DINCH metabolites were observed for the sum of ∑DEHP metabolites [56.4 (35.6, 93.4)], followed by MiBP [33.9 (21.5, 57.7)], MnBP [26.3 (15.5, 47.2)] and MEP [21.8 (10.4, 54.4)], while MBzP [4.14 (1.80, 7.99)] and ∑DINCH [3.30 (1.68, 6.68)] showed the lowest concentrations (Table 1). Phthalate metabolite concentrations, including MEP, MnBP, ∑DEHP and ∑DINCH, differed across the cohorts, highlighting substantial variability linked to different countries and lifestyles (Table 1). The Spearman's correlation coefficients among phthalate metabolites were in general low to moderate, with the exception of metabolites belonging to the same parent compound, which, as expected, showed strong correlations among them (Supplementary Figure S1).

Table 1. Characteristics and data availability of included studies.

Variables	NACII-IT	PCB-SK	InAirQ-HU	NEBII-NO	CRP-SLO	Combined [b]
	Cohort characteristics [Mean (SD) or frequency (percentage)]					
Country	Italy	Slovakia	Hungary	Norway	Slovenia	-
Initial sample size	300	294	257	297	124	1148
Age	7.0 (0.2)	11.1 (0.4)	9.3 (0.7)	9.83 (1.17)	8.85 (0.97)	9.32 (1.68)
BMI (Kg/m^2)	16.9 (2.26)	20.2 (4.3)	17.4 (3.3)	17.4 (2.46)	17.2 (3.09)	18.0 (3.50)
BMI z-score	0.49 (1.15)	0.92 (1.22)	0.27 (1.35)	0.32 (0.81)	0.34 (1.44)	0.51 (1.17)
Child sex = boys	150 (50%)	130 (44%)	130 (51%)	160 (54%)	55 (44%)	570 (50%)
Maternal education						
Low	42 (14%)	33 (11%)	9 (4%)	0 (0%)	11 (9%)	84 (8%)
Medium	139 (46%)	235 (80%)	122 (52%)	39 (13%)	30 (24%)	535 (47%)
High	119 (40%)	25 (9%)	103 (44%)	258 (87%)	83 (67%)	505 (45%)
Maternal IQ	120 (10.7) [68–128]	108 (15.1) [63–132]	-	-	-	114.1 (14.5) [63–132]
	Behavior CBCL assessment (Mean (SD)/[range] and BDNF levels (Median, P25, P75; µg/g creatinine)					
Total CBCL score	23.2 (15.0) [0–87]	-	-	-	-	-
Externalizing score	5.4 (4.6) [0–27]	-	-	-	-	-
Internalizing score	6.3 (5.4) [0–34]	-	-	-	-	-
Urinary BDNF [a]	2.87 (1.98, 4.21)	2.65 (1.85, 3.83)	2.52 (1.70, 3.54)	1.52 (1.08, 1.98)	-	2.31 (1.52, 3.46)
Serum BDNF	-	-	-	-	45.1 (38.9, 52.6)	-
	Phthalate concentrations (Median, P25, P75; µg/g creatinine)					
MEP	63.5 (33.3, 131.9)	23.9 (12.9, 54.9)	9.64 (5.82, 18.1)	14.3 (9.70, 25.6)	25.3 (15.7, 51.2)	21.8 (10.4, 54.4)
MnBP	21.9 (14.5, 33.0)	58.8 (35.9, 102.8)	18.6 (11.5, 30.0)	23.9 (15.1, 35.2)	19.9 (14.0, 26.9)	26.3 (15.5, 47.2)
MiBP	34.2 (24.6, 50.2)	48.3 (25.9, 81.0)	34.0 (19.9, 51.7)	26.7 (16.1, 41.3)	27.0 (17.5, 43.5)	33.9 (21.5, 57.7)
MBzP	6.69 (3.89, 11.1)	3.94 (0.77, 7.58)	1.59 (0.93, 2.84)	4.88 (2.95, 8.61)	3.47 (2.12, 5.52)	4.14 (1.80, 7.99)
∑DEHP	69.4 (51.8, 108.1)	88.2 (57.8, 136.6)	36.9 (23.8, 56.3)	40.8 (29.3, 62.3)	55.4 (38.4, 87.4)	56.4 (35.6, 93.4)
MEHHP	19.2 (14.3, 28.8)	20.3 (14.0, 31.8)	10.8 (6.85, 16.6)	9.29 (6.35, 14.5)	14.6 (9.11, 21.9)	14.8 (8.56, 23.5)
MEOHP	9.82 (6.89, 15.0)	18.1 (12.2, 25.6)	5.23 (3.43, 8.42)	5.74 (3.89, 8.66)	11.1 (7.27, 17.4)	8.62 (4.91, 15.4)
MECPP	24.2 (16.8, 37.3)	28.2 (18.0, 46.8)	11.4 (7.32, 18.3)	17.2 (12.3, 24.8)	17.7 (11.7, 25.8)	19.6 (12.5, 31.6)
∑DINCH [c]	6.17 (3.40, 10.5)	2.52 (1.43, 4.97)	1.97 (1.25, 4.11)	-	4.27 (2.43, 7.05)	3.30 (1.68, 6.68)
MHINCH	3.65 (2.03, 6.68)	1.65 (0.90, 3.35)	1.20 (0.74, 2.42)	2.94 (1.91, 5.52)	2.61 (1.41, 4.52)	2.28 (1.23, 4.54)
MCOCH [c]	2.24 (1.26, 4.13)	0.86 (0.46, 1.62)	0.90 (0.49, 1.68)	-	1.55 (0.89, 2.70)	1.21 (0.58, 2.49)

Abbreviations: BDNF (brain-derived neurotrophic factor); BMI (body mass index); CBCL (Child Behavior Checklist); ISCED (International Standard Classification of Education); IQ (intelligence quotient); NACII-IT (Italian Northern Adriatic Cohort II); PCB-SK (Slovakian PCB-SK cohort); InAirQ-HU (Hungarian Integrated Indoor Air Quality Management cohort); MEP (mono-ethyl phthalate); MnBP (mono-n-butyl phthalate); MiBP (mono-isobutyl phthalate); MBzP (mono-benzyl phthalate); DEHP (di(2-ethylhexyl) phthalate); MEHHP (mono(2-ethyl-5-hydroxyhexyl) phthalate); MEOHP (mono(2-ethyl-5-oxohexyl) phthalate); MECPP (mono(2-ethyl-5-carboxypentyl) phthalate); DINCH (di(isononyl)cyclohexane-1,2-dicarboxylate); MHINCH (mono-hydroxy isononyl cyclohexane-1,2-dicarboxylate); MCOCH (mono-oxo isononyl cyclohexane-1,2-dicarboxylate). [a] (µg/g of creatinine). [b] Combined data from NACII-IT, PCB-SK, InAirQ-HU and NEBII-NO cohorts, excluding CRP-SLO since the later cohort did not had data on urinary BDNF levels. [c] The ∑DINCH measure was estimated without considering the NEBII-NO cohort, which only had data on the MHINCH but not the MCOCH metabolite.

3.2. Phthalate–BDNF–Behavior Associations in the NACII-IT Cohort

No associations were observed between phthalate metabolites and CBCL scores when boys and girls were analyzed together (Supplementary Table S1). However, among boys, each natural-log-unit increase in urinary MnBP and DEHP metabolite concentrations was associated with higher externalizing behavior scores (Figure 1, Supplementary Table S2): MnBP (IRR: 1.20; 95% CI: 1.02, 1.42), ∑DEHP (IRR: 1.23; 95% CI: 1.00, 1.53), MEHHP (IRR: 1.20; 95% CI: 0.98, 1.48), MEOHP (IRR: 1.26; 95% CI: 1.03, 1.55) and MECPP (IRR: 1.23; 95% CI: 1.00, 1.51). Regarding the MnBP–externalizing behavior association, evidence of effect modification by sex was found (p-interaction: 0.01), with boys showing higher externalizing scores (IRR: 1.20; 95% CI: 1.02, 1.42) as abovementioned, and girls showing a nonsignificant opposite trend toward lower externalizing scores (IRR: 0.89; 95% CI: 0.74, 1.08) (Figure 1, Supplementary Table S2).

Each natural-log-unit increase in several urinary phthalate metabolites was associated with higher urinary BDNF levels among both boys and girls from the NACII-IT cohort (Figure 2 and Supplementary Table S3): MnBP (PC: 21.2%; 95% CI: 8.74, 35.1), MiBP (PC: 27.8%; 95% CI: 11.8, 46.2), MBzP (PC: 14.4%; 95% CI: 3.93, 25.9) and MEOHP (PC: 14.0%; 95% CI: 0.11, 29.9). When sex-stratified, boys showed stronger and significant associations with higher urinary BDNF levels for MEP, MnBP, MiBP, MBzP and MEOHP, while only MnBP was borderline-associated with higher BDNF levels among girls (Figure 2, Supplementary Table S3).

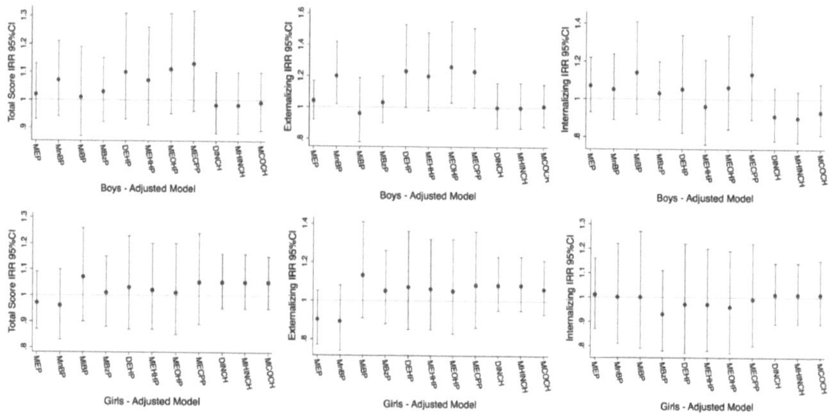

Figure 1. Adjusted cross-sectional associations between concurrent childhood urinary phthalate metabolite concentrations and CBCL scores in boys and girls from the NACII-IT cohort (n = 298). Note: Phthalate concentrations were natural log transformed, so the incidence rate ratio (IRR) and 95% confidence interval (95% CI) correspond to the multiplicative change in the probability of CBCL scores for each 2.7-fold increase in exposure biomarkers. Models adjusted for maternal education (low (0–2), middle (3–4) or high (>=5) ISCED classification), child body mass index (BMI) z-score, creatinine (log transformed), child age at CBCL assessment, season of urine collection (summer/autumn vs. winter/spring) and maternal IQ. The numeric results corresponding to this figure are available in Supplementary Table S2.

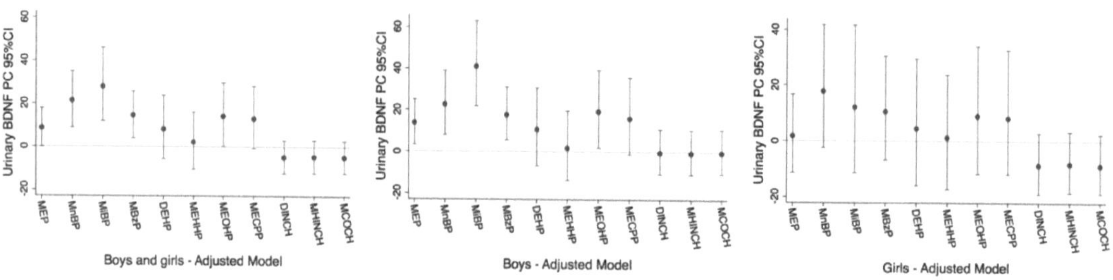

Figure 2. Adjusted cross-sectional associations between concurrent childhood urinary phthalate metabolite concentrations and urinary BDNF levels in children from the NACII-IT cohort (n = 299). Note: Phthalate concentrations were natural log transformed, so the percent change (PC) and 95% confidence interval (95% CI) correspond to the mean change in the urinary BDNF levels for each 2.7-fold increase in exposure biomarkers. Models adjusted for maternal education (low (0–2), middle (3–4) or high (>=5) ISCED classification), child body mass index (BMI) z-score, creatinine (log transformed), child sex, child age at urine collection and season of urine collection (summer/autumn vs. winter/spring). The numeric results corresponding to this figure are available in Supplementary Table S3.

When boys and girls from the NACII-IT cohort were considered together, each natural-log-unit increase in urinary BDNF levels was associated with higher internalizing problems (IRR: 1.15; 95% CI: 1.00, 1.32) but not with externalizing or total CBCL scores (Figure 3, Supplementary Table S4).

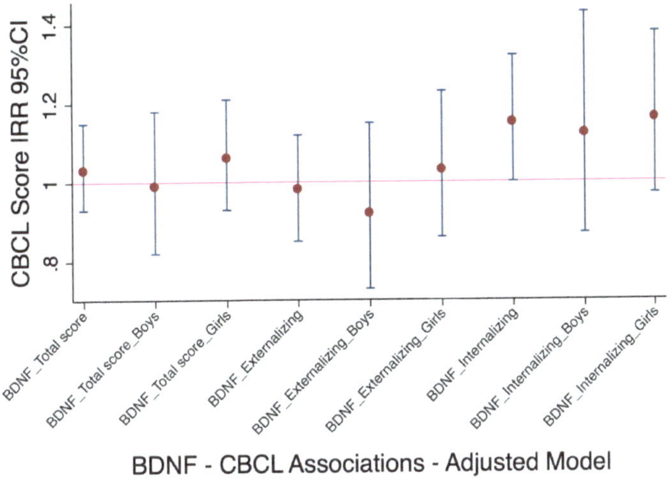

Figure 3. Adjusted cross-sectional associations between concurrent childhood urinary BDNF levels and CBCL scores in children from the NACII-IT cohort (n = 299). Note: BDNF levels were natural log transformed, so the incidence rate ratio (IRR) and 95% confidence interval (95% CI) correspond to the multiplicative change in the probability of CBCL scores for each 2.7-fold increase in urinary BDNF levels. Models adjusted for maternal education (low (0–2), middle (3–4) or high (>=5) ISCED classification), child body mass index (BMI) z-score, creatinine (log transformed), child age at CBCL assessment, season of urine collection (summer/autumn vs. winter/spring) and maternal IQ. The numeric results corresponding to this figure are available in Supplementary Table S4.

3.3. Phthalate–Urinary BDNF Associations in NACII-IT, PCB-SK, InAirQ-HU and NEBII-NO Cohorts (N = 1148)

When the four cohorts with available urinary BDNF measures were analyzed together, with the exception of DINCH metabolites, each natural-log-unit increase in most urinary phthalate metabolites was strongly associated with higher urinary BDNF levels among both boys and girls together (Figure 4 and Supplementary Table S5): MEP (PC: 4.2%; 95% CI: 0.20, 8.3), MnBP (PC: 19.0%; 95% CI: 12.7, 25.6), MiBP (PC: 14.4%; 95% CI: 8.6, 20.5), ∑DEHP (PC: 13.0%; 95% CI: 6.8, 19.4), MEHHP (PC: 13.5%; 95% CI: 7.7, 19.7), MEOHP (PC: 17.5%; 95% CI: 11.6, 23.8) and MECPP (PC: 9.4%; 95% CI: 3.5, 15.7). Sex-stratified models showed that, among boys, all the previous associations were strengthened (Figure 4 and Supplementary Table S5), while in girls, only MnBP and MEOHP were positively associated. Among girls, inverse associations were also observed between higher urinary DINCH metabolites and lower urinary BDNF levels (Figure 4 and Supplementary Table S5): DINCH (PC: −8.4%; 95% CI: −14.3, −2.2), MHINCH (PC: −11.3%; 95% CI: −16.6, −5.6) and MCOCH (PC: −8.4%; 95% CI: −14.3, −2.2).

3.4. Phthalate–Serum BDNF Associations in the CRP-SLO Cohort (N = 124)

In boys but not girls of the CRP-SLO cohort, higher urinary concentrations of ∑DINCH and their metabolites were associated with lower serum BDNF levels: ∑DINCH (PC: −8.8%; 95% CI: −16.7, −0.3), MHINCH (PC: −7.9; 95% CI: −15.6, 0.4) and MCOCH (PC: −10.1%; 95% CI: −18.0, −1.5) (Supplementary Table S6). In girls, a borderline-significant trend toward lower serum BDNF levels (PC: −6.83%; 95% CI: −14.0, 0.98) was also noted in relation to higher urinary MiBP concentrations (Supplementary Table S6).

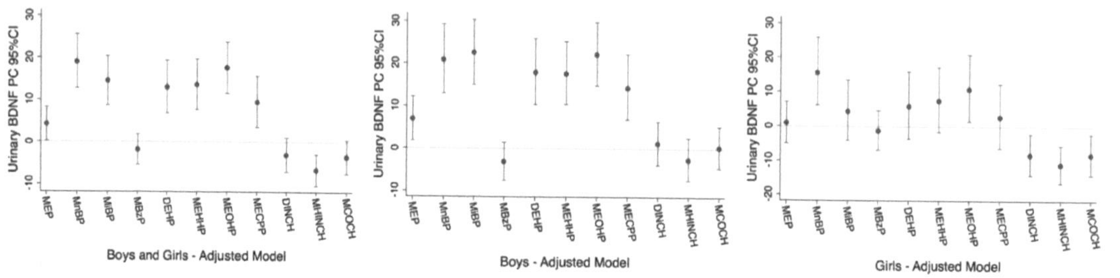

Figure 4. Adjusted cross-sectional associations between concurrent childhood urinary phthalate metabolite concentrations and urinary BDNF levels in all children from the NACII-IT, PCB-SK, InAirQ-HU and NEBII-NO cohorts (N = 1148). Note: Phthalate concentrations were natural log transformed, so the percent change (PC) and 95% confidence interval (95% CI) correspond to the mean change in the urinary BDNF levels for each 2.7-fold increase in exposure biomarkers. Models adjusted for maternal education (low (0–2), middle (3–4) or high (>=5) ISCED classification), child body mass index (BMI) z-score, creatinine (log transformed), child sex, child age at urine collection and season of urine collection (summer/autumn vs. winter/spring). The ∑DINCH measure was estimated without considering the NEBII-NO cohort, which only had data on the MHINCH but not the MCOCH metabolite. The numeric results corresponding to this figure are available in Supplementary Table S4.

3.5. Phthalate's Mixture Associations

Regarding CBCL scores in the NACII-IT cohort, WQS regression models showed that each tertile increase in the phthalate mixture was suggestively associated with higher externalizing scores in the whole population (IRR = 1.15; 95% CI: 0.97, 1.36), mostly driven by boys (IRR = 1.20; 95% CI: 0.96, 1.49) (Table 2). Both in the overall population and among boys, MEOHP, MnBP and MCOCH metabolites were identified as the main contributors to the mixture effect (Supplementary Figure S2).

Table 2. Mixture associations using weighted quantile sum (WQS) regression.

Outcome	All	Boys	Girls
	IRR or PC (95% CI)	IRR or PC (95% CI)	IRR or PC (95% CI)
Total CBCL score	1.09 (0.96, 1.23)	1.07 (0.91, 1.25)	1.11 (0.90, 1.35)
Externalizing score	1.15 (0.97, 1.36)	1.20 (0.96, 1.49)	1.11 (0.88, 1.39)
Internalizing score	1.03 (0.90, 1.17)	1.04 (0.85, 1.29)	1.00 (0.76, 1.30)
Urinary BDNF levels	25.9 (17.6, 34.7)	36.0 (24.3, 48.9)	18.6 (7.92, 30.5)

Note: Sample size for CBCL was 298, 149 for boys and 149 for girls, respectively. Sample size for urinary BDNF was 1148 in total, 570 and 578 for boys and girls, respectively (since NEBII-NO participants had data on MHINCH but not MCOCH, mixture models only included the MHINCH metabolite). Mixture models between phthalate exposure and serum BDNF in the CRP-SLO cohort were not computed due to the limited sample size available. The incidence rate ratio (IRR) or percentage change (PC) represents the mean change in CBCL score or BDNF levels per tertile increase of the mixture, with its related 95% confidence interval (95% CI). The weight contributed by each chemical compound to the most relevant mixture associations can be consulted in Supplementary Figure S2. Mixture models were adjusted for the same set of covariates considered in single-pollutant analyses.

Regarding urinary BDNF levels in the four cohorts together, WQS regression models showed that each tertile increase in the phthalate mixture was significantly associated with higher BDNF levels in the whole population (PC = 25.9%; 95% CI: 17.6, 34.7), also in girls (PC = 18.6%; 95% CI: 7.92, 30.5) and especially among boys (PC = 36.0%; 95% CI: 24.3, 48.9). The main contributors to the mixture effect on urinary BDNF levels were the metabolites MEOHP, MEP, MnBP and MiBP (Supplementary Figure S2).

3.6. Sensitivity Analysis

When phthalate–BDNF associations were stratified by cohort, we observed higher urinary BDNF levels in response to higher phthalate metabolites in the NACII-IT cohort, followed by boys of the PCB-SK cohort (Figure 2). Mostly null associations were observed for the NEBII-NO and InAirQ-HU cohorts (Supplementary Figure S3).

Very similar results were observed for the five approaches used to account for urinary creatinine levels (Supplementary Figure S4), indicating that the positive associations observed between urinary phthalates and BDNF concentrations in children from the four European cohorts combined were robust to all types of creatinine adjustments. Indeed, the only associations that varied were the negative associations between DINCH metabolites and urinary BDNF levels, which were only observed in models that accounted for creatinine as a covariate.

4. Discussion

This cross-sectional study in European children including five HBM4EU aligned studies showed that (1) in the NACII-IT cohort, higher urinary concentrations of MnBP and DEHP metabolites were associated with externalizing problems among boys but not girls, while the phthalate mixture was suggestively associated with higher externalizing scores in both the whole population and boys; (2) most urinary phthalate metabolites, individually or as a mixture, were associated with higher urinary BDNF protein levels in NACII-IT, PCB-SK, InAirQ-HU and NEB-NO cohorts combined, with stronger associations found in boys; (3) higher urinary DINCH concentrations were associated with lower serum BDNF levels in boys of the CRP-SLO cohort and (4) higher urinary BDNF levels were associated with more internalizing—but not externalizing—problems among both boys and girls of the NACII-IT cohort.

4.1. Phthalate–CBCL Associations

Our observation that urinary concentrations of MnBP and DEHP metabolites, individually and as a mixture, were associated with higher externalizing problems in boys is consistent with previous epidemiological studies. For example, higher prenatal exposure to DEHP metabolites was associated with more externalizing problems among boys in one epidemiological study [68], and in both boys and girls in others [69,70]. Studies have also observed that higher maternal urinary DEHP concentrations during pregnancy were associated with more attention-deficit hyperactivity disorder (ADHD) symptoms among boys [71]. Breastmilk DEHP exposure was also associated with higher infant inattention and hyperactivity symptoms [72]. Higher prenatal exposure to MnBP was also associated with more externalizing [73], as well as with social and emotional problems among boys [74] and total behavioral problems in boys [75]. Additionally, the few postnatal studies performed to date have shown that concurrent exposure to DnBP metabolites was associated with total, emotional and hyperactivity behavioral problems in 7-year-old boys and girls [76] and that MnBP and DEHP metabolites were cross-sectionally associated with more social problems in boys [77]. In children, DEHP exposure has also been cross-sectionally associated with a higher risk of ADHD [78]. Notwithstanding, other studies reported null or contradictory associations [79,80]. Despite heterogeneity in exposure timing, sampling protocols and the different neurobehavioral tests and domains assessed, the epidemiological literature generally supports that developmental exposure to MnBP and DEHP metabolites is linked to poorer neurobehavioral scores among children.

Toxicological studies in rodents have also revealed effects of perinatal DEHP exposure on exploratory, social and anxiety behaviors [81–87]. Toxicological data for the effect of MnBP and its parent compound on the brain and behavior are scarce but also point in a similar direction [88].

Regarding possible modes of action, both MnBP and DEHP have shown anti-androgenic activities in both toxicological [87–91] and epidemiological studies [92,93], providing a potential clue as to why associations were observed in boys but not girls. Notwithstanding, it should be noted that these phthalates can also act through other modes of action including thyroid disruption, oxidative stress, peroxisome proliferator-activated receptors (PPARs) and estrogen pathways [25,94]. Overall, scientific findings including ours support that MnBP and DEHP metabolites, individually and as a mixture, may have a deleterious impact on children's neurodevelopment, calling for further policies aiming at reducing children's exposure to phthalates [19,95].

4.2. Urinary Phthalates and Serum BDNF Levels

The inverse associations found between exposure to DINCH metabolites and serum BDNF levels among boys of the CRP-SLO cohort suggest that exposure to this non-phthalate replacement may be related to lower systemic levels of BDNF. Since lower serum BDNF levels have been linked to a variety of neuropsychiatric disorders including depression and anxiety [96,97], further investigation of this phthalate substitute is warranted. Given the limited sample size and that this is the first human study investigating the DINCH–BDNF association, additional studies are needed to validate these results.

4.3. Urinary Phthalate–Urinary BDNF Associations—Possible "Urine Concentration Bias"?

This is the first epidemiological study that investigated phthalate exposure in relation to the neurological effect biomarker BDNF. Our results indicate consistently higher BDNF protein levels in urine in response to most phthalate metabolites. This association was additionally observed in phthalate mixture models, especially driven by boys. Since the analysis in the NACII-IT cohort also showed that higher urinary BDNF levels were predictive of more internalizing problems among boys and girls analyzed together, urinary BDNF concentrations may be linked to poorer neurobehavior. However, it is not clear whether higher urinary BDNF concentrations reflect higher or lower systemic BDNF levels, and even more complex, both low and high serum BDNF levels have been linked to altered neurobehavior, with lower levels associated with anxiety and depression [96,97], while high levels are linked to ADHD and autism spectrum disorders [98,99]. Caution is needed when interpreting associations with urinary BDNF levels because of the fact that most—although not all—phthalate metabolites showed strong and consistent associations toward higher urinary BDNF levels (a result not frequently observed in the environmental epidemiology field), that could also reflect a potential and unintentional noncausal artifact.

An alternative noncausal explanation could be that urine samples with a higher concentration (due to longer storage in the bladder or lower hydration status) in general would tend to correlate with higher phthalate metabolites and with higher physiological urinary waste products, including urinary BDNF levels. To test this hypothesis, we performed a post hoc analysis checking Spearman's correlations between urinary creatinine in relation to phthalate/DINCH metabolites and BDNF levels (Supplementary Table S7). Results showed that urinary creatinine levels were correlated positively and significantly ($p < 0.001$) with urinary BDNF and all urinary phthalate/DINCH metabolites examined, with rho values ranging from 0.26 to 0.59 (Supplementary Table S6). This post hoc analysis supports the hypothesis that, in spot urine samples, urinary concentration tends to be correlated with higher exposure (phthalate/DINCH metabolites) and effect biomarker (BDNF) levels. This led us to formulate the following question: in studies measuring exposure and effect biomarkers in the same spot urine sample, could urine concentration favor the appearance of noncausal associations toward a positive direction, even when standardizing and adjusting models for creatinine such as in the current work?

Until now, most previous studies in the field have assessed exposure to phthalates, phenols and metals in relation to oxidative stress and kidney damage biomarkers measured in urine, in most cases finding positive associations. Examples include the review by Steffensen et al. reporting that urinary bisphenol A exposure was associated with higher urinary oxidative stress biomarkers in most of the studies evaluated [100]. Urinary metals have also been consistently linked to higher urinary oxidative stress and renal biomarkers [101–103], even showing contradictory associations between antioxidant essential metals such as selenium and higher urinary oxidative stress biomarkers [103]. Urinary phthalate metabolites have also been associated with higher oxidative stress in many studies [104–107]. Noteworthy, Davalos et al. were able to compare associations between urinary phthalates and oxidative stress biomarkers in urine samples collected at the same time or some days after exposure was assessed. In the cross-sectional evaluation, they found that many phthalate metabolites were associated with increased oxidative biomarkers. In the short-term longitudinal analysis, they still found some associations toward increased oxidative stress, but the magnitude of associations was importantly attenuated: for example, percent increase per interquartile range (IQR) difference in individual phthalate metabolites and the 8-iso-PGF2α metabolite ranged from 16% to 63% in cross-sectional models but only 2.9% to 9.8% in longitudinal models [104]. Therefore, these results suggest that although phthalate exposure may increase oxidative stress as shown by experimental laboratory models [108,109], human studies assessing phthalate metabolites and oxidative stress biomarkers in the same urine sample may considerably overestimate the magnitude of associations or even produce spurious positive associations. Against this alternative noncausal hypothesis is the fact that, in the current study, positive associations were not systematically observed in all cohorts, and even a few negative associations were observed in one of the studies examined (InAirQ-HU, Supplementary Figure S3), although this result was an exception compared to findings in the remaining cohorts.

Overall, our study raises a concern when exposure and effect biomarkers are measured in the same urine sample because urine concentration may tend to exert an effect that statistical methods are not able to disentangle, as shown in our dataset by a comparison of five different methods to account for urinary creatinine levels (Supplementary Figure S4). While additional methodological studies are needed to achieve a final response to this question, it is advisable that future biomarker studies, when possible, measure urinary exposure and effect biomarkers in urine samples separated in time [104] or measure the effect biomarkers of interest in blood or alternative matrices to urine in order to avoid this potential "urine concentration bias".

4.4. BDNF Biomarkers in Relation to CBCL

Regardless of whether the observed urinary phthalate–urinary BDNF associations are causal or not, higher urinary BDNF levels appeared as a predictor of internalizing problems in both boys and girls of the NACII-IT cohort. This is relevant since current evidence supports that (a) BDNF can be reliably measured in urine [53], and (b) it has a physiological or pathophysiological meaning worthy of further examination in future epidemiological studies. Of note, we initially aimed to assess serum protein BDNF levels and blood BDNF DNA methylation in addition to urinary levels, but it was not possible due to the lack of availability of these biological matrices in the HBM4EU aligned studies in children (with the exception of serum BDNF levels in the CRP-SLO cohort). Measuring BDNF in urine has the advantage of being a noninvasive matrix widely available in human biomonitoring studies. Notwithstanding, in a previous HBM4EU pilot study using the INMA (Environment and Childhood) cohort from Granada (Southern Spain), we showed that blood BDNF DNA methylation was altered in response to bisphenol A, heavy metals and nonpersistent pesticide exposure [28,30,31], being apparently more useful compared to serum or urine BDNF protein levels. In the case of bisphenol A, BDNF DNA methylation levels even mediated about a third of the association observed between urinary bisphenol A concentrations measured at 9 years and behavior assessed using the CBCL test at 15 years of age [28]. More evidence is needed to identify which BDNF biomarkers show the best predictive potential for neurobehavioral outcomes; meanwhile, a combination of blood- and urine-based BDNF biomarkers may be advisable as this strategy offers the highest sensitivity to detect the potential effects of environmental chemicals on the nervous system. Finally, given that in our study urinary BDNF levels were associated with internalizing but not externalizing problems (the scale associated with MnBP and DEHP metabolites in boys), we did not perform mediation analyses.

4.5. Strengths and Limitations

Among the strengths of this study is the inclusion of five cohorts belonging to different European countries, which better reflected the differences in sociodemographic and exposure profiles in Europe. Chemical measurements were performed under the rigorous quality controls and interlaboratory comparisons within the HBM4EU project [50], promoting an optimal exposure measurement. Since urinary BDNF measurements were also previously validated and performed in the same laboratory [53], the acquisition of exposure and effect data in this study was comparable and pre-harmonized. Furthermore, this is the first study to evaluate associations between phthalate metabolites and BDNF in children, using urine as a noninvasive and readily available matrix. Among the limitations, we must highlight the cross-sectional design that prevented us from establishing the temporality of associations and is more prone to possible reverse causality issues. In addition, we did not correct for multiple comparisons since this approach may be counterproductive due to a disproportionate increase in type 2 errors [110] and because our interpretation was not based solely on statistical significance but also considered previous toxicological and epidemiological data. Although the possibility of some chance findings cannot be fully excluded, the observed associations between DnBP/DEHP metabolites and externalizing problems were in line with the weight of evidence, which increases our confidence in these results. Although data for phthalates and BDNF were available for 1148 children, behavioral data were only available for 298 children from the NACII-IT cohort. Phthalate and DINCH metabolites are short-lived compounds with rapid excretion and high variability over days/weeks. Therefore, the use of one spot urine sample may have led to exposure misclassification. However, in general, this would tend to result in attenuation bias [111–114], that is, a bias toward null findings, meaning that we would be underestimating rather than overestimating the associations with phthalate/DINCH metabolites. Finally, despite models were controlled for relevant covariates, we cannot discard possible unmeasured or residual confounders such as diet quality or physical exercise.

5. Conclusions

In European children aged 7–12 years, exposure to urinary concentrations of DnBP and DEHP metabolites were cross-sectionally associated with higher externalizing problems in the NACII-IT cohort, especially among boys. Higher urinary concentrations of the non-phthalate replacement DINCH were associated with lower serum BDNF levels among children of the CRP-SLO cohort. Although most urinary phthalate metabolites were associated with higher urinary BDNF protein levels when four out of the five cohorts were examined together, we identified the possibility of a "urine concentration bias" that could also explain the associations observed in noncausal terms, so these results should be carefully interpreted. Finally, higher urinary BDNF levels were associated with more internalizing problems among boys and girls from the NACII-IT cohort, suggesting that urinary BDNF levels could be used as a noninvasive biomarker to predict behavioral alterations in children. Given this is the first observational study investigating phthalate exposure in relation to BDNF biomarkers, future studies are needed to validate the associations observed.

Supplementary Materials: The following supporting information can be downloaded at https://www.mdpi.com/article/10.3390/toxics12090642/s1, Figure S1: Spearman's correlation heatplot of creatinine-standardized phthalate metabolite concentrations, Figure S2: Relative weights of phthalates for mixture associations, Figure S3: Sensitivity analysis showing adjusted cross-sectional associations between concurrent childhood urinary phthalate metabolite concentrations and urinary BDNF levels in children of the PCB-SK, InAirQ-HU and NEBII-NO cohorts, Figure S4: Sensitivity analyses comparing five different methods of creatinine adjustment, Table S1: Adjusted cross-sectional associations between concurrent childhood urinary phthalate metabolite concentrations and CBCL scores among all children in the NACII-IT cohort (n = 298), Table S2: Adjusted cross-sectional associations between concurrent childhood urinary phthalate metabolite concentrations and CBCL scores in boys and girls separately in the NACII-IT cohort (n = 298), Table S3: Adjusted cross-sectional associations between concurrent childhood urinary phthalate metabolite concentrations and urinary BDNF levels in children of the NACII-IT cohort (n = 299), Table S4: Adjusted cross-sectional associations between concurrent childhood urinary BDNF levels and CBCL scores in children of the NACII-IT cohort (n = 299), Table S5: Adjusted cross-sectional associations between concurrent childhood urinary phthalate metabolite concentrations and urinary BDNF levels in children of the NACII-IT, PCB-SK and InAirQ cohorts (N = 1148), Table S6: Adjusted cross-sectional associations between concurrent childhood urinary phthalate metabolite concentrations and serum BDNF levels in children of the CRP-SLO cohort (N = 124), Table S7: Spearman's correlations between urinary creatinine and BDNF and phthalate/DINCH metabolite levels in NACII-IT, PCB-SK and InAirQ cohorts combined (N = 1148).

Author Contributions: Conceptualization, V.M.; data curation, M.M., L.R., Ľ.P.M., L.F., T.S., R.K., L.S.H., B.L., J.S.T., T.K., S.P.-D., and S.R.; formal analysis, L.E.-M., A.O.-M., R.M.P., A.K.S., L.S.H., G.J., S.V., H.J., R.G., I.P., B.K., and M.E.; funding acquisition, M.K.-G. and M.F.F.; investigation, A.P.-C., J.L.M.-R., and M.F.F.; methodology, E.S.-F., L.E.-M., V.R., and V.M.; project administration, E.G. and S.R.; resources, F.B. and S.P.-D.; supervision, G.S., M.F.F., and V.M.; writing—original draft, E.S.-F., L.E.-M., and V.M.; writing—review and editing, A.O.-M., A.P.-C., J.L.M.-R., R.M.P., F.B., V.R., M.M., L.R., Ľ.P.M., L.F., T.S., R.K., A.K.S., L.S.H., B.L., J.S.T., T.K., G.J., S.V., H.J., R.G., I.P., B.K., M.E., S.P.-D., M.K.-G., S.R., E.G., G.S., and M.F.F. All authors have read and agreed to the published version of the manuscript.

Funding: This work was carried out under the framework of the European Partnership for the Assessment of Risks from Chemicals (PARC) and Human Biomonitoring for Europe (HBM4EU) projects and has received funding from the European Union's Horizon Europe research and innovation programs under grant agreement no. 101057014 and no. 733032, respectively. The views and opinions expressed are, however, those of the authors only and do not necessarily reflect those of the European Union or the Health and Digital Executive Agency. Neither the European Union nor the granting authority can be held responsible for them. This work also received co-funding from the authors' organizations. PCB cohort: The PCB cohort was funded by the Slovak Research and Development Agency, project no. APVV-0571–12, and by the Ministry of Health of the Slovak Republic, project no. 2014/47-SZU-11. The development of methods and laboratory measurements were also co-funded by the Slovak Research and Development Agency, project no. APVV-20-0462. The Norwegian Institute of Public Health (NIPH) has contributed to the funding of the Norwegian Environmental Biobank (NEB). The laboratory measurements have partly been funded by the Research Council of Norway through research projects (275903 and 268465). Vicente Mustieles was supported by a Sara Borrell postdoctoral research contract (CD22/00176), granted by Instituto de Salud Carlos III (Spain) and NextGenerationEU funds. Alicia Olivas-Martínez was supported by a predoctoral contract (grant no. FI21/00236) granted by Instituto de Salud Carlos III (Spain).

Institutional Review Board Statement: The specific ethics committee approvals of each of the cohorts included in our study are as follows: NEBII cohort: approved by The Regional Committees for Medical and Health Research Ethics in Norway, Reference: 2015/1340; InAirQ cohort: approved by the Medical Research Council of Hungary (registration number: 46399-4/2017/EKU); PCB cohort: approved by the ethical committee of the Slovak, Medical University in Bratislava (No. 01/2018 and 02/2019); SLO-CRP cohort: approved by the National Medical Ethics Committee, Republic of Slovenia (NMEC, number of accordance: 0120-118/2017/3); NACII cohort: approved by the Ethics Committees of the University of Udine and of the Institute for Maternal and Child Health—IRCCS Burlo Garofolo, Trieste (Italy), (CE/V-70—05/02/2007; CE/V-109-12/04/2010).

Informed Consent Statement: All the participant studies obtained the respective ethic committee authorizations before recruiting subjects, and informed consent was obtained from the caregivers of all subjects involved in the current work. Studies were conducted in accordance with the Declaration of Helsinki.

Data Availability Statement: Metadata of the five cohorts are available from IPCHEM, the European Commission's Information Platform for Chemical Monitoring. The summary statistics (percentiles P5, P10, P25, P50, P75, P90, P95) of the exposure biomarkers are made available on the openly accessible online European HBM dashboard (https://hbm.vito.be/eu-hbm-dashboard, accessed on 28 August 2024) and IPCHEM (https://ipchem.jrc.ec.europa.eu/, accessed on 28 August 2024), where they can be accessed. Data sharing of individual-level data is possible upon request (https://hbm.vito.be/peh-data-platform, accessed on 28 August 2024).

Acknowledgments: The authors acknowledge all of the participating families. We also acknowledge the laboratory technical assistance provided by Raquel Quesada-Jiménez at the Centre for Biomedical Research, University of Granada, 18016 Granada, Spain. This article will form part of the doctoral thesis developed by Lydia Espín-Moreno in the context of the "Clinical Medicine and Public Health Program" of the University of Granada.

Conflicts of Interest: The authors declare they have no conflicts of interest.

References

1. Koch, H.M.; Calafat, A.M. Human Body Burdens of Chemicals Used in Plastic Manufacture. *Philos. Trans. R. Soc. B Biol. Sci.* **2009**, *364*, 2063–2078. [CrossRef] [PubMed]
2. Coltro, L.; Pitta, J.B.; da Costa, P.A.; Fávaro Perez, M.Â.; de Araújo, V.A.; Rodrigues, R. Migration of Conventional and New Plasticizers from PVC Films into Food Simulants: A Comparative Study. *Food Control* **2014**, *44*, 118–129. [CrossRef]
3. Wittassek, M.; Koch, H.M.; Angerer, J.; Brüning, T. Assessing Exposure to Phthalates—The Human Biomonitoring Approach. *Mol. Nutr. Food Res.* **2011**, *55*, 7–31. [CrossRef] [PubMed]
4. Qadeer, A.; Kirsten, K.L.; Ajmal, Z.; Xingru, Z. Rebuttal to Comment on "Alternative Plasticizers as Emerging Global Environmental and Health Threat: Another Regrettable Substitution?" Focus on DINCH as an Example. *Environ. Sci. Technol.* **2022**, *56*, 5294–5297. [CrossRef]
5. Mustieles, V.; Arrebola, J.P.; Porta, M. From Old Pollutants to the Regulation of Bisphenol A: Lessons Learned for Health Promotion and Disease Prevention. *Prev. Med.* **2023**, *169*, 107460. [CrossRef] [PubMed]
6. Lemke, N.; Murawski, A.; Lange, R.; Weber, T.; Apel, P.; Dębiak, M.; Koch, H.M.; Kolossa-Gehring, M. Substitutes Mimic the Exposure Behaviour of REACH Regulated Phthalates—A Review of the German HBM System on the Example of Plasticizers. *Int. J. Hyg. Environ. Health* **2021**, *236*, 113780. [CrossRef]
7. Schütze, A.; Kolossa-Gehring, M.; Apel, P.; Brüning, T.; Koch, H.M. Entering Markets and Bodies: Increasing Levels of the Novel Plasticizer Hexamoll® DINCH® in 24 h Urine Samples from the German Environmental Specimen Bank. *Int. J. Hyg. Environ. Health* **2014**, *217*, 421–426. [CrossRef]
8. Schettler, T.; Skakkebæk, N.E.; De Kretser, D.; Leffers, H. Human Exposure to Phthalates via Consumer Products. *Int. J. Androl.* **2006**, *29*, 134–139. [CrossRef]
9. Philippat, C.; Rolland, M.; Lyon-Caen, S.; Pin, I.; Sakhi, A.K.; Sabaredzovic, A.; Thomsen, C.; Slama, R. Pre- and Early Post-Natal Exposure to Phthalates and DINCH in a New Type of Mother-Child Cohort Relying on within-Subject Pools of Repeated Urine Samples. *Environ. Pollut.* **2021**, *287*, 117650. [CrossRef]
10. Weise, P.; Apel, P.; Kolossa-Gehring, M. [Human Biomonitoring for Europe (HBM4EU)-First Insights into the Results of the Initiative]. *Bundesgesundheitsblatt Gesundheitsforschung Gesundheitsschutz* **2022**, *65*, 936–939. [CrossRef]
11. Shu, H.; Jönsson, B.A.; Gennings, C.; Svensson, Å.; Nånberg, E.; Lindh, C.H.; Knutz, M.; Takaro, T.K.; Bornehag, C.G. Temporal Trends of Phthalate Exposures during 2007-2010 in Swedish Pregnant Women. *J. Expo. Sci. Environ. Epidemiol.* **2018**, *28*, 437–447. [CrossRef] [PubMed]
12. Rodriguez Martin, L.; Gilles, L.; Helte, E.; Åkesson, A.; Tägt, J.; Covaci, A.; Sakhi, A.K.; Van Nieuwenhuyse, A.; Katsonouri, A.; Andersson, A.M.; et al. Time Patterns in Internal Human Exposure Data to Bisphenols, Phthalates, DINCH, Organophosphate Flame Retardants, Cadmium and Polyaromatic Hydrocarbons in Europe. *Toxics* **2023**, *11*, 819. [CrossRef] [PubMed]
13. Vogel, N.; Frederiksen, H.; Lange, R.; Jørgensen, N.; Koch, H.M.; Weber, T.; Andersson, A.M.; Kolossa-Gehring, M. Urinary Excretion of Phthalates and the Substitutes DINCH and DEHTP in Danish Young Men and German Young Adults between 2000 and 2017—A Time Trend Analysis. *Int. J. Hyg. Environ. Health* **2023**, *248*, 114080. [CrossRef] [PubMed]
14. Legler, J. Are Brominated Flame Retardants Endocrine Disruptors? *Environ. Int.* **2003**, *29*, 879–885. [CrossRef]
15. Breous, E.; Wenzel, A.; Loos, U. The Promoter of the Human Sodium/Iodide Symporter Responds to Certain Phthalate Plasticisers. *Mol. Cell. Endocrinol.* **2005**, *244*, 75–78. [CrossRef] [PubMed]
16. Dong, J.; Cong, Z.; You, M.; Fu, Y.; Wang, Y.; Wang, Y.; Fu, H.; Wei, L.; Chen, J. Effects of Perinatal Di (2-Ethylhexyl) Phthalate Exposure on Thyroid Function in Rat Offspring. *Environ. Toxicol. Pharmacol.* **2019**, *67*, 53–60. [CrossRef]
17. Wenzel, A.; Franz, C.; Breous, E.; Loos, U. Modulation of Iodide Uptake by Dialkyl Phthalate Plasticisers in FRTL-5 Rat Thyroid Follicular Cells. *Mol. Cell. Endocrinol.* **2005**, *244*, 63–71. [CrossRef]
18. La Merrill, M.A.; Vandenberg, L.N.; Smith, M.T.; Goodson, W.; Browne, P.; Patisaul, H.B.; Guyton, K.Z.; Kortenkamp, A.; Cogliano, V.J.; Woodruff, T.J.; et al. Consensus on the Key Characteristics of Endocrine-Disrupting Chemicals as a Basis for Hazard Identification. *Nat. Rev. Endocrinol.* **2020**, *16*, 45–57. [CrossRef]
19. Engel, S.M.; Patisaul, H.B.; Brody, C.; Hauser, R.; Zota, A.R.; Bennet, D.H.; Swanson, M.; Whyatt, R.M. Neurotoxicity of Ortho-Phthalates: Recommendations for Critical Policy Reforms to Protect Brain Development in Children. *Am. J. Public Health* **2021**, *111*, 687–695. [CrossRef]
20. Bakoyiannis, I.; Kitraki, E.; Stamatakis, A. Endocrine-Disrupting Chemicals and Behaviour: A High Risk to Take? *Best Pract. Res. Clin. Endocrinol. Metab.* **2021**, *35*, 101517. [CrossRef]

21. Palanza, P.; Paterlini, S.; Brambilla, M.M.; Ramundo, G.; Caviola, G.; Gioiosa, L.; Parmigiani, S.; vom Saal, F.S.; Ponzi, D. Sex-Biased Impact of Endocrine Disrupting Chemicals on Behavioral Development and Vulnerability to Disease: Of Mice and Children. *Neurosci. Biobehav. Rev.* **2021**, *121*, 29–46. [CrossRef]
22. Ejaredar, M.; Nyanza, E.C.; Ten Eycke, K.; Dewey, D. Phthalate Exposure and Childrens Neurodevelopment: A Systematic Review. *Environ. Res.* **2015**, *142*, 51–60. [CrossRef] [PubMed]
23. Radke, E.G.; Braun, J.M.; Nachman, R.M.; Cooper, G.S. Phthalate Exposure and Neurodevelopment: A Systematic Review and Meta-Analysis of Human Epidemiological Evidence. *Environ. Int.* **2020**, *137*, 105408. [CrossRef] [PubMed]
24. Diamanti-Kandarakis, E.; Bourguignon, J.P.; Giudice, L.C.; Hauser, R.; Prins, G.S.; Soto, A.M.; Zoeller, R.T.; Gore, A.C. Endocrine-Disrupting Chemicals: An Endocrine Society Scientific Statement. *Endocr. Rev.* **2009**, *30*, 293–342. [CrossRef] [PubMed]
25. Benjamin, S.; Masai, E.; Kamimura, N.; Takahashi, K.; Anderson, R.C.; Faisal, P.A. Phthalates Impact Human Health: Epidemiological Evidences and Plausible Mechanism of Action. *J. Hazard. Mater.* **2017**, *340*, 360–383. [CrossRef] [PubMed]
26. Hlisníková, H.; Petrovičová, I.; Kolena, B.; Šidlovská, M.; Sirotkin, A. Effects and Mechanisms of Phthalates' Action on Neurological Processes and Neural Health: A Literature Review. *Pharmacol. Rep.* **2021**, *73*, 386–404. [CrossRef] [PubMed]
27. Mustieles, V.; D'Cruz, S.C.; Couderq, S.; Rodríguez-Carrillo, A.; Fini, J.-B.; Hofer, T.; Steffensen, I.-L.; Dirven, H.; Barouki, R.; Olea, N.; et al. Bisphenol A and Its Analogues: A Comprehensive Review to Identify and Prioritize Effect Biomarkers for Human Biomonitoring. *Environ. Int.* **2020**, *144*, 105811. [CrossRef]
28. Mustieles, V.; Rodríguez-Carrillo, A.; Vela-Soria, F.; D'Cruz, S.C.; David, A.; Smagulova, F.; Mundo-López, A.; Olivas-Martínez, A.; Reina-Pérez, I.; Olea, N.; et al. BDNF as a Potential Mediator between Childhood BPA Exposure and Behavioral Function in Adolescent Boys from the INMA-Granada Cohort. *Sci. Total Environ.* **2022**, *803*, 150014. [CrossRef]
29. Klein, A.B.; Williamson, R.; Santini, M.A.; Clemmensen, C.; Ettrup, A.; Rios, M.; Knudsen, G.M.; Aznar, S. Blood BDNF concentrations reflect brain-tissue BDNF levels across species. *Int. J. Neuropsychopharmacol.* **2011**, *14*, 347–353. [CrossRef]
30. Rodríguez-Carrillo, A.; DĆruz, S.C.; Mustieles, V.; Suárez, B.; Smagulova, F.; David, A.; Peinado, F.; Artacho-Cordón, F.; López, L.C.; Arrebola, J.P.; et al. Exposure to Non-Persistent Pesticides, BDNF, and Behavioral Function in Adolescent Males: Exploring a Novel Effect Biomarker Approach. *Environ. Res.* **2022**, *211*, 113115. [CrossRef]
31. Rodríguez-Carrillo, A.; Mustieles, V.; D'Cruz, S.C.; Legoff, L.; Gil, F.; Olmedo, P.; Reina-Pérez, I.; Mundo, A.; Molina, M.; Smagulova, F.; et al. Exploring the Relationship between Metal Exposure, BDNF, and Behavior in Adolescent Males. *Int. J. Hyg. Environ. Health* **2022**, *239*, 113877. [CrossRef] [PubMed]
32. Lee, S.M.; Jeon, S.; Jeong, H.J.; Kim, B.N.; Kim, Y. Dibutyl Phthalate Exposure during Gestation and Lactation in C57BL/6 Mice: Maternal Behavior and Neurodevelopment in Pups. *Environ. Res.* **2020**, *182*, 109025. [CrossRef]
33. Ponsonby, A.L.; Symeonides, C.; Vuillermin, P.; Mueller, J.; Sly, P.D.; Saffery, R. Epigenetic Regulation of Neurodevelopmental Genes in Response to in Utero Exposure to Phthalate Plastic Chemicals: How Can We Delineate Causal Effects? *Neurotoxicology* **2016**, *55*, 92–101. [CrossRef] [PubMed]
34. HBM4EU—Science and Policy for a Healthy Future. Available online: https://www.hbm4eu.eu/ (accessed on 19 January 2024).
35. Gilles, L.; Govarts, E.; Martin, L.R.; Andersson, A.M.; Appenzeller, B.M.R.; Barbone, F.; Castaño, A.; Coertjens, D.; Den Hond, E.; Dzhedzheia, V.; et al. Harmonization of Human Biomonitoring Studies in Europe: Characteristics of the HBM4EU-Aligned Studies Participants. *Int. J. Environ. Res. Public Health* **2022**, *19*, 6787. [CrossRef]
36. Gilles, L.; Govarts, E.; Rambaud, L.; Vogel, N.; Castaño, A.; Esteban López, M.; Rodriguez Martin, L.; Koppen, G.; Remy, S.; Vrijheid, M.; et al. HBM4EU Combines and Harmonises Human Biomonitoring Data across the EU, Building on Existing Capacity—The HBM4EU Survey. *Int. J. Hyg. Environ. Health* **2021**, *237*, 113809. [CrossRef] [PubMed]
37. Govarts, E.; Gilles, L.; Rodriguez Martin, L.; Santonen, T.; Apel, P.; Alvito, P.; Anastasi, E.; Andersen, H.R.; Andersson, A.M.; Andryskova, L.; et al. Harmonized Human Biomonitoring in European Children, Teenagers and Adults: EU-Wide Exposure Data of 11 Chemical Substance Groups from the HBM4EU Aligned Studies (2014–2021). *Int. J. Hyg. Environ. Health* **2023**, *249*, 114119. [CrossRef]
38. Kolossa-Gehring, M.; Pack, L.K.; Hülck, K.; Gehring, T. HBM4EU from the Coordinator's Perspective: Lessons Learnt from Managing a Large-Scale EU Project. *Int. J. Hyg. Environ. Health* **2023**, *247*, 114072. [CrossRef]
39. Zare Jeddi, M.; Hopf, N.B.; Viegas, S.; Price, A.B.; Paini, A.; van Thriel, C.; Benfenati, E.; Ndaw, S.; Bessems, J.; Behnisch, P.A.; et al. Towards a Systematic Use of Effect Biomarkers in Population and Occupational Biomonitoring. *Environ. Int.* **2021**, *146*, 106257. [CrossRef]
40. Rodríguez-Carrillo, A.; Mustieles, V.; Salamanca-Fernández, E.; Olivas-Martínez, A.; Suárez, B.; Bajard, L.; Baken, K.; Blaha, L.; Bonefeld-Jørgensen, E.C.; Couderq, S.; et al. Implementation of Effect Biomarkers in Human Biomonitoring Studies: A Systematic Approach Synergizing Toxicological and Epidemiological Knowledge. *Int. J. Hyg. Environ. Health* **2023**, *249*, 114140. [CrossRef]
41. PARC Partnership for the Assessment of Risks from Chemicals | Parc. Available online: https://www.eu-parc.eu/ (accessed on 19 January 2024).
42. Valent, F.; Horvat, M.; Sofianou-Katsoulis, A.; Spiric, Z.; Mazej, D.; Little, D.; Prasouli, A.; Mariuz, M.; Tamburlini, G.; Nakou, S.; et al. Neurodevelopmental Effects of Low-Level Prenatal Mercury Exposure from Maternal Fish Consumption in a Mediterranean Cohort: Study Rationale and Design. *J. Epidemiol.* **2013**, *23*, 146–152. [CrossRef]
43. Brumatti, L.V.; Rosolen, V.; Mariuz, M.; Piscianz, E.; Valencic, E.; Bin, M.; Athanasakis, E.; D'adamo, P.; Fragkiadoulaki, E.; Calamandrei, G.; et al. Impact of Methylmercury and Other Heavy Metals Exposure on Neurocognitive Function in Children Aged 7 Years: Study Protocol of the Follow-Up. *J. Epidemiol.* **2021**, *31*, 157–163. [CrossRef] [PubMed]

44. Simeone, R.M.; Howards, P.P.; Anderson, E.; Jusko, T.A.; Drobná, B.; Kočan, A.; Čonka, K.; Fabišiková, A.; Murínová, Ľ.P.; Canfield, R.L.; et al. Pre- and Postnatal Polychlorinated Biphenyl Exposure and Cognitive and Behavioral Development at Age 45 Months in a Cohort of Slovak Children. *Chemosphere* **2022**, *287*, 132375. [CrossRef] [PubMed]
45. Szabados, M.; Kakucs, R.; Páldy, A.; Kotlík, B.; Kazmarová, H.; Dongiovanni, A.; Di Maggio, A.; Kozajda, A.; Jutraz, A.; Kukec, A.; et al. Association of Parent-Reported Health Symptoms with Indoor Air Quality in Primary School Buildings—The InAirQ Study. *Build. Environ.* **2022**, *221*, 109339. [CrossRef]
46. Szabados, M.; Csákó, Z.; Kotlík, B.; Kazmarová, H.; Kozajda, A.; Jutraz, A.; Kukec, A.; Otorepec, P.; Dongiovanni, A.; Di Maggio, A.; et al. Indoor Air Quality and the Associated Health Risk in Primary School Buildings in Central Europe—The InAirQ Study. *Indoor Air* **2021**, *31*, 989–1003. [CrossRef]
47. Magnus, P.; Birke, C.; Vejrup, K.; Haugan, A.; Alsaker, E.; Daltveit, A.K.; Handal, M.; Haugen, M.; Høiseth, G.; Knudsen, G.P.; et al. Cohort Profile Update: The Norwegian Mother and Child Cohort Study (MoBa). *Int. J. Epidemiol.* **2016**, *45*, 382–388. [CrossRef]
48. Runkel, A.A.; Stajnko, A.; Snoj Tratnik, J.; Mazej, D.; Horvat, M.; Přibylová, P.; Kosjek, T. Exposure of Children and Adolescents from Northeastern Slovenia to Per- and Polyfluoroalkyl Substances. *Chemosphere* **2023**, *321*, 138096. [CrossRef]
49. Stajnko, A.; Snoj Tratnik, J.; Kosjek, T.; Mazej, D.; Jagodic, M.; Eržen, I.; Horvat, M. Seasonal Glyphosate and AMPA Levels in Urine of Children and Adolescents Living in Rural Regions of Northeastern Slovenia. *Environ. Int.* **2020**, *143*, 105985. [CrossRef]
50. Esteban López, M.; Göen, T.; Mol, H.; Nübler, S.; Haji-Abbas-Zarrabi, K.; Koch, H.M.; Kasper-Sonnenberg, M.; Dvorakova, D.; Hajslova, J.; Antignac, J.P.; et al. The European Human Biomonitoring Platform—Design and Implementation of a Laboratory Quality Assurance/Quality Control (QA/QC) Programme for Selected Priority Chemicals. *Int. J. Hyg. Environ. Health* **2021**, *234*, 113740. [CrossRef]
51. Mol, H.G.J.; Elbers, I.; Pälmke, C.; Bury, D.; Göen, T.; López, M.E.; Nübler, S.; Vaccher, V.; Antignac, J.P.; Dvořáková, D.; et al. Proficiency and Interlaboratory Variability in the Determination of Phthalate and DINCH Biomarkers in Human Urine: Results from the HBM4EU Project. *Toxics* **2022**, *10*, 57. [CrossRef]
52. Vogel, N.; Schmidt, P.; Lange, R.; Gerofke, A.; Sakhi, A.K.; Haug, L.S.; Jensen, T.K.; Frederiksen, H.; Szigeti, T.; Csákó, Z.; et al. Current Exposure to Phthalates and DINCH in European Children and Adolescents—Results from the HBM4EU Aligned Studies 2014 to 2021. *Int. J. Hyg. Environ. Health* **2023**, *249*, 114101. [CrossRef] [PubMed]
53. Olivas-Martinez, A.; Suarez, B.; Salamanca-Fernandez, E.; Reina-Perez, I.; Rodriguez-Carrillo, A.; Mustieles, V.; Olea, N.; Freire, C.; Fernández, M.F. Development and Validation of Brain-Derived Neurotrophic Factor Measurement in Human Urine Samples as a Non-Invasive Effect Biomarker. *Front. Mol. Neurosci.* **2023**, *15*, 1075613. [CrossRef]
54. Achenbach, T.; Rescorla, L. *Manual for the ASEBA School-Age Forms and Profiles: An Integrated System of Multi-Informant Assessment*; University of Vermont, Research Center for Children, Youth and Families: Burlington, VT, USA, 2001.
55. Rosolen, V.; Giordani, E.; Mariuz, M.; Parpinel, M.; Ronfani, L.; Vecchi Brumatti, L.; Bin, M.; Calamandrei, G.; Mustieles, V.; Gilles, L.; et al. Concurrent Assessment of Phthalates/HEXAMOLL® DINCH Exposure and Wechsler Intelligence Scale for Children Performance in Three European Cohorts of the HBM4EU Aligned Studies. *Toxics* **2022**, *10*, 538. [CrossRef] [PubMed]
56. *ISCED 2011*; International Standard Classification of Education. UNESCO Institute for Statistics: Montreal, QC, Canada, 2011.
57. de Onis, M. Development of a WHO Growth Reference for School-Aged Children and Adolescents. *Bull. World Health Organ.* **2007**, *85*, 660–667. [CrossRef]
58. Lubin, J.H.; Colt, J.S.; Camann, D.; Davis, S.; Cerhan, J.R.; Severson, R.K.; Bernstein, L.; Hartge, P. Epidemiologic Evaluation of Measurement Data in the Presence of Detection Limits. *Environ. Health Perspect.* **2004**, *112*, 1691–1696. [CrossRef]
59. Textor, J.; van der Zander, B.; Gilthorpe, M.S.; Liskiewicz, M.; Ellison, G.T. Robust Causal Inference Using Directed Acyclic Graphs: The R Package "Dagitty". *Int. J. Epidemiol.* **2016**, *45*, 1887–1894. [CrossRef]
60. Lozano, M.; Murcia, M.; Soler-Blasco, R.; González, L.; Iriarte, G.; Rebagliato, M.; Lopez-Espinosa, M.J.; Esplugues, A.; Ballester, F.; Llop, S. Exposure to Mercury among 9-Year-Old Children and Neurobehavioural Function. *Environ. Int.* **2021**, *146*, 106173. [CrossRef] [PubMed]
61. Maitre, L.; Julvez, J.; López-Vicente, M.; Warembourg, C.; Tamayo-Uria, I.; Philippat, C.; Gützkow, K.B.; Guxens, M.; Andrusaityte, S.; Basagaña, X.; et al. Early-Life Environmental Exposure Determinants of Child Behavior in Europe: A Longitudinal, Population-Based Study. *Environ. Int.* **2021**, *153*, 106523. [CrossRef] [PubMed]
62. Dalsager, L.; Jensen, T.K.; Nielsen, F.; Grandjean, P.; Bilenberg, N.; Andersen, H.R. No Association between Maternal and Child PFAS Concentrations and Repeated Measures of ADHD Symptoms at Age $2\frac{1}{2}$ and 5 Years in Children from the Odense Child Cohort. *Neurotoxicol. Teratol.* **2021**, *88*, 107031. [CrossRef]
63. Carrico, C.; Gennings, C.; Wheeler, D.C.; Factor-Litvak, P. Characterization of Weighted Quantile Sum Regression for Highly Correlated Data in a Risk Analysis Setting. *J. Agric. Biol. Environ. Stat.* **2015**, *20*, 100–120. [CrossRef]
64. Guilbert, A.; Rolland, M.; Pin, I.; Thomsen, C.; Sakhi, A.K.; Sabaredzovic, A.; Slama, R.; Guichardet, K.; Philippat, C. Associations between a Mixture of Phenols and Phthalates and Child Behaviour in a French Mother–Child Cohort with Repeated Assessment of Exposure. *Environ. Int.* **2021**, *156*, 106697. [CrossRef]
65. Tanner, E.M.; Bornehag, C.G.; Gennings, C. Repeated Holdout Validation for Weighted Quantile Sum Regression. *MethodsX* **2019**, *6*, 2855–2860. [CrossRef] [PubMed]
66. O'Brien, K.M.; Upson, K.; Cook, N.R.; Weinberg, C.R. Environmental chemicals in urine and blood: Improving methods for creatinine and lipid adjustment. *Environ. Health Perspect.* **2016**, *124*, 220–227. [CrossRef]

67. Amrhein, V.; Greenland, S.; McShane, B. Scientists Rise up against Statistical Significance. *Nature* **2019**, *567*, 305–307. [CrossRef] [PubMed]
68. Cohen-Eliraz, L.; Ornoy, A.; Ein-Mor, E.; Bar-Nitsan, M.; Pilowsky Peleg, T.; Calderon-Margalit, R. Prenatal exposure to phthalates and emotional/behavioral development in young children. *Neurotoxicology* **2023**, *98*, 39–47. [CrossRef] [PubMed]
69. Huang, H.B.; Kuo, P.H.; Su, P.H.; Sun, C.W.; Chen, W.J.; Wang, S.L. Prenatal and childhood exposure to phthalate diesters and neurobehavioral development in a 15-year follow-up birth cohort study. *Environ. Res.* **2019**, *172*, 569–577. [CrossRef]
70. Dewey, D.; Martin, J.W.; MacDonald, A.M.; Kinniburgh, D.W.; Letourneau, N.; Giesbrecht, G.F.; Field, C.J.; Bell, R.C.; England-Mason, G.; APrON Study Team. Sex-specific associations between maternal phthalate exposure and neurodevelopmental outcomes in children at 2 years of age in the APrON cohort. *Neurotoxicology* **2023**, *98*, 48–60. [CrossRef]
71. Munk Andreasen, S.; Frederiksen, H.; Bilenberg, N.; Andersson, A.M.; Juul, A.; Kyhl, H.B.; Kold Jensen, T. Maternal Concentrations of Phthalates and Attention-Deficit Hyperactivity Disorder (ADHD-) Related Symptoms in Children Aged 2 to 4 Years from Odense Child Cohort. *Environ. Int.* **2023**, *180*, 108244. [CrossRef]
72. Kim, J.H.; Moon, N.; Ji, E.; Moon, H.B. Effects of Postnatal Exposure to Phthalate, Bisphenol a, Triclosan, Parabens, and per- and Poly-Fluoroalkyl Substances on Maternal Postpartum Depression and Infant Neurodevelopment: A Korean Mother-Infant Pair Cohort Study. *Environ. Sci. Pollut. Res. Int.* **2023**, *30*, 96384–96399. [CrossRef]
73. Jedynak, P.; Maitre, L.; Guxens, M.; Gützkow, K.B.; Julvez, J.; López-Vicente, M.; Sunyer, J.; Casas, M.; Chatzi, L.; Gražulevičienė, R.; et al. Prenatal Exposure to a Wide Range of Environmental Chemicals and Child Behaviour between 3 and 7 Years of Age—An Exposome-Based Approach in 5 European Cohorts. *Sci. Total Environ.* **2021**, *763*, 144115. [CrossRef]
74. Philippat, C.; Nakiwala, D.; Calafat, A.M.; Botton, J.; De Agostini, M.; Heude, B.; Slama, R. EDEN Mother–Child Study Group Prenatal Exposure to Nonpersistent Endocrine Disruptors and Behavior in Boys at 3 and 5 Years. *Environ. Health Perspect.* **2017**, *125*, 097014. [CrossRef]
75. England-Mason, G.; Martin, J.W.; MacDonald, A.; Kinniburgh, D.; Giesbrecht, G.F.; Letourneau, N.; Dewey, D. Similar Names, Different Results: Consistency of the Associations between Prenatal Exposure to Phthalates and Parent-Ratings of Behavior Problems in Preschool Children. *Environ. Int.* **2020**, *142*, 105892. [CrossRef] [PubMed]
76. Jankowska, A.; Polańska, K.; Koch, H.M.; Pälmke, C.; Waszkowska, M.; Stańczak, A.; Wesołowska, E.; Hanke, W.; Bose-O'Reilly, S.; Calamandrei, G.; et al. Phthalate Exposure and Neurodevelopmental Outcomes in Early School Age Children from Poland. *Environ. Res.* **2019**, *179*, 108829. [CrossRef]
77. Won, E.K.; Kim, Y.; Ha, M.; Burm, E.; Kim, Y.S.; Lim, H.; Jung, D.E.; Lim, S.; Kim, S.Y.; Kim, Y.M.; et al. Association of Current Phthalate Exposure with Neurobehavioral Development in a National Sample. *Int. J. Hyg. Environ. Health* **2016**, *219*, 364–371. [CrossRef]
78. Engel, S.M.; Villanger, G.D.; Nethery, R.C.; Thomsen, C.; Sakhi, A.K.; Drover, S.S.M.; Hoppin, J.A.; Zeiner, P.; Knudsen, G.P.; Reichborn-Kjennerud, T.; et al. Prenatal Phthalates, Maternal Thyroid Function, and Risk of Attention-Deficit Hyperactivity Disorder in the Norwegian Mother and Child Cohort. *Environ. Health Perspect.* **2018**, *126*, 057004. [CrossRef]
79. Barrett, E.S.; Day, D.B.; Szpiro, A.; Peng, J.; Loftus, C.T.; Ziausyte, U.; Kannan, K.; Trasande, L.; Zhao, Q.; Nguyen, R.H.N.; et al. Prenatal Exposures to Phthalates and Life Events Stressors in Relation to Child Behavior at Age 4–6: A Combined Cohort Analysis. *Environ. Int.* **2024**, *183*, 108425. [CrossRef] [PubMed]
80. Arbuckle, T.E.; Davis, K.; Boylan, K.; Fisher, M.; Fu, J. Bisphenol A, Phthalates and Lead and Learning and Behavioral Problems in Canadian Children 6-11 Years of Age: CHMS 2007–2009. *Neurotoxicology* **2016**, *54*, 89–98. [CrossRef]
81. Quinnies, K.M.; Harris, E.P.; Snyder, R.W.; Sumner, S.S.; Rissman, E.F. Direct and Transgenerational Effects of Low Doses of Perinatal Di-(2-Ethylhexyl) Phthalate (DEHP) on Social Behaviors in Mice. *PLoS ONE* **2017**, *12*, e0171977. [CrossRef] [PubMed]
82. Lee, K.I.; Chiang, C.W.; Lin, H.C.; Zhao, J.F.; Li, C.T.; Shyue, S.K.; Lee, T.S. Maternal Exposure to Di-(2-Ethylhexyl) Phthalate Exposure Deregulates Blood Pressure, Adiposity, Cholesterol Metabolism and Social Interaction in Mouse Offspring. *Arch. Toxicol.* **2016**, *90*, 1211–1224. [CrossRef]
83. Shen, R.; Zhao, L.-L.; Yu, Z.; Zhang, C.; Chen, Y.-H.; Wang, H.; Zhang, Z.-H.; Xu, D.-X. Maternal Di-(2-Ethylhexyl) Phthalate Exposure during Pregnancy Causes Fetal Growth Restriction in a Stage-Specific but Gender-Independent Manner. *Reprod. Toxicol.* **2017**, *67*, 117–124. [CrossRef]
84. Carbone, S.; Ponzo, O.J.; Gobetto, N.; Samaniego, Y.A.; Reynoso, R.; Scacchi, P.; Moguilevsky, J.A.; Cutrera, R. Antiandrogenic Effect of Perinatal Exposure to the Endocrine Disruptor Di-(2-Ethylhexyl) Phthalate Increases Anxiety-like Behavior in Male Rats during Sexual Maturation. *Horm. Behav.* **2013**, *63*, 692–699. [CrossRef]
85. Martinez-Arguelles, D.B.; McIntosh, M.; Rohlicek, C.V.; Culty, M.; Zirkin, B.R.; Papadopoulos, V. Maternal in Utero Exposure to the Endocrine Disruptor Di-(2-Ethylhexyl) Phthalate Affects the Blood Pressure of Adult Male Offspring. *Toxicol. Appl. Pharmacol.* **2013**, *266*, 95–100. [CrossRef] [PubMed]
86. Barakat, R.; Lin, P.C.; Park, C.J.; Best-Popescu, C.; Bakry, H.H.; Abosalem, M.E.; Abdelaleem, N.M.; Flaws, J.A.; Ko, C.M. Prenatal Exposure to DEHP Induces Neuronal Degeneration and Neurobehavioral Abnormalities in Adult Male Mice. *Toxicol. Sci.* **2018**, *164*, 439–452. [CrossRef] [PubMed]
87. Xu, X.; Yang, Y.; Wang, R.; Wang, Y.; Ruan, Q.; Lu, Y. Perinatal Exposure to Di-(2-Ethylhexyl) Phthalate Affects Anxiety- and Depression-like Behaviors in Mice. *Chemosphere* **2015**, *124*, 22–31. [CrossRef]

88. Mao, G.; Zhou, Z.; Chen, Y.; Wang, W.; Wu, X.; Feng, W.; Cobbina, S.J.; Huang, J.; Zhang, Z.; Xu, H.; et al. Neurological Toxicity of Individual and Mixtures of Low Dose Arsenic, Mono and Di (n-Butyl) Phthalates on Sub-Chronic Exposure to Mice. *Biol. Trace Elem. Res.* **2016**, *170*, 183–193. [CrossRef]
89. Rider, C.V.; Wilson, V.S.; Howdeshell, K.L.; Hotchkiss, A.K.; Furr, J.R.; Lambright, C.R.; Gray, L.E. Cumulative Effects of in Utero Administration of Mixtures of "Antiandrogens" on Male Rat Reproductive Development. *Toxicol. Pathol.* **2009**, *37*, 100–113. [CrossRef] [PubMed]
90. Rider, C.V.; Furr, J.R.; Wilson, V.S.; Gray, L.E. Cumulative Effects of in Utero Administration of Mixtures of Reproductive Toxicants That Disrupt Common Target Tissues via Diverse Mechanisms of Toxicity. *Int. J. Androl.* **2010**, *33*, 443–462. [CrossRef] [PubMed]
91. Varshavsky, J.R.; Zota, A.R.; Woodruff, T.J. A Novel Method for Calculating Potency-Weighted Cumulative Phthalates Exposure with Implications for Identifying Racial/Ethnic Disparities among U.S. Reproductive-Aged Women in NHANES 2001–2012. *Environ. Sci. Technol.* **2016**, *50*, 10616–10624. [CrossRef] [PubMed]
92. Swan, S.H.; Liu, F.; Hines, M.; Kruse, R.L.; Wang, C.; Redmon, J.B.; Sparks, A.; Weiss, B. Prenatal Phthalate Exposure and Reduced Masculine Play in Boys. *Int. J. Androl.* **2010**, *33*, 259–269. [CrossRef]
93. Swan, S.H.; Sathyanarayana, S.; Barrett, E.S.; Janssen, S.; Liu, F.; Nguyen, R.H.N.; Redmon, J.B. First Trimester Phthalate Exposure and Anogenital Distance in Newborns. *Hum. Reprod.* **2015**, *30*, 963–972. [CrossRef]
94. Baken, K.A.; Lambrechts, N.; Remy, S.; Mustieles, V.; Rodríguez-Carrillo, A.; Neophytou, C.M.; Olea, N.; Schoeters, G. A Strategy to Validate a Selection of Human Effect Biomarkers Using Adverse Outcome Pathways: Proof of Concept for Phthalates and Reproductive Effects. *Environ. Res.* **2019**, *175*, 235–256. [CrossRef]
95. Gerofke, A.; Lange, R.; Vogel, N.; Schmidt, P.; Weber, T.; David, M.; Frederiksen, H.; Baken, K.; Govarts, E.; Gilles, L.; et al. Phthalates and Substitute Plasticizers: Main Achievements from the European Human Biomonitoring Initiative HBM4EU. *Int. J. Hyg. Environ. Health* **2024**, *259*, 114378. [CrossRef]
96. Suliman, S.; Hemmings, S.M.; Seedat, S. Brain-Derived Neurotrophic Factor (BDNF) Protein Levels in Anxiety Disorders: Systematic Review and Meta-Regression Analysis. *Front. Integr. Neurosci.* **2013**, *7*, 55. [CrossRef]
97. Shi, Y.; Luan, D.; Song, R.; Zhang, Z. Value of Peripheral Neurotrophin Levels for the Diagnosis of Depression and Response to Treatment: A Systematic Review and Meta-Analysis. *Eur. Neuropsychopharmacol.* **2020**, *41*, 40–51. [CrossRef] [PubMed]
98. Qin, X.Y.; Feng, J.C.; Cao, C.; Wu, H.T.; Loh, Y.P.; Cheng, Y. Association of Peripheral Blood Levels of Brain-Derived Neurotrophic Factor with Autism Spectrum Disorder in Children: A Systematic Review and Meta-Analysis. *JAMA Pediatr.* **2016**, *170*, 1079–1086. [CrossRef] [PubMed]
99. Zhang, J.; Luo, W.; Li, Q.; Xu, R.; Wang, Q.; Huang, Q. Peripheral Brain-Derived Neurotrophic Factor in Attention-Deficit/Hyperactivity Disorder: A Comprehensive Systematic Review and Meta-Analysis. *J. Affect. Disord.* **2018**, *227*, 298–304. [CrossRef] [PubMed]
100. Steffensen, I.L.; Dirven, H.; Couderq, S.; David, A.; D'cruz, S.C.; Fernández, M.F.; Mustieles, V.; Rodríguez-Carillo, A.; Hofer, T. Bisphenols and Oxidative Stress Biomarkers—Associations Found in Human Studies, Evaluation of Methods Used, and Strengths and Weaknesses of the Biomarkers. *Int. J. Environ. Res. Public Health* **2020**, *17*, 3609. [CrossRef]
101. Choi, S.H.; Choi, K.H.; Won, J.U.; Kim, H. Impact of Multi-Heavy Metal Exposure on Renal Damage Indicators in Korea: An Analysis Using Bayesian Kernel Machine Regression. *Medicine* **2023**, *102*, E35001. [CrossRef]
102. Cavalheiro Paulelli, A.C.; Cruz, J.C.; Rocha, B.A.; Souza, M.C.O.; de Oliveira, R.S.; Cesila, C.A.; Devoz, P.P.; dos Reis Pedreira Filho, W.; Soares de Campos, M.; Domingo, J.L.; et al. Association between Urinary Concentrations of Toxic Metals/Metalloids and Oxidative Stress in Brazilians Living in Areas Affected by the Fundão Dam Failure. *Environ. Res.* **2023**, *238*, 117307. [CrossRef]
103. Kim, S.S.; Meeker, J.D.; Keil, A.P.; Aung, M.T.; Bommarito, P.A.; Cantonwine, D.E.; McElrath, T.F.; Ferguson, K.K. Exposure to 17 Trace Metals in Pregnancy and Associations with Urinary Oxidative Stress Biomarkers. *Environ. Res.* **2019**, *179*, 108854. [CrossRef]
104. Davalos, A.D.; Mínguez-Alarcón, L.; van t' Erve, T.J.; Keil, A.P.; Williams, P.L.; Meeker, J.D.; Milne, G.L.; Zhao, S.; Hauser, R.; Ferguson, K.K. Associations between Mixtures of Urinary Phthalate Metabolite Concentrations and Oxidative Stress Biomarkers among Couples Undergoing Fertility Treatment. *Environ. Res.* **2022**, *212*, 113342. [CrossRef]
105. Zhang, Y.J.; Wu, L.H.; Wang, F.; Liu, L.Y.; Zeng, E.Y.; Guo, Y. DNA Oxidative Damage in Pregnant Women upon Exposure to Conventional and Alternative Phthalates. *Environ. Int.* **2021**, *156*, 106743. [CrossRef]
106. Liao, Q.; Huang, H.; Zhang, X.; Ma, X.; Peng, J.; Zhang, Z.; Chen, C.; Lv, Y.; Zhu, X.; Zheng, J.; et al. Assessment of Health Risk and Dose-Effect of DNA Oxidative Damage for the Thirty Chemicals Mixture of Parabens, Triclosan, Benzophenones, and Phthalate Esters. *Chemosphere* **2022**, *308*, 136394. [CrossRef]
107. Ferguson, K.K.; Chen, Y.-H.; VanderWeele, T.J.; McElrath, T.F.; Meeker, J.D.; Mukherjee, B. Mediation of the Relationship between Maternal Phthalate Exposure and Preterm Birth by Oxidative Stress with Repeated Measurements across Pregnancy. *Environ. Health Perspect.* **2017**, *125*, 488–494. [CrossRef]
108. Li, L.; Huang, L.; Lei, R.; Zhang, P.; Yang, Y.; Liu, H.; Zhang, Y. DEHP and DBP, Common Phthalates, Induce Glucose Metabolism Disorders in Rats via Oxidative Damage of PI3K/Akt/GLUT4 Signaling. *Environ. Pollut.* **2024**, *341*, 122948. [CrossRef] [PubMed]
109. Chen, L.; Qi, M.; Zhang, L.; Yu, F.; Tao, D.; Xu, C.; Xu, S. Di(2-Ethylhexyl) Phthalate and Microplastics Cause Necroptosis and Apoptosis in Hepatocytes of Mice by Inducing Oxidative Stress. *Environ. Toxicol.* **2023**, *38*, 1226–1238. [CrossRef] [PubMed]
110. Rothman, K.J. Six persistent research misconceptions. *J. Gen. Intern. Med.* **2014**, *29*, 1060–1064. [CrossRef]
111. Perrier, F.; Giorgis-Allemand, L.; Slama, R.; Philippat, C. Within-Subject Pooling of Biological Samples to Reduce Exposure Misclassification in Biomarker-Based Studies. *Epidemiology* **2016**, *27*, 378–388. [CrossRef]

112. Vernet, C.; Philippat, C.; Agier, L.; Calafat, A.M.; Ye, X.; Lyon-Caen, S.; Hainaut, P.; Siroux, V.; Schisterman, E.F.; Slama, R. An Empirical Validation of the Within-Subject Biospecimens Pooling Approach to Minimize Exposure Misclassification in Biomarker-Based Studies. *Epidemiology* **2019**, *30*, 756–767. [CrossRef]
113. Mustieles, V.; Lascouts, A.; Pozo, O.J.; Haro, N.; Lyon-Caen, S.; Jedynak, P.; Bayat, S.; Thomsen, C.; Sakhi, A.K.; Sabaredzovic, A.; et al. Longitudinal Associations between Prenatal Exposure to Phthalates and Steroid Hormones in Maternal Hair Samples from the SEPAGES Cohort. *Environ. Sci. Technol.* **2023**, *57*, 19202–19213. [CrossRef] [PubMed]
114. Mustieles, V.; Rolland, M.; Pin, I.; Thomsen, C.; Sakhi, A.; Sabaredzovic, A.; Muckle, G.; Guichardet, K.; Slama, R.; Philippat, C. Early-Life Exposure to a Mixture of Phenols and Phthalates in Relation to Child Social Behavior: Applying an Evidence-Based Prioritization to a Cohort with Improved Exposure Assessment. *Environ. Health Perspect.* **2023**, *131*, 087006. [CrossRef] [PubMed]

toxics

MDPI

Article

Analysis of Mercury in Aquifers in Gold Mining Areas in the Ecuadorian Amazon and Its Associated Risk for Human Health

Irene Passarelli [1], Michelle Vanessa Villacis Verdesoto [2], Samantha Jiménez-Oyola [3], Ana Gabriela Flores Huilcapi [4], Demmy Mora-Silva [5], Giorgio Anfuso [6], Jose Fernando Esparza Parra [7], Mirian Jimenez-Gutierrez [2], Luis Santiago Carrera Almendáriz [2], Victor Gabriel Avalos Peñafiel [2], Salvatore Straface [1] and Carlos Mestanza-Ramón [1,*]

1. Department of Environmental Engineering, University of Calabria, 87036 Rende, Italy; pssrni88t62i854p@studenti.unical.it (I.P.)
2. Green Amazon, Research Center, Nueva Loja 210150, Ecuador; mivavill@espol.edu.ec (M.V.V.V.); luissantiago.carrera@espoch.edu.ec (L.S.C.A.)
3. Escuela Superior Politécnica del Litoral, ESPOL, Facultad de Ingeniería en Ciencias de la Tierra, ESPOL Polytechnic University, Campus Gustavo Galindo km 30.5 Vía Perimetral, P.O. Box 09-01-5863, Guayaquil 090101, Ecuador
4. Faculty of Food Science and Engineering and Biotechnology, Universidad Técnica de Ambato, Av. Colombia y Chile, Ambato 180104, Ecuador
5. Research Group YASUNI-SDC, Escuela Superior Politécnica de Chimborazo, Sede Orellana, El Coca 220001, Ecuador
6. Department of Earth Sciences, Faculty of Marine and Environmental Sciences, University of Cádiz, 11510 Puerto Real, Spain; giorgio.anfuso@uca.es
7. Escuela Superior Politécnica de Chimborazo (ESPOCH), Faculty of Natural Resources, Panamericana Sur km $1\frac{1}{2}$, Riobamba 060155, Ecuador

* Correspondence: cmestanza@ug.uchile.cl

Citation: Passarelli, I.; Villacis Verdesoto, M.V.; Jiménez-Oyola, S.; Flores Huilcapi, A.G.; Mora-Silva, D.; Anfuso, G.; Esparza Parra, J.F.; Jimenez-Gutierrez, M.; Carrera Almendáriz, L.S.; Avalos Peñafiel, V.G.; et al. Analysis of Mercury in Aquifers in Gold Mining Areas in the Ecuadorian Amazon and Its Associated Risk for Human Health. *Toxics* **2024**, *12*, 162. https://doi.org/10.3390/toxics12020162

Academic Editor: Carina Ladeira

Received: 25 January 2024
Revised: 12 February 2024
Accepted: 15 February 2024
Published: 19 February 2024

Abstract: Gold mining activity is a source of supply in many areas of the world, and especially in developing countries, it is practiced illegally and by applying unsafe techniques. Particularly in Ecuador, artisanal and small-scale gold mining (ASGM) is widespread, and it is based on the use of toxic substances, such as mercury (Hg), in gold recovery. Hg is a heavy metal that is water-insoluble, which, once mobilized, poses a threat to both the environment and human health. This study analyzes Hg concentrations in the six provinces of Napo, Sucumbíos, Orellana, Pastaza, Morona Santiago, and Zamora Chinchipe of the Ecuadorian Amazon region to conduct a human health risk assessment. Significant differences in Hg levels were found between provinces, but concentrations were below MPL imposed by Ecuadorian regulations everywhere. Nevertheless, a worrisome picture emerges, especially with regard to the most vulnerable receptors represented by the child population. There are multiple factors of incidence that may affect the possible future development of the phenomenon, and with reference to the social, economic, and environmental context of the region, it can be concluded that it may be appropriate to plan further investigation to arrive at a more comprehensive assessment. The results of this study can be used by decision makers to plan further investigation and to implement monitoring networks, risk mitigation strategies, and groundwater protection measures.

Keywords: artisanal small-scale mining; groundwater pollution; environmental and health risks; Ecuadorian Amazon; risk evaluation

1. Introduction

The present investigations were conducted to determine the concentration of mercury (Hg) in the coastal region of Ecuador to assess the resulting risks to the environment and human health using both deterministic and probabilistic methods.

Artisanal and small-scale gold mining (ASGM) is an important source of economic livelihood in many areas of the world (where mineral deposits exist), especially in rural areas where employment opportunities are very limited [1]. In these areas, in most cases, this activity is carried out illegally and in violation of present environmental regulations. The most serious associated problem is the extensive use of mercury (Hg) in gold recovery processes [2]. In recent decades, the problem has worsened as global demand for gold has increased, driving criminal groups to finance illegal gold mining and, to increase production, processes have shifted from rudimentary to mechanized, resulting in even greater use of Hg [3]. In Ecuador, mining can be described as a historical activity dating back to colonial and precolonial times. However, in the 1990s, it increased significantly when the existence of precious metal deposits was established through geological studies financed by the Central Bank [4]. In its beginnings, mining was carried out with rudimentary techniques; in the last decades, an important enhancement of the methods used, which vary from place to place, has been recorded. It can be carried out at different scales and in different ways; it often continues to be carried out illegally or informally, and in some cases, local authorities tend to tolerate it [5]. Currently, medium- and large-scale mining projects have developed and consolidated alongside artisanal and small-scale activities [6].

Hg is considered one of the most hazardous heavy metals for both human health and the environment [7]. It can enter the atmosphere as inorganic mercury from geogenic and anthropogenic sources and then be converted to organic mercury by natural processes. The organic forms of Hg mercury, such as methylmercury and dimethylmercury, are the most toxic and harmful [8,9]. The effects of Hg on human health can vary depending on the duration of exposure and the level of toxicity. The toxicity of a metal is influenced by several factors: chemical form, rate of absorption and elimination from an organism, and tissue concentration [10]. The main effects of short-term exposure include nervous system abnormalities and disorders of the respiratory and gastrointestinal tract [5,11]. In cases where exposure is more prolonged or toxicity levels are higher, the effects can be much more severe. These include cancer, DNA mutations, cardiovascular system dysfunction, and damage to the central nervous system, blood cells, and vital organs such as the lungs, liver, and kidneys [10,12].

There are multiple migration pathways by which a contaminant can reach human receptors, but in general, these depend on the particular environment under consideration and the species prevalent there [13]. For residential and recreational uses, the most frequent mode of exposure appears to be accidental ingestion of contaminated water. Added to this is the consumption of fish caught near mining areas, which can introduce organic mercury (methylmercury) into the human body [13,14]. In workplaces where mining is improperly conducted and where amalgamation processes involving mercury are carried out, inhalation of elemental Hg vapors is the most frequent type of exposure [15,16]. A final mode of exposure that deserves mention is dermal contact. This phenomenon is not simply a contaminant but plays a crucial role in exposure to harmful substances [17]. According to this scenario, people living near areas where mining is practiced are subject to numerous exposure pathways. Therefore, they have a high probability of developing adverse health effects.

In the Amazon region of Ecuador, mining has traditionally been carried out by applying rudimentary techniques based on the use of Hg in amalgamation processes, although this has been banned since 2015 [18]. These methods were widely used, especially in the southern region, i.e., in the jurisdiction of Chinapintza (Zamora Chinchipe province). In this region, the impacts of Hg use and inadequate management of tailing dumps have mainly resulted in deforestation and water contamination. Several studies have shown a significant increase in gold mining activities in four provinces of the Ecuadorian Amazon: Napo, Orellana, Sucumbíos, and Zamora [19–22].

A large part of the Amazonian population lives from activities such as agriculture, so mining also emerges as a key and attractive activity. Recent studies have found an alarming expansion of illegal mining in the provinces of Napo, Orellana, Sucumbíos, Pastaza, Morona Santiago, and Zamora Chinchipe, with evident effects of deforestation and contamination of surface and subsurface waters [23,24]. The present investigation is focused on determining mercury concentrations in groundwater near mining areas and assessing the resulting risk to the human health of exposed populations using deterministic and probabilistic methods. To carry out the risk assessment, 147 water samples were taken from wells near the mining areas. The deterministic approach performs a point-type analysis, while the probabilistic approach is based on modeling a probability distribution. This will allow the spatial pattern of social and environmental effects to be determined to fully understand and better manage the risk, providing decision makers with a tool to implement the pollution mitigation strategy.

The aim of this study is to examine Hg concentrations in groundwater in the Ecuadorian Amazon and assess the corresponding health risk. To date, there are no further studies on groundwater in the Amazon region. There are interactions between different environmental compartments, so groundwater contaminants can easily reach receptors. The guardianship of groundwater is extremely important for the protection of human health, considering that it accounts for almost all drinking water resources. In addition, it feeds the spring and, especially in lean periods, the flow of surface water courses. Therefore, water contamination can quickly spread to humans, the fish population, and vegetation. In the case of Hg, the situation is even more worrying, considering that this metal is water-insoluble. The results of this study can be used as a guideline for planning more detailed investigation plans, risk mitigation actions, and groundwater protection interventions.

2. Materials and Methods

2.1. Study Area

The study covers six provinces in the Amazon region of Ecuador: Sucumbíos, Orellana, Napo, Pastaza, Morona Santiago, and Zamora Chinchipe (Figure 1). This area includes six watersheds and a population of 739,814 inhabitants [25]. This area has been characterized mainly by economic activities such as cattle ranching, agriculture, and, since the end of the 1960s, oil and gold mining activities. These last two activities have intensified in recent decades, causing serious impacts to the biotic, abiotic, and social environment. These impacts are directly related to the release of heavy metals and other pollutants residues from mining processes [13]. In the Amazon Region, there are 225 mining concessions, of which two are of the general regime, 164 correspond to small mining, five to medium, and 54 to large mining, of which 50 are in the exploitation stage, mainly in the south of the region in the provinces of Morona Santiago and Zamora Chinchipe. In addition, in the last ten years, the region has seen a notable increase in socio-environmental conflicts caused by illegal gold mining. Moreover, the impacts caused by gold mining activity have significantly affected surface water bodies due to the indiscriminate discharge of mining waste and the use of Hg in amalgamation processes [21,26].

2.2. Sampling and Laboratory Analisys

The field work was carried out between March and July 2022, a period characterized by rainfall between 200 and 350 mm [27]. Samples were collected from domestic water consumption wells in hamlets or rural communities located near the mining concessions (Figure 1). This study focused on analyzing the concentration of Hg in water samples from aquifers. Regarding the procedure for taking and transporting the samples, 250 mL amber flasks were used and subjected to acidification with 0.10 mL of nitric acid. All the samples were transported under a rigorous chain of custody to the Science Laboratory of the Escuela Superior Politécnica de Chimborazo, located in Orellana, Ecuador. The procedure to determine the Hg concentration was based on atomic absorption and hydride generation (atomic absorption spectrophotometry; PG Instruments, Leicestershire, UK). The application of the method for the determination of mercury (Hg) concentration has been preceded by a procedure for quality assurance and quality control protocol in accordance with the EPA method 7473 [25]. Analyses were conducted by operators wearing safety equipment to avoid any potential contact (ingestion, inhalation, and dermal absorption) with mercury (Hg) compounds. In addition, a quality check was carried out on water samples to detect samples with a high organic content. Such samples require special care during analysis, and the sample size should be reduced to avoid ignition of the samples in the decomposition tube.

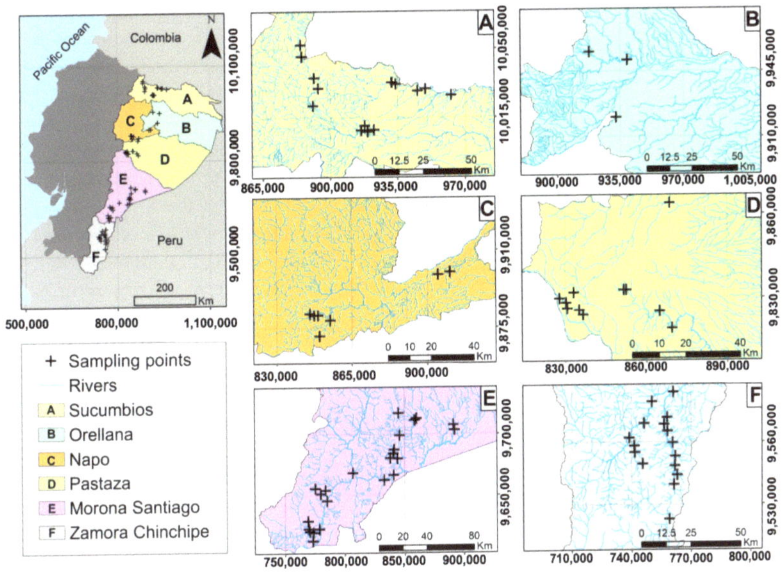

Figure 1. Study area and location of the sampled sites.

First of all, the laboratory demonstrated its initial proficiency by following the sample preparations and analysis procedures according to the method. For each batch of samples processed, both a blank method and a laboratory control sample were carried out. A blank method is prepared by using a reagent of water at the specified volume and carried through the appropriate steps. If the method blank does not contain the target analyte at a level that interferes with the quality of the project, it is considered acceptable. In the absence of specific indications, this means that a maximum level of 10% of the lowest concentration of the sample is considered acceptable. The laboratory control samples must be carried out for each analyte of interest at the project level. Acceptance criteria can be set based on historical data, or in the absence of these, the limit can be set at ±20% of the spiked value.

The Hg measurement range was from 0.0005 to 10 mg/L, and the reference method used was Standard Methods, Ed. 23. 2017, 3112B-Acid Digestion: EPA Method 3015, 2007. Samples were prepared according to the nitric acid digestion procedure described in EPA Method 7473 [28]. The entire process, from sample collection to processing in the laboratory, complied with international quality, confidentiality, and code of ethics policies.

2.2.1. Brief Description of EPA Standard Method 3015

An aqueous sample undergoes extraction with concentrated nitric acid or a combination of concentrated nitric acid and hydrochloric acid, using microwave heating in a suitable oven. The acidified sample is then placed in a microwave-friendly container made of fluorocarbon polymer (DuPont, Wilmington, DE, USA) (such as PFA or TFM) or quartz or placed in a container liner. The container is closed and heated in the microwave oven for a time. After cooling, the contents of the container are filtered, centrifuged, or allowed to settle, then diluted to volume and ready for analysis (Environmental Protection Agency).

2.2.2. Brief Description of EPA Standard Method 7473

EPA Method 7473 is a suitable standard for the analysis of mercury (in organic and inorganic forms) in sediment and aqueous samples or in digested solutions. Application of this method requires prior full calibration of the instrument's working range. This is performed at the initial stage by setting the significant instrumental parameters and is called primary calibration. The primary calibration should be performed after the replacement of the decomposition tube, amalgamator, or oxygen tank. This is followed by a daily calibration to ensure that the primary calibration is valid. The products and instruments used are in accordance with the indications of this standard; thus, the suitability of performance has been documented for them (Environmental Protection Agency).

Controlled heating in an oxygen-containing decomposition oven releases mercury from solid and aqueous samples into the instrument. The sample is dried and then thermally and chemically decomposed in the decomposition oven. The decomposition products are transported to the catalytic part of the furnace by a stream of oxygen. The oxidation is complete, and halogens and nitrogen/sulfur oxides are captured. The remaining decomposition products are transported to an amalgamator, which selectively captures the mercury. After the system is oxygenated to remove any remaining gases or decomposition products, the amalgamator heats up rapidly and releases mercury vapors. The oxygen stream transports mercury vapor through absorption cells located in the light path of a single-wavelength atomic absorption spectrophotometer (Thermo Fisher Scientific, Waltham, MA, USA). The absorbance (peak height or peak area) is measured at 253.7 nm as a function of mercury concentration (Environmental Protection Agency).

2.3. *Risk Assessment and Characterization*

For the human health risk assessment, the residential scenario was considered. The main routes of exposure were water ingestion and dermal contact. The potential risk associated with human health was assessed by the hazard quotient (HQ) calculated for non-carcinogenic type substances, defined as the ratio between the average daily dose (ADD) and the reference dose (RfD). The average daily dose (ADD) was determined by Equations (1) and (2), USEPA 2001 and 2004, respectively. The RfD-Hg was obtained from the Risk Assessment Information System website. The risk was evaluated in terms of cumulative risk through the parameter known as the hazard index (HI), given by the sum of the HQs related to the two exposure routes. If HQs or HI have values above 1, it means that the safe exposure threshold has been exceeded and, therefore, adverse effects related to Hg exposure may occur. Both the traditional deterministic method and the probabilistic method were applied to perform the risk assessment.

$$\mathrm{ADD}_{\mathrm{ingestion}} = \frac{\mathrm{C_{sw}} \cdot \mathrm{EF} \cdot \mathrm{IR} \cdot \mathrm{ED}}{\mathrm{AT} \cdot \mathrm{BW} \cdot \mathrm{CF}} \tag{1}$$

Equation (1)—Average daily dose by the route of ingestion, USEPA 2001.

$$\mathrm{ADD}_{\mathrm{dermal\ contact}} = \frac{\mathrm{C_{sw}} \cdot \mathrm{EF} \cdot \mathrm{ET} \cdot \mathrm{ED} \cdot \mathrm{SA} \cdot \mathrm{k_p}}{\mathrm{AT} \cdot \mathrm{BW} \cdot \mathrm{CF}} \tag{2}$$

Equation (2)—Average daily dose by the route of dermal contact, USEPA 2004, where the meaning of the parameters is as follows:

$$\begin{cases} \mathrm{C_{sw}} = \text{Hg concentration in water } \left(\frac{\mathrm{mg}}{\mathrm{L}}\right) \\ \mathrm{EF} = \text{exposure time } \left(\frac{\mathrm{days}}{\mathrm{year}}\right) \\ \mathrm{IR} = \text{ingestion rate of water } \left(\frac{\mathrm{L}}{\mathrm{day}}\right) \\ \mathrm{ET} = \text{exposure time } \left(\frac{\mathrm{hours}}{\mathrm{event}}\right) \\ \mathrm{ED} = \text{lifetime exposure duration (years)} \\ \mathrm{SA} = \text{skin surface area exposed } (\mathrm{cm}^2) \\ \mathrm{k_p} = \text{skin permeability constant } \left(\frac{\mathrm{cm}}{\mathrm{hour}}\right) \\ \mathrm{AT} = \text{averaging time (days)} \\ \mathrm{BW} = \text{body weight (kg)} \\ \mathrm{CF} = \text{conversion factor} \end{cases}$$

Insight into the significance of parameters:

- C_{SW} (Concentration): It represents the concentration of mercury detected at selected sampling points.
- EF (Exposure Frequency): It is a representative parameter of the average number of days per year the receptor is considered to be exposed to contamination.
- IR (Ingestion Rate): It represents, on average, the amount of contaminated water ingested daily by the receptor. Clearly, this quantity varies depending on the type of receptor.
- ET (Exposure Time): It represents the duration of exposure with reference to the individual contamination event.
- ED (Exposure Duration): It represents the duration, expressed in years, over which, on average, the receptor is considered to be exposed to contamination. Therefore, the ED value varies depending on the type of receptor: adult or child.
- SA (Skin Area): It is the average area of skin considered to be exposed to contamination through dermal contact. It varies depending on the type of receptor considered.
- Kp (Skin Permeability Constant): The amount of contaminant absorbed per centimeter of skin exposed per hour.
- AT (Averaging Time): It represents the period over which the exposure is averaged.
- BW (Body Weight): It is the average body weight of the receptor, so it has a different value for adult and child receptors.
- CF (Conversion Factor): It is used to standardize units of measurement.
- RfD (Chronic Reference Dose): It represents the maximum dose of toxic contaminant that can be accepted. In essence, it is the concentration value of the pollutant for which no adverse effects on human health have been found in the literature.

Regarding the values to be used for each parameter, depending on the receptor type, the values recommended by international standards are considered. Table 1 presents the values assigned to each parameter and the probability distributions used in the human risk assessment.

Table 1. Parameters used in the risk assessment.

	Deterministic	Probabilistic	
Parameters	Value	Distribution	Values
EF	350	Triangular	(26–260)
ED (adults)	120	LogNormal	11.36 ± 13.72
ED (children)	0.22	Uniform	(1–6)
ET (adults and children)	2.6	Triangular	(0.5–6)
SA (adults)	23,000	Normal	18,400 ± 2300
SA (children)	7380	Normal	6800 ± 600
BW (adults)	72	Normal	72 ± 15.9
BW (children)	15.6	Normal	15.6 ± 3.7
IR (adults)	2.04	-	2.04
IR (children)	1.28	-	1.28
Kp	0.001	-	0.001

EF = Exposure Frequency; ED = Exposure Duration; ET = Exposure Time; SA = Skin Area; BW = Body Weight; IR = Ingestion rate; Kp = Skin Permeability Constant [29–32].

3. Results

3.1. Hg Concentration in Water

The summary statistics of Hg concentration in the groundwater samples for each province are presented in Table 2. The Hg concentration for the 50th percentile (p50) decreased in the following order: Napo > Sucumbíos > Morona Santiago > Pastaza >

Zamora Chinchipe > Orellana. Forty-one percent of the samples were below the detection limit of the measuring equipment (LoD = 0.0005 mg/L). This could indicate that the concentrations are below the instrument's detection threshold. However, undetected values could also be caused by malfunctions in the instrumentation or accidental errors on the part of the operators. In such cases, it may be appropriate to repeat the measurements several times. Moreover, it is possible to consider constructing a new data set cleaned of non-detected values. In the present study, a Hg concentration of half the MPL was assumed at the points where the values were not detected.

Table 2. Hg concentration (mg/L) in groundwater samples.

Province	*n*	Min–Max	p50	S.D.
Sucumbíos	14	0.0009–0.0044	0.0030	0.0013
Orellana	3	*	*	*
Napo	7	0.0029–0.0045	0.0037	0.0011
Pastaza	12	0.0009–0.0045	0.0022	0.0013
Morona Santiago	23	0.0008–0.0056	0.0026	0.0015
Zamora Chinchipe	16	0.0007–0.0037	0.0015	0.0010

n = number of samples; min = minimum; max = maximum; p50 = percentile 50; S.D. = Standard Deviation. * Values below the detection limit of the measuring equipment (LoD = 0.0005 mg/L) 0.006 mg/L = limit of Hg for human health protection in drinking water (INEN 1108). In this case, a value of half the LoD (0.00025 mg/L) is assumed for the purpose of the analysis.

With the exception of Orellana province, generally, the Hg concentration was in the range of 0.0007 mg/L and 0.0056 mg/L; therefore, 100% of the samples presented Hg concentration below the MPL established in the Ecuadorian regulation for water quality for human consumption (0.006 mg/L) (INEN 1108).

As a background, it should be noted that there is currently no information about the mercury concentration in groundwater in the Amazon region of Ecuador. Studies to date have reported Hg concentration in surface water, sediments [13,20], soils [21], biological samples [33], and fish [34]. Thus, our study constitutes a baseline for future research concerning groundwater quality in the region.

The Hg concentration in groundwater detected in our study (0.0007 and 0.0056 mg/L) was in the range of Hg values reported in surface waters in the Ecuadorian Amazon; Mestanza-Ramón et al. [13] reported Hg concentrations between 0.0005 and 0.0070 mg/L in several rivers and streams of the Amazon region; and Capparelli et al. [20] reported Hg concentrations between 0.0005 and 0.0112 mg/L in 14 tributaries of the Napo River, however, the observations only partially match these values. This relationship between Hg concentrations in surface and groundwater is evidence of the interaction between these two environments and may be related to infiltration processes that occur when Hg present in surface water bodies is mobilized to the aquifer [35]. These findings highlight the importance of determining the sources of aquifer recharge in order to identify potential sources of contamination that may be affecting the quality of groundwater resources [36].

Several studies have reported the presence of various anthropogenic sources of contamination in the Ecuadorian Amazon, which can affect water quality, both surface and groundwater; among the main ones are artisanal and small-scale gold mining (mainly carried out illegally), inadequate management of mining environmental liabilities [5], fish farming [20], and natural Hg leaching from deforestation [37]. Therefore, assessing water quality in these environments is essential to ensure the protection of water resources and, thus, the safety of users.

Additionally, the results of our study were compared with studies that evaluated Hg concentration in groundwater worldwide. Ghana [38] reported Hg values between 0.00001 mg/L and 0.0132 mg/L in well water samples in areas with auriferous mining influence. In Saudi Arabia [39], the presence of Hg in groundwater in areas near the Mah Adh Dhahab gold mine, where Hg concentrations of up to 9.990 mg/L were recorded, was evaluated. In Nigeria, ref. [40] analyzed Hg concentrations and detected average concentrations of 0.0001 mg/L. In this context, the Hg concentrations detected in the Ecuadorian Amazon are within the range of the values reported in the aforementioned studies but tend to be lower than the maximum values detected in Saudi Arabia. Despite this, it is important to note that Hg levels measured in the Ecuadorian Amazon may represent a significant long-term risk to both health and the environment and should, therefore, be monitored regularly.

3.2. Human Health Risk Assessment

3.2.1. Deterministic Approach

The point risk maps obtained from the deterministic assessment are presented in Figure 2. In addition, Table 2 summarizes the deterministic non-carcinogenic risk values (HI) for the 50th percentile (p50) of Hg exposure for each province assessed and for each receptor. HI values > 1 were reported in the provinces of Sucumbíos, Napo, Pastaza, and Morona Santiago. As shown in Figure 2, six of the sampled sites are above the acceptable risk threshold (HI > 1) for minor receptors, which are the most vulnerable to Hg exposure.

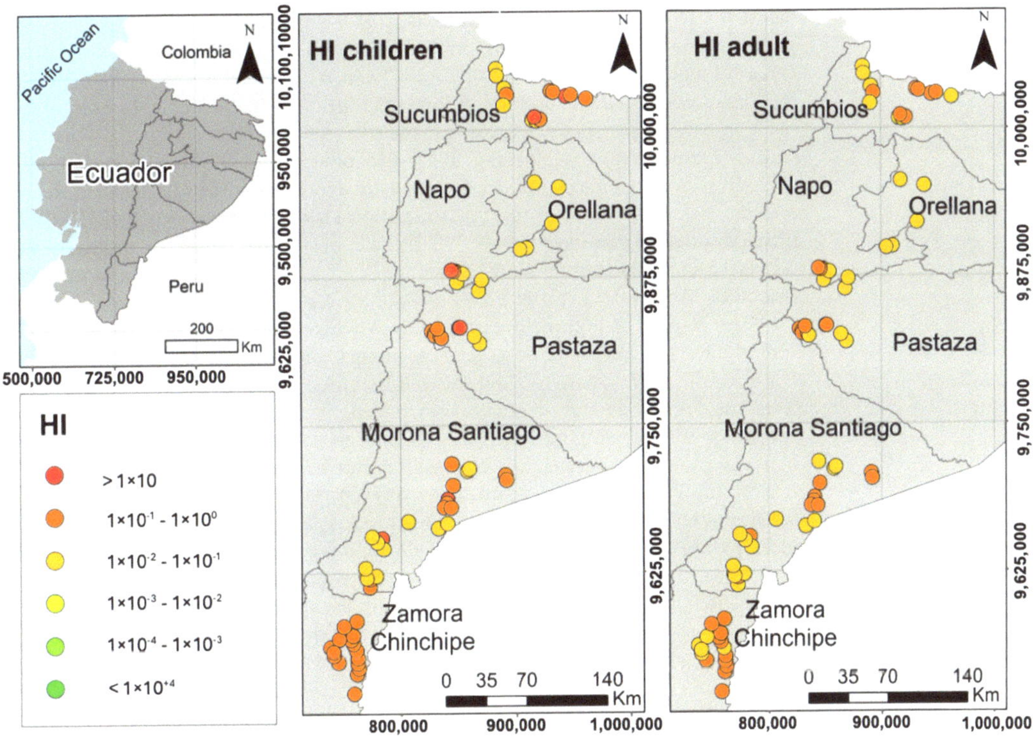

Figure 2. Point risk map of hazard index (HI) for receptors exposed to polluted ground waters in the Ecuadorian Amazon.

The HQ values for water ingestion were below 1 for adults (in the order of 10^{-1} and 10^{-2}) (HI < 1) but above 1 in children in four of the six provinces analyzed; in Sucumbíos, two sites were recorded with HI values slightly above the safe exposure threshold; in Napo and Pastaza, a single sampling point was recorded where the HI was greater than 1; and in Morona Santiago, two sites were observed that represent risk of exposure. Morona Santiago had the highest non-carcinogenic risk value in children (HI = 1.49). In contrast, in Orellana and Zamora Chinchipe, the systemic risk values do not exceed the safe exposure limit for both receptors.

The results of our investigation agree with the findings reported by Mestanza-Ramón et al. [13], who evaluated the systemic health risk from exposure to surface water in the Amazon region through ingestion, obtaining deterministic risk values (p95) in children of HI = 2.51 in Sucumbíos, HI = 2.34 in Orellana, HI = 2.56 in Napo, HI = 2.90×10^{-1} in Pastaza, HI = 8.80×10^{-1} in Morona, and HI = 1.58 in Zamora Chinchipe. As expected, the risk values for surface water exposure, mainly in children, were significantly higher than those reported for groundwater exposure. Likewise, Jiménez-Oyola et al. [29] evaluated the deterministic risk to human health from exposure to surface water for recreational purposes in the province of Napo, obtaining values that exceeded the safe exposure threshold, with an HI = 1.85 for adults and an HI = 4.83 for children, values significantly higher than those determined for exposure to groundwater in our study.

The most relevant route of exposure with the highest contribution to risk was water ingestion, with HI values between 2.26×10^{-2} and 5.07×10^{-1} for adults and in the range of 6.56×10^{-2} to 1.47×10^{1} for children. Regarding the dermal contact route, this presented very low HQ values, in the range of 8.02×10^{-4} and 1.80×10^{-2} for adults and between 1.17×10^{-3} and 2.63×10^{-2} for children; for both cases, the risk from dermal contact was below the acceptable limit for the receptors, so it is considered that this route of exposure presents a negligible contribution in the evaluation of the total risk. Similar results were reported by Mestanza et al. and Jiménez-Oyola et al. [13,29] in risk studies conducted in the Ecuadorian Amazon, with ingestion being the route that contributes most to the entry of Hg into the human body, according to the risk scenarios evaluated in the aforementioned studies. Despite this, it is important to consider other routes of exposure, such as inhalation of Hg in sites where illegal mining is carried out and Hg is used in amalgamation processes, and the risk assessment of food consumption should also be considered, as these can bioaccumulate Hg in their tissues. All these exposure scenarios can increase the risk to which the residents of potentially contaminated areas are exposed. Hence, a comprehensive risk study is essential to protect the health of the communities.

The impact of Hg on human health is widely known. The consumption of Hg-contaminated groundwater poses a risk to living beings, even in minute concentrations [41]. For example, in Pakistan, ref. [42] showed that the groups most affected by Hg exposure were young children and pregnant women using well water for domestic and agricultural purposes. Even during gestation, Hg in its organic form can cross the placental barrier between mother and fetus, causing brain damage and mental retardation [43]. For this reason, the World Health Organization (WHO) [44] recommends a maximum value of 0.006 mg/L Hg for drinking water consumption; Hg values above this limit can cause kidney damage, especially in children through long-term ingestion [45]. On this basis, the wells sampled in this study, despite having Hg concentrations slightly below the limit established by WHO, are not extensive enough to cause long-term health risks to the child population, so periodic sampling of groundwater quality is recommended, in addition to taking biological samples to assess the presence of Hg in the most vulnerable populations.

3.2.2. Probabilistic Approach

The results of the probabilistic risk analysis yielded values below the safe exposure limit for adults and children; thus, according to the probabilistic analysis, there are no exposure levels that require attention for the receptors (Figure 3 and Table 3). The hazard quotient (HQ) values for children ranged between 1.2×10^{-2} and 8.66×10^{-1} for water ingestion and between 2.65×10^{-4} and 2.71×10^{-2} for dermal contact. For adults, the risk was lower, in the range of 5.14×10^{-3} and 4.39×10^{-1} for ingestion and 8.55×10^{-5} and 7.42×10^{-3} for dermal contact. In this sense, it is evident that, although the quantified risk for children is higher than the risk detected in adults, in both cases, the safe exposure limit is not exceeded.

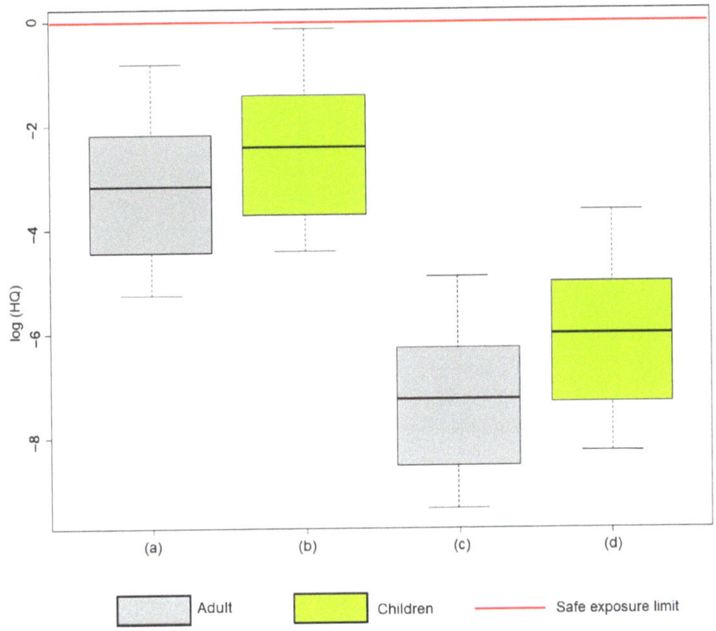

Figure 3. Boxplot of hazard quotient for (**a**) adult ingestion, (**b**) child ingestion, (**c**) adult dermal contact, and (**d**) child dermal contact. HQ values are given in logarithm.

Table 3. Deterministic HI from exposure to Hg in groundwaters for both receptors by province.

Province	Adults				Children			
	Min–Max	p50	p95	S.D.	Min–Max	p50	p95	S.D.
Sucumbíos	2.34×10^{-2}–4.13×10^{-1}	9.85×10^{-2}	4.07×10^{-1}	1.53×10^{-1}	6.67×10^{-2}–1.17	2.80×10^{-1}	1.16	4.36×10^{-1}
Orellana	2.34×10^{-2}–2.34×10^{-2}	2.34×10^{-2}	2.34×10^{-2}	1.0×10^{-3}	6.67×10^{-2}–6.67×10^{-2}	6.67×10^{-2}	6.67×10^{-2}	1.0×10^{-3}
Napo	2.34×10^{-2}–4.22×10^{-1}	2.34×10^{-2}	3.77×10^{-1}	1.64×10^{-1}	6.67×10^{-2}–1.20	6.67×10^{-2}	1.07	4.66×10^{-1}
Pastaza	2.34×10^{-2}–4.22×10^{-1}	9.38×10^{-2}	3.60×10^{-1}	1.39×10^{-1}	6.67×10^{-2}–1.20	2.67×10^{-1}	1.03	3.97×10^{-1}
Morona Santiago	2.34×10^{-2}–5.25×10^{-1}	2.34×10^{-2}	3.95×10^{-1}	1.47×10^{-1}	2.34×10^{-2}–1.49	6.67×10^{-2}	1.12	4.18×10^{-1}
Zamora Chinchipe	6.56×10^{-2}–3.47×10^{-1}	1.36×10^{-1}	3.26×10^{-1}	9.27×10^{-2}	1.87×10^{-1}–9.88×10^{-1}	3.87×10^{-1}	9.28×10^{-1}	2.64×10^{-1}

n = number of samples; min = minimum; max = maximum; p50 = percentile 50; p95 = percentile 95; S.D. = Standard Deviation.

The results of the deterministic analysis differ from the results of the probabilistic analysis (Table 4); according to the probabilistic methodology, the residential scenario does not generate risk for adult and child receptors, whereas, according to the deterministic methodology, there are sites where exposure generates risk for minor receptors. Therefore, based on a conservative criterion, the use of groundwater for domestic consumption is not recommended, mainly in sites located in areas of mining activity that show signs of contamination, since prolonged exposure of vulnerable populations to Hg could affect their health.

To have more conclusive results on the risks for the population, a detailed investigation should be carried out that includes in situ data on exposure and more Hg determinations in different matrices (soil, sediments, surface water, food, human tissue, etc.), with the aim of making a comprehensive risk assessment and minimizing the uncertainty in the risk assessment. It is also recommended to analyze other potentially toxic elements since other contaminants may be present in illegal mining operations that could affect the environment and the population.

Table 4. Probabilistic HQ and HI from exposure to Hg in groundwaters for both receptors. Values in bold exceed the safe exposure threshold.

Risk	Adults				Children			
	Min–Max	p50	p95	S.D.	Min–Max	p50	p95	S.D.
$HQ_{ingestion}$	5.14×10^{-3}–4.39×10^{-1}	4.18×10^{-2}	2.05×10^{-1}	6.89×10^{-2}	1.20×10^{-2}–8.66×10^{-1}	8.94×10^{-2}	4.22×10^{-1}	1.42×10^{-1}
HQ_{dermal}	8.55×10^{-5}–7.42×10^{-3}	7.02×10^{-4}	3.58×10^{-3}	1.17×10^{-3}	2.65×10^{-4}–2.71×10^{-2}	2.54×10^{-3}	1.19×10^{-2}	4.13×10^{-3}
HI	5.23×10^{-3}–4.47×10^{-1}	4.26×10^{-2}	2.08×10^{-1}	7.01×10^{-2}	1.23×10^{-2}–8.93×10^{-1}	9.20×10^{-2}	4.33×10^{-1}	1.47×10^{-1}

min = minimum; max = maximum; p50 = percentile 50; p95 = percentile 95; S.D. = Standard Deviation.

4. Discussion

The results of the study carried out show a worrying scenario regarding Hg concentrations in the Amazon region of Ecuador. The samples to be analyzed were collected from groundwater bodies close to gold mining activities, so some provinces, being more affected, have a denser network of measurements, such as Morona Santiago, Zamora Chinchipe, and Sucumbíos. It can be concluded that the most affected areas in the region are the south and the north, where the average concentration obtained is between 0.0007 mg/L and 0.0056 mg/L. This value is below the maximum permissible limit imposed by Ecuadorian regulations for water intended for human consumption. However, the real risk to the population also depends on the vulnerability of the exposed receptors. In fact, the analysis showed that the child population is at greater risk.

To gain a more objective overview, the measured concentrations were also compared with the maximum permissible limits imposed by regulations in other countries, again with regard to water intended for human consumption. In particular, with respect to the European environmental quality standard (EQS) for groundwater bodies, all analyzed samples exceeded the maximum admissible limit of 0.07 μg/L. By contrast, with reference to the Canadian water quality guidelines, about 35% of the samples exceeded the maximum permissible limit of 0.001 mg/L. Therefore, although not exceeding legal limits, the detected Hg concentrations must be interpreted as a wake-up call. Firstly, it may be necessary to conduct more in-depth analyses in certain areas. Furthermore, it would be advisable to keep these areas under constant observation to ensure that Hg levels do not rise further.

According to many experts in the field, the high levels of Hg found in the analyzed samples can be directly attributed to gold mining, especially illegal gold mining, which is conducted with disregard for both environmental safety and the safety of the workers themselves [5,20,46]. The effects on the environment are manifold and, among others, include deforestation and contamination of groundwater bodies [5,22]. When a contaminant is released into the environment, it interacts with the environment and, depending on its chemical and physical properties, can move from one environmental compartment to another [47]. This increases the likelihood that the contaminant itself will spread and come into contact with human receptors. Thus, from the deep soil, the contaminant can reach the groundwater, flow back up to the soil, and contaminate surface water bodies and vice versa. Moreover, polluted water poisons fish fauna [48]. In addition, it can be used in agriculture for irrigation and contaminate agricultural products for human consumption. Finally, contaminated water can come into direct contact with humans through the known exposure routes of ingestion and dermal contact. The growth in Hg use in Ecuador has been proportional to the increase in gold mining activities [49]. Although the state has sanction instruments at its disposal, it has not yet used them to counter this problem [10,44]. This is partly due to the difficulty in controlling all the illegal activities taking place in the territory and partly due to inefficiency on the part of the administrations. Similar situations are widespread in other developing countries, such as Venezuela [50], Brazil [51,52], Colombia [53,54], and Peru [7]. In these countries, the indiscriminate use of mercury has caused severe impacts on the environment and human health [55]. To reverse the trend, the state of Ecuador should first increase control and monitoring activities in order to unmask illegal gold mining activities. Secondly, the state could create fiscal instruments to foster the creation and growth of mining activities that work within the legal framework. In this way, on the one hand, the state would alleviate the current problem of contamination by Hg, and on the other hand, it would favor the economic growth of the country [56].

5. Conclusions

The present study detected Hg concentrations in groundwater in the Amazon region of Ecuador. The study conducted found that the most affected provinces are Napo, Sucumbíos, and Morona Santiago. In contrast, in the province of Orellana, all samples analyzed exhibited Hg concentrations below the detection threshold of the measuring instruments used. Overall, the totality of samples analyzed (100%) had Hg concentrations below the MPL established by Ecuadorian regulations for water for human consumption. Although below legal limits, Hg levels detected in some provinces are significantly high. In addition, taking into account the fact that the mining activity is constantly expanding, the Hg level may increase at a later date. The overall picture suggests that the most affected provinces should be subjected to more control and possibly constant monitoring.

In any case, the results of the risk analysis vary depending on the two approaches used: the analysis carried out with a deterministic approach suggests a more severe scenario. This is because the deterministic approach is based on point-by-point analysis and, therefore, requires a significant amount of data. Instead, the probabilistic approach is based on the synthetic generation of a larger sample of data. This allows statistical uncertainties to be taken into account and provides a more complete view.

Nevertheless, a more accurate risk analysis is required to reconstruct a model that takes into account the following aspects: social, ecological, and geological. In fact, the risk is proportional to the actual vulnerability of the receptors, which in turn is influenced by several factors, such as the uses and habits of the exposed populations, the properties of the environmental compartments, and the geological and geomorphological properties of the soil. Knowledge of the properties of environmental compartments provides information on the essential interactions between them and the pollutant. Similarly, the geological characteristics of the subsurface provide a better understanding of the transport mechanism of the pollutant that may develop. Finally, knowing the social habits of the exposed population makes it possible to trace the most likely routes of exposure. The results obtained from this study can be used, together with other research, to plan monitoring and risk mitigation activities resulting from the use of Hg.

Author Contributions: Conceptualization, I.P., S.S. and C.M.-R.; methodology, I.P., S.S., C.M.-R. and S.J.-O.; software, I.P., M.J.-G. and S.J.-O.; validation, S.S., C.M.-R. and S.J.-O.; formal analysis, I.P., S.S., C.M.-R. and S.J.-O.; investigation, L.S.C.A., V.G.A.P., A.G.F.H., D.M.-S., M.J.-G., G.A., J.F.E.P. and C.M.-R.; writing—original draft preparation, I.P., S.S., C.M.-R., M.V.V.V., D.M.-S. and S.J.-O.; writing—review and editing, S.S., C.M.-R., S.J.-O. and D.M.-S.; visualization, S.S., C.M.-R., M.V.V.V. and S.J.-O.; supervision, S.S. and C.M.-R. All authors have read and agreed to the published version of the manuscript.

Funding: The researchers are grateful for the financial support of the Escuela Superior Politécnica de Chimborazo and the University of Calabria.

Institutional Review Board Statement: Not applicable.

Informed Consent Statement: Informed consent was obtained from all subjects involved in the study.

Data Availability Statement: The data presented in this study are available upon request from the corresponding author.

Conflicts of Interest: The authors declare no conflicts of interest.

References

1. Verbrugge, B.; Geenen, S. The Gold Commodity Frontier: A Fresh Perspective on Change and Diversity in the Global Gold Mining Economy. *Extr. Ind. Soc.* **2019**, *6*, 413–423. [CrossRef]
2. Niane, B.; Guédron, S.; Feder, F.; Legros, S.; Ngom, P.M.; Moritz, R. Impact of Recent Artisanal Small-Scale Gold Mining in Senegal: Mercury and Methylmercury Contamination of Terrestrial and Aquatic Ecosystems. *Sci. Total Environ.* **2019**, *669*, 185–193. [CrossRef]
3. Bengtsson, U.G.; Hylander, L.D. Increased Mercury Emissions from Modern Dental Amalgams. *Biometals* **2017**, *30*, 277–283. [CrossRef]
4. Capparelli, V.M.; Cabrera, M.; Rico, A.; Lucas-Solis, O.; Alvear-S, D.; Vasco, S.; Galarza, E.; Shiguango, L.; Pinos-Velez, V.; Pérez-González, A. An Integrative Approach to Assess the Environmental Impacts of Gold Mining Contamination in the Amazon. *Toxics* **2021**, *9*, 149. [CrossRef]
5. Mestanza-Ramón, C.; Mora-Silva, D.; D'Orio, G.; Tapia-Segarra, E.; Gaibor, I.D.; Esparza Parra, J.F.; Chávez Velásquez, C.R.; Straface, S. Artisanal and Small-Scale Gold Mining (ASGM): Management and Socioenvironmental Impacts in the Northern Amazon of Ecuador. *Sustainability* **2022**, *14*, 6854. [CrossRef]
6. Chamba Quizhpe, A.V. Diseño de Un Sistema de Gestión Ambiental Aplicado a La Agencia de Regulación y Control de Energía y Recursos Naturales No Renovables En Las Áreas Administrativas y Laboratorio Ubicados En El Edificio de La Armenia. Master's Thesis, Universidad de las Fuerzas Armadas, Sangolquí, Ecuador, 2022.
7. World Health Organization. *COVID-19: Operational Guidance for Maintaining Essential Health Services during an Outbreak: Interim Guidance*; World Health Organization: Geneva, Switzerland, 2021.
8. Fuentes-Gandara, F.; Herrera-Herrera, C.; Pinedo-Hernández, J.; Marrugo-Negrete, J.; Díez, S. Assessment of Human Health Risk Associated with Methylmercury in the Imported Fish Marketed in the Caribbean. *Environ. Res.* **2018**, *165*, 324–329. [CrossRef] [PubMed]
9. Rodrigues, S.M.; Coelho, C.; Cruz, N.; Monteiro, R.J.R.; Henriques, B.; Duarte, A.C.; Römkens, P.; Pereira, E. Oral Bioaccessibility and Human Exposure to Anthropogenic and Geogenic Mercury in Urban, Industrial and Mining Areas. *Sci. Total Environ.* **2014**, *496*, 649–661. [CrossRef] [PubMed]
10. Briffa, J.; Sinagra, E.; Blundell, R. Heavy Metal Pollution in the Environment and Their Toxicological Effects on Humans. *Heliyon* **2020**, *6*, e04691. [CrossRef] [PubMed]
11. Fernandes Azevedo, B.; Barros Furieri, L.; Peçanha, F.M.; Wiggers, G.A.; Frizera Vassallo, P.; Ronacher Simões, M.; Fiorim, J.; Rossi de Batista, P.; Fioresi, M.; Rossoni, L. Toxic Effects of Mercury on the Cardiovascular and Central Nervous Systems. *Biomed Res. Int.* **2012**, *2012*, 949048. [CrossRef] [PubMed]
12. Khan, F.; Momtaz, S.; Abdollahi, M. The Relationship between Mercury Exposure and Epigenetic Alterations Regarding Human Health, Risk Assessment and Diagnostic Strategies. *J. Trace Elem. Med. Biol.* **2019**, *52*, 37–47. [CrossRef] [PubMed]
13. Mestanza-Ramón, C.; Jiménez-Oyola, S.; Montoya, A.V.G.; Vizuete, D.D.C.; D'Orio, G.; Cedeño-Laje, J.; Straface, S. Assessment of Hg Pollution in Stream Waters and Human Health Risk in Areas Impacted by Mining Activities in the Ecuadorian Amazon. *Environ. Geochem. Health* **2023**, *45*, 7183–7197. [CrossRef] [PubMed]
14. Abercrombie, M.I. Mercury in Sediment, Fish and Hair: An Ecosystem Study of Bioaccumulation of Mercury in Mexico. Ph.D. Thesis, Rensselaer Polytechnic Institute, Troy, NY, USA, 2010.
15. De Miguel, E.; Clavijo, D.; Ortega, M.F.; Gómez, A. Probabilistic Meta-Analysis of Risk from the Exposure to Hg in Artisanal Gold Mining Communities in Colombia. *Chemosphere* **2014**, *108*, 183–189. [CrossRef]
16. Zhao, H.; Wang, Q.; Groves, D.I.; Santosh, M.; Zhang, J.; Fan, T. Genesis of Orogenic Gold Systems in the Daduhe Belt: Evidence of Long-Lived Fertile Mantle Lithosphere as a Source of Diverse Metallogeny on the Western Margin of the Yangtze Craton, China. *Ore Geol. Rev.* **2022**, *145*, 104861. [CrossRef]
17. Tran, T.S.; Dinh, V.C.; Nguyen, T.A.H.; Kim, K.-W. Soil Contamination and Health Risk Assessment from Heavy Metals Exposure near Mining Area in Bac Kan Province, Vietnam. *Environ. Geochem. Health* **2022**, *44*, 1189–1202. [CrossRef] [PubMed]
18. Mestanza-Ramón, C.; D'ORIO, G.; Straface, S. Gold Mining in Ecuador: Innovative Recommendations for the Management and Remediation of Mercury-Contaminated Waters. *Green World J.* **2021**, *4*, 11. [CrossRef]
19. Melo, C.; Mena, C.F.; Arsel, M.; Pellegrini, L.; CoCooN, N. The State Is Dead, Long Live the State: Re-Inserting the State in the Gold-Mining Industry in Zamora-Chinchipe, Ecuador. *CoCooN NEBE* **2013**.
20. Capparelli, M.V.; Moulatlet, G.M.; de Souza Abessa, D.M.; Lucas-Solis, O.; Rosero, B.; Galarza, E.; Tuba, D.; Carpintero, N.; Ochoa-Herrera, V.; Cipriani-Avila, I. An Integrative Approach to Identify the Impacts of Multiple Metal Contamination Sources on the Eastern Andean Foothills of the Ecuadorian Amazonia. *Sci. Total Environ.* **2020**, *709*, 136088. [CrossRef] [PubMed]
21. López-Blanco, C.; Collahuazo, L.; Torres, S.; Chinchay, L.; Ayala, D.; Benítez, P. Mercury Pollution in Soils from the Yacuambi River (Ecuadorian Amazon) as a Result of Gold Placer Mining. *Bull. Environ. Contam. Toxicol.* **2015**, *95*, 311–316. [CrossRef]
22. Mestanza-Ramón, C.; Cuenca-Cumbicus, J.; D'Orio, G.; Flores-Toala, J.; Segovia-Cáceres, S.; Bonilla-Bonilla, A.; Straface, S. Gold Mining in the Amazon Region of Ecuador: History and a Review of Its Socio-Environmental Impacts. *Land* **2022**, *11*, 221. [CrossRef]
23. Hecht, S.; Schmink, M.; Abers, R.; Assad, E.; Humphreys Bebbington, D.; Eduaro, B.; Costa, F.; Durán Calisto, A.M.; Fearnside, P.M.; Garrett, R. The Amazon in Motion: Changing Politics, Development Strategies, Peoples, Landscapes, and Livelihoods. In *Amazon Assessment Report 2021, Part II*; United Nations Development Solutions Network: New York, NY, USA, 2021; pp. 12–14.

24. Martinez-Alier, J.; Walter, M. Social Metabolism and Conflicts over Extractivism. *Environ. Gov. Lat. Am.* **2016**, 58–85.
25. Villacís, B.; Carrillo, D.; Martínez, A.G. *Estadística Demográfica En El Ecuador: Diagnóstico y Propuesta*; Instituto Nacional de Estadística y Censos: Quito, Ecuador, 2011; pp. 1–74.
26. Requelme, M.E.R.; Ramos, J.F.F.; Angélica, R.S.; Brabo, E.S. Assessment of Hg-Contamination in Soils and Stream Sediments in the Mineral District of Nambija, Ecuadorian Amazon (Example of an Impacted Area Affected by Artisanal Gold Mining). *Appl. Geochem.* **2003**, *18*, 371–381. [CrossRef]
27. Ilbay-Yupa, M.; Lavado-Casimiro, W.; Rau, P.; Zubieta, R.; Castillón, F. Updating Regionalization of Precipitation in Ecuador. *Theor. Appl. Climatol.* **2021**, *143*, 1513–1528. [CrossRef]
28. Mensh, M. Direct Mercury Analysis of Soil, Sediments and Waste Waters Using Method 7473. *Soil Sediment Contam.* **2004**, *13*, 150.
29. Jiménez-Oyola, S.; Escobar Segovia, K.; García-Martínez, M.-J.J.; Ortega, M.; Bolonio, D.; García-Garizabal, I.; Salgado, B.; Segovia-Escobar, K.; García-Martínez, M.-J.J.; Ortega, M.; et al. Human Health Risk Assessment for Exposure to Potentially Toxic Elements in Polluted Rivers in the Ecuadorian Amazon. *Water* **2021**, *13*, 613. [CrossRef]
30. Spence, L.R.; Walden, T.; Sunbury, U.K. *RISC4. Risk Integrated Software for Clean-Ups*; Bp-Amoco Oil: Sunbury, UK, 2001; p. 464.
31. Israeli, M.; Nelson, C.B. Distribution and Expected Time of Residence for U.S. Households. *Risk Anal.* **1992**, *12*, 65–72. [CrossRef]
32. Anderson, M.R.; Scruton, D.A.; Williams, U.P.; Payne, J.F. Mercury in Fish in the Smallwood Reservoir, Labrador, Twenty One Years after Impoundment. *Water. Air. Soil Pollut.* **1995**, *80*, 927–930. [CrossRef]
33. González-Merizalde, M.V.; Menezes-Filho, J.A.; Cruz-Erazo, C.T.; Bermeo-Flores, S.A.; Sánchez-Castillo, M.O.; Hernández-Bonilla, D.; Mora, A. Manganese and Mercury Levels in Water, Sediments, and Children Living near Gold-Mining Areas of the Nangaritza River Basin, Ecuadorian Amazon. *Arch. Environ. Contam. Toxicol.* **2016**, *71*, 171–182. [CrossRef]
34. Webb, J.; Coomes, O.T.; Mainville, N.; Mergler, D. Mercury Contamination in an Indicator Fish Species from Andean Amazonian Rivers Affected by Petroleum Extraction. *Bull. Environ. Contam. Toxicol.* **2015**, *95*, 279–285. [CrossRef]
35. Yan, J.; Li, R.; Ali, M.U.; Wang, C.; Wang, B.; Jin, X.; Shao, M.; Li, P.; Zhang, L.; Feng, X. Mercury Migration to Surface Water from Remediated Mine Waste and Impacts of Rainfall in a Karst Area–Evidence from Hg Isotopes. *Water Res.* **2023**, *230*, 119592. [CrossRef] [PubMed]
36. Abdelaal, A.; Sultan, M.; Abotalib, A.Z.; Bedair, M.; Krishnamurthy, R.V.; Elhebiry, M. Emerging Mercury and Methylmercury Contamination from New Artisanal and Small-Scale Gold Mining along the Nile Valley, Egypt. *Environ. Sci. Pollut. Res.* **2023**, *30*, 52514–52534. [CrossRef] [PubMed]
37. Mainville, N.; Webb, J.; Lucotte, M.; Davidson, R.; Betancourt, O.; Cueva, E.; Mergler, D. Decrease of Soil Fertility and Release of Mercury Following Deforestation in the Andean Amazon, Napo River Valley, Ecuador. *Sci. Total Environ.* **2006**, *368*, 88–98. [CrossRef]
38. Dorleku, M.K.; Nukpezah, D.; Carboo, D. Effects of Small-Scale Gold Mining on Heavy Metal Levels in Groundwater in the Lower Pra Basin of Ghana. *Appl. Water Sci.* **2018**, *8*, 1–11. [CrossRef]
39. Al-Hobaib, A.S.; Al-Jaseem, Q.K.; Baioumy, H.M.; Ahmed, A.H. Heavy Metals Concentrations and Usability of Groundwater at Mahd Adh Dhahab Gold Mine, Saudi Arabia. *Arab. J. Geosci.* **2013**, *6*, 259–270. [CrossRef]
40. Taiwo, A.M.; Awomeso, J.A. Assessment of Trace Metal Concentration and Health Risk of Artisanal Gold Mining Activities in Ijeshaland, Osun State Nigeria—Part 1. *J. Geochem. Explor.* **2017**, *177*, 1–10. [CrossRef]
41. Feng, Z.; Deng, L.; Guo, Y.; Guo, G.; Wang, L.; Zhou, G.; Huan, Y.; Liang, T. The Spatial Analysis, Risk Assessment and Source Identification for Mercury in a Typical Area with Multiple Pollution Sources in Southern China. *Environ. Geochem. Health* **2023**, *45*, 4057–4069. [CrossRef]
42. Khattak, S.A.; Rashid, A.; Tariq, M.; Ali, L.; Gao, X.; Ayub, M.; Javed, A. Potential Risk and Source Distribution of Groundwater Contamination by Mercury in District Swabi, Pakistan: Application of Multivariate Study. *Environ. Dev. Sustain.* **2021**, *23*, 2279–2297. [CrossRef]
43. Grandjean, P.; Budtz-Jørgensen, E.; Jørgensen, P.J.; Weihe, P. Umbilical Cord Mercury Concentration as Biomarker of Prenatal Exposure to Methylmercury. *Environ. Health Perspect.* **2005**, *113*, 905–908. [CrossRef] [PubMed]
44. World Health Organization. *The World Health Report 2005: Make Every Mother and Child Count*; World Health Organization: Geneva, Switzerland, 2005; ISBN 9241562900.
45. World Health Organization. *World Health Statistics 2008*; World Health Organization: Geneva, Switzerland, 2008; ISBN 9241563591.
46. Cano Polania, A.J.; Malagón Sanchez, M.C. *Evaluación de Trazas de Mercurio En El Tramo Caña Brava-Buenos Aires Del Río Cotuhé, Asociados a La Actividad Minera Artesanal de Oro En El Amazonas Colombiano*; Universidad de La Salle: Boston, MA, USA, 2016.
47. Manahan, S.E. *Introducción a La Química Ambiental*; Reverté: Barcelona, Spain, 2006; ISBN 8429179070.
48. Toro Ruiz, E. *Impactos Socioambientales Derivados Del Uso Del Mercurio En La Minería Aurífera Ilegal En El Bajo Cauca Antioqueño*; Tecnológico de Antioquia: Medellín, Colombia, 2021.
49. Hammond, D.S.; Rosales, J.; Ouboter, P.E. Gestión Del Impacto de La Explotación Minera a Cielo Abierto Sobre El Agua Dulce En América Latina. *Banco Interam. Desarro.* **2013**, *40*.
50. Santos-Francés, F.; García-Sánchez, A.; Alonso-Rojo, P.; Contreras, F.; Adams, M. Distribution and Mobility of Mercury in Soils of a Gold Mining Region, Cuyuni River Basin, Venezuela. *J. Environ. Manag.* **2011**, *92*, 1268–1276. [CrossRef] [PubMed]
51. Castilhos, Z.; Rodrigues-Filho, S.; Cesar, R.; Rodrigues, A.P.; Villas-Bôas, R.; de Jesus, I.; Lima, M.; Faial, K.; Miranda, A.; Brabo, E. Human Exposure and Risk Assessment Associated with Mercury Contamination in Artisanal Gold Mining Areas in the Brazilian Amazon. *Environ. Sci. Pollut. Res.* **2015**, *22*, 11255–11264. [CrossRef] [PubMed]

52. Lino, A.S.; Kasper, D.; Guida, Y.S.; Thomaz, J.R.; Malm, O. Total and Methyl Mercury Distribution in Water, Sediment, Plankton and Fish along the Tapajós River Basin in the Brazilian Amazon. *Chemosphere* **2019**, *235*, 690–700. [CrossRef] [PubMed]
53. Cordy, P.; Veiga, M.M.; Salih, I.; Al-Saadi, S.; Console, S.; Garcia, O.; Mesa, L.A.; Velásquez-López, P.C.; Roeser, M. Mercury Contamination from Artisanal Gold Mining in Antioquia, Colombia: The World's Highest per Capita Mercury Pollution. *Sci. Total Environ.* **2011**, *410*, 154–160. [CrossRef]
54. Restrepo, E.; Ciro, E. Residuos Mineros, Una Fuente Alternativa de Materias Primas. *Rev. Colomb. Mater.* **2021**, 1–2. [CrossRef]
55. Villegas Rosas, C.A. *Impacto Ambiental Por El Uso de Mercurio En Minería Aurífera: Una Revisión de La Literatura Científica Entre Los Años 2009–2019*; Universidad Privada del Norte: Trujillo, Peru, 2020.
56. Fernández-Martínez, R.; Loredo, J.; Ordóñez, A.; Rucandio, I. Mercury Availability by Operationally Defined Fractionation in Granulometric Distributions of Soils and Mine Wastes from an Abandoned Cinnabar Mine. *Environ. Sci. Process. Impacts* **2014**, *16*, 1069–1075. [CrossRef] [PubMed]

toxics

MDPI

Article

K252a Prevents Microglial Activation Induced by Anoxic Stimulation of Carotid Bodies in Rats

Ricardo Cuéllar-Pérez [1], Fernando Jauregui-Huerta [1], Yaveth Ruvalcaba-Delgadillo [1], Sergio Montero [2], Mónica Lemus [3], Elena Roces de Álvarez-Buylla [3], Joaquín García-Estrada [4] and Sonia Luquín [1,*]

1 Microscopía de Alta Resolución, Depto, de Neurociencias, Universidad de Guadalajara, Guadalajara 44340, Mexico; ricardo.cuellarp@academicos.udg.mx (R.C.-P.)
2 Facultad de Medicina, Universidad de Colima, Colima 28040, Mexico
3 Centro Universitario de Investigaciones Biomédicas, Universidad de Colima, Colima 28040, Mexico
4 División de Neurociencias, Centro de Investigación Biomédica de Occidente (CIBO), Instituto Mexicano del Seguro Social, Guadalajara 44340, Mexico
* Correspondence: sonialuqin@gmail.com

Abstract: Inducing carotid body anoxia through the administration of cyanide can result in oxygen deprivation. The lack of oxygen activates cellular responses in specific regions of the central nervous system, including the Nucleus Tractus Solitarius, hypothalamus, hippocampus, and amygdala, which are regulated by afferent pathways from chemosensitive receptors. These receptors are modulated by the brain-derived neurotrophic factor receptor TrkB. Oxygen deprivation can cause neuroinflammation in the brain regions that are activated by the afferent pathways from the chemosensitive carotid body. To investigate how microglia, a type of immune cell in the brain, respond to an anoxic environment resulting from the administration of NaCN, we studied the effects of blocking the TrkB receptor on this cell-type response. Male Wistar rats were anesthetized, and a dose of NaCN was injected into their carotid sinus to induce anoxia. Prior to the anoxic stimulus, the rats were given an intracerebroventricular (icv) infusion of either K252a, a TrkB receptor inhibitor, BDNF, or an artificial cerebrospinal fluid (aCSF). After the anoxic stimulus, the rats were perfused with paraformaldehyde, and their brains were processed for microglia immunohistochemistry. The results indicated that the anoxic stimulation caused an increase in the number of reactive microglial cells in the hypothalamic arcuate, basolateral amygdala, and dentate gyrus of the hippocampus. However, the infusion of the K252a TrkB receptor inhibitor prevented microglial activation in these regions.

Keywords: chemosensitive receptors; cyanide; microglia; BDNF; hypothalamus; amygdala; hippocampus

Citation: Cuéllar-Pérez, R.; Jauregui-Huerta, F.; Ruvalcaba-Delgadillo, Y.; Montero, S.; Lemus, M.; Roces de Álvarez-Buylla, E.; García-Estrada, J.; Luquín, S. K252a Prevents Microglial Activation Induced by Anoxic Stimulation of Carotid Bodies in Rats. *Toxics* **2023**, *11*, 871. https://doi.org/10.3390/toxics11100871

Academic Editors: Carina Ladeira and Pamela Lein

Received: 28 July 2023
Revised: 13 September 2023
Accepted: 16 October 2023
Published: 20 October 2023

1. Introduction

Recent studies have found that the carotid body chemoreceptors (CBCs) can modulate the activity of other physiological systems [1]. This modulation can lead, for example, to changes in blood pressure, heart rate, and respiratory rate [2,3]. Moreover, studies have shown that the CBCs are intimately involved in glucose homeostasis and energy metabolism. Specifically, they have been shown to sense changes in glucose levels in the bloodstream and signal to the brain to regulate food intake and energy expenditure. CBCs can also detect the levels of oxygen (O_2), carbon dioxide (CO_2), and pH in arterial blood [4–6].

Sodium cyanide (NaCN) causes oxidative stress in CBCs by blocking cytochrome C in the mitochondria, inducing a metabolic crisis and activating a chemosensory reflex [7]. This reflex involves the activation of multiple signaling pathways that are initiated by the CBCs and travel through the carotid sinus nerve to the brainstem's nucleus of the solitary tract (NTS). The NTS is a key site for the integration of sensory information from the CBCs as it receives inputs from various afferent fibers that are involved in the regulation of cardiorespiratory and autonomic function [8,9]. The activation of these pathways is crucial for the survival of the organism, as they trigger a series of responses that increase oxygen delivery to the tissues and maintain cellular homeostasis under hypoxic conditions. From there, the signals are transmitted to other regions of the brain, such as the hypothalamus, hippocampus, and amygdala, where they modulate a wide range of physiological processes, including stress responses, emotional regulation, and memory formation [10–13].

However, the activation of the chemosensory reflex also leads to an increase in reactive oxygen species (ROS) production and oxidative stress, which can cause cellular damage and trigger an inflammatory response in the brain. Moreover, traces of cyanide can reach the brain and stimulate microglia cells by hypoxia [12]. These responses are mainly mediated by microglia, the resident immune cells of the brain, which become activated, produce proinflammatory cytokines, and transform from a "resting" to an "amoeboid" phase [14,15]. The activation of microglia is a double-edged sword, as it is essential for clearing damaged cells and debris from the brain [16] but can also lead to neuroinflammation and neurodegeneration [17,18]. Moreover, it has been suggested that the activation of microglia and the subsequent inflammatory response may contribute to the development and progression of several neurological diseases, such as Alzheimer's disease and Parkinson's disease [19,20]. In fact, studies have shown that the activation of microglia can lead to neuronal damage and death, as well as the production of pro-inflammatory cytokines and reactive oxygen species (ROS) that can damage the surrounding tissue [21,22].

Recent studies have investigated the role of brain-derived neurotrophic factor (BDNF) and its receptor TrkB in the activation of microglia after hypoxic stimulation [23,24]. BDNF is a neurotrophic factor that is involved in the regulation of neuronal survival, differentiation, and synaptic plasticity, and it has been shown to be upregulated in response to hypoxic stimulation [25,26]. It is also a survival and proliferation factor [27–30] that can modulate the shift between acute and chronic neuroinflammation [31]. It is released by chemo-reflex pathways of the carotid body [32,33], expressed in microglial cells [27,28], and overexpressed in response to nerve damage [34]. TrkB is the high-affinity receptor for BDNF and is expressed in microglia, as well as in neurons and other glial cells [23,35]. The activation of the TrkB receptor in microglia has been shown to promote their proliferation, migration, and phagocytosis, as well as the production of pro-inflammatory cytokines [36,37], and K252a is a potent Trk pan-inhibitor phosphorylator [38,39]. However, it is still unclear how the activation of TrkB in microglia is regulated in response to hypoxic stimulation.

To further investigate the response of the dentate gyrus of the hippocampus (HDG), arcuate nucleus of the hypothalamus (Arc), and basolateral amygdala (BLA) to anoxic stimuli, we examined microglia transformation after NaCN administration in carotid bodies. We also assessed the role of BDNF and its TrkB receptor as possible mediators of these responses in order to better understand how carotid-body-induced inflammation is regulated.

2. Materials and Methods

2.1. Animals and Experimental Design

Male Swiss Wistar rats (N = 12, 280–300 g) were housed in individual transparent polycarbonate boxes with 12:12 h light–dark cycles at a temperature of 22–23 °C with normal laboratory rodent diet and water ad libitum. Until 15 days before the start of the study, they were handled for adaptation and to reduce stress. The rats were grouped, randomly, into the following four groups: (a) *Sham-operated*, received 10 μL/4 min of artificial cerebrospinal fluid (aCSF) within the left lateral ventricle (LV), followed by 100 μL of saline in the isolated carotid sinus (n = 3); (b) aCSF, received 10 μL/4 min of aCSF in the LV four minutes before the injection of NaCN (5 μg/100 g diluted in 100 μL of saline) in the isolated carotid sinus (n = 3); (c) BDNF, received a dose of BDNF (0.25 μg/10 μL of aCSF for 4 min) in the LV four minutes before the injection of NaCN in the isolated carotid sinus (n = 3); and (d) TrkB receptor inhibitor, K252a, received a dose of K252a (25 μg/10 μL of aCSF for 4 min) in the LV four minutes before the injection of NaCN in the isolated carotid sinus (n = 3). Afterwards, the animals were sacrificed by intracardiac perfusion 48 h after the NaCN stimulation; the protocols were performed according to the Guide for the Care and Use of Laboratory Animals of the United States National Institutes of Health.

2.2. Substances and Compounds

Substances and compounds used: (a) sodium cyanide (NaCN) (Fluka Biochemika, Sigma, Mexico. Dose: 5 µg/100 g/100 µL saline [5]; icv route, (b) freshly prepared artificial cerebrospinal fluid (aCSF containing: 145 mM NaCl, 2.7 mM KCl, 1.0 mM MgCl, 1.2 mM $CaCl_2$, 2.0 mM ascorbate, 2 mM NaH2PO4 (pH 7.3–7.4) [40]; (c) brain-derived neurotrophic factor (BDNF, 0.25 µg/10 µL of aCSF) [41]; (d) TrkB receptor inhibitor, K252a (25 µg/10 µL of aCSF) [42]. The same volume of aCSF was injected to the control rats.

2.3. Surgery

The rats were anesthetized with an intraperitoneal (ip) injection of sodium pentobarbital (Pisabental, PiSA Agropecuaria, Hidalgo, Mexico) diluted 1:1 in saline solution (5 mg/100 g); this anesthesia was implemented by an ip infusion of Nembutal in saline (0.063 mg/100 g/min/saline). To ensure deep anesthesia, blink and pinch reflexes were monitored frequently, and the animals were kept artificially ventilated with a tracheal cannula.

2.4. Intracerebroventricular Injection in the Left Lateral Ventricle (LV)

Unilateral microinjections of drugs were made using the stellar rat stereotaxic frame (Stoelting Co., Wood Dale, IL, USA). A craniotomy in the left LV was performed with a trephination bur of 1/32 inches to expose the surface of the skull. A glass micropipette (tip diameter 40–50 µm) was inserted using the bregma coordinates: AP = −1.3 mm, L = −1.8 mm, V = 3.6 mm [43]. Drugs were injected through a microinfusion pump (BASI Syringe Pump, MD-1000, West Lafayette, IN, USA) at a rate of 5 µL/min for 2 min, with a connected syringe of a polyethylene tube to the injection cannula. Finally, the cannula was removed, and the craniotomy was sealed with bone wax (Ethicon, Livingstone, UK).

2.5. Microglia (Griffonia simplicifolia)

The α-D-galactosil-specific isolectin B4 derived from *Griffonia simplicifolia* seed has shown selective staining in rat microglial cells in both normal and pathologically altered brains [44,45]. To do this, 48 h after icv infusion, rats were transcardially perfused with 37 °C isotonic saline, followed by ice-cold paraformaldehyde ((2.5%) in 0.1 M phosphate-buffered saline (PBS)). After the perfusion, the brains were immediately removed from the skulls and placed overnight in a fixative containing 1% paraformaldehyde in PBS and 30% sucrose at 4 °C. After two days of washing in PBS, the brainstem was sectioned coronally (30 µm) on a vibratome (Leica VT-1000E). Brain sections were incubated in a solution containing PBS 0.1M/Triton 0.1%/Albumin 0.1% (PBS-T-A) for 15 min at room temperature then transferred to a cation solution (PBS 0.1M, $MgCl_2$, $MnCl_2$, $CaCl_2$) for 10 min. After that, the tissues were washed with PBS and left to incubate in free-floating with Griffonia *Simplicifolia* Lectin I (GSL I) non-conjugated goat isolectin B4 (1:100, Vector L1104, Burlingame, CA, USA) for 13 h at 4 °C. Then, the tissues were washed and placed in a Tris-buffered saline solution, 0.1 M (TBS), with Triton X-100 0.1% (TBS-T). To block non-specific reactivity in the tissues, they were incubated in TBS-T with 2% normal rabbit serum for 60 min at room temperature. After washing with PBS-T-A, the brain slices were incubated with the primary antibody, non-conjugated anti-griffonia (Bandeiraea) simplicifolia lectin (GSL, BSL) from goat, at a dilution of 1:100, Vector AS-2104, Burlingame, CA, USA, for 15 h at 4 °C. Then, the slices were washed with PBS-T-A and incubated with the secondary antibody for 2 h at room temperature. Subsequently, the slices were incubated with the ABC complex (Vectastain kit, Vector Labs, Burlingame, CA, USA) for 2 h at room temperature, washed with PBS-T-A, followed by a wash with PBS, and a final wash with Tris-HCl 0.1 M. The peroxidase reaction product was visualized by incubating the slices for 8 min with a chromogenic solution to obtain a brown color (3,3′-diaminobencidina 0.07% and H_2O_2 0.01%) (Aldrich, St. Louis, MO, USA). Then, the slices were washed with PBS, 0.1M; this procedure is summarized in Table 1. Finally, the slices were mounted on slides, dried and dehydrated, and covered with Entellan (Merck, Darmstadt, Germany). The slices were analyzed with a Leica microscope (DMi8, Leica Microsystem, Wetzlar, Germany), and the images were recorded with the LAS AF Lite program (v4.6, Leica, Microsystem, Wetzlar, Germany). Morphometric analysis was performed by an investigator who was unaware of the identity of the experimental groups. For quantitative investigation, 10 equidistant pairs of adjacent sections were used. The numerical density of microglia cells was estimated by using the optical disector method in the brain regions mentioned above, using the total thickness of the slices for a disector height of 20X [46], and a two-dimensional point intersect template was calibrated so that the area associated with each point was known (125 × 75 μm). Microglia cells for which the clearest profiles fell within the disector volume and did not touch the left and bottom borders of the superimposed counting template nor the superior were counted. The estimated volume was indirectly calculated by multiplying the template × slice thickness × number of analyzed slices by the numerical density value using the disector method; this gives us an indirect estimation of the total number of microglia cells in each analyzed region [47]. In addition, the percentage of microglia cells with different morphologies was also evaluated. The cells were classified into five morphological types: Type I, cells with few cellular processes (two or less); Type II, cells displaying three to five extensions; Type III, cells with numerous (more than five) and long extensions and with a small cell body; Type IV, cells with large somas and thick and retracted extensions; and Type V, ameboid-shaped cells with numerous short extensions. For each animal, 120 cells were analyzed in each region ipsilateral to the CBC stimulation [48].

Table 1. Summarizes the general procedure of the immunohistochemical process.

(1) Chemical Union of GRIFFONIA to Glycocalix.	(2) Incubation with Primary Antibody (Ab).	(3) Incubation with Secondary Ab.	(4) Incubation with Avidin–Biotin Complex (ABC)
Griffonia simplicifolia Lectin I non-conjugated goat isolectin B4 1:100	Anti-griffonia (Bandeiraea) simplicifolia lectin goat 1:100	Biotinylated rabbit anti-goat IgG 1:150	Sandwich-type reaction and color development in chromogenic solution

2.6. Statistical Analysis

Values are presented as means ± SEM, and the number and morphology of microglia cells were compared among the four conditions (aCSF, BDNF, K252a, and *Sham-operated*) using a one-way ANOVA followed by Tukey's post hoc test. Values of $p < 0.05$ were considered statistically significant. The statistical analysis was performed using GraphPad Prism 5 (GraphPad Software Inc., San Diego, CA, USA).

3. Results

3.1. Stimulation of the CBC with NaCN Increases the Density of Microglia in the Amygdala, Dentate Gyrus Hilus, and Arcuate

The rats infused with aCSF prior to CBC stimulation with cyanide had an increase density of microglial cells in the basolateral amygdala (BLA), dentate gyrus hilus (HDG), and arcuate (Arc) (7229 ± 574, 6882 ± 706, 5647 ± 874 microglia cells/mm^3, respectively) compared to the reactivity displayed in the BLA, HDG, and Arc in the *Sham-operated* group (2743 ± 711, 3879 ± 927, 3048 ± 880 microglia cells/mm^3, respectively; p = 0.0007, p = 0.027 and p = 0.048, respectively) (Table 2, Figure 1A,B). Additionally, we found that the microglial morphology observed in the aCSF + NaCN group in greater quantity was type II in BLA (58%), type III in HDG (36%), and type II in Arc (54%), while the *Sham-operated* group mostly presented type I cells in the three regions analyzed (BLA 89%, HDG 57%, and

Arc 57%) in Table 3. These results indicate that anoxic stimulation of the CBC with NaCN activates the immune response in brain regions involved in the chemosensory response.

Table 2. Multiple comparisons on the number of microglia cells in response to each treatment.

	Area		
Drug	**BLA**	**HDG**	**Arc**
aCSF	7209 ± 488 *	6882 ± 706 *	5647 ± 873 *
BDNF	6095 ± 408 *	3694 ± 491 #	5012 ± 621
K252a	3556 ± 536	2241 ± 433 #	3288 ± 450 #
Sham-operated	2743 ± 711	3879 ± 927	3048 ± 880 *

These values are presented as means ± M.E.S. The asterisk sign (*) indicates a significant difference with the *Sham-operated* group, and the number sign (#) indicates a significant difference with the aCSF group (ANOVA, post hoc Tukey test).

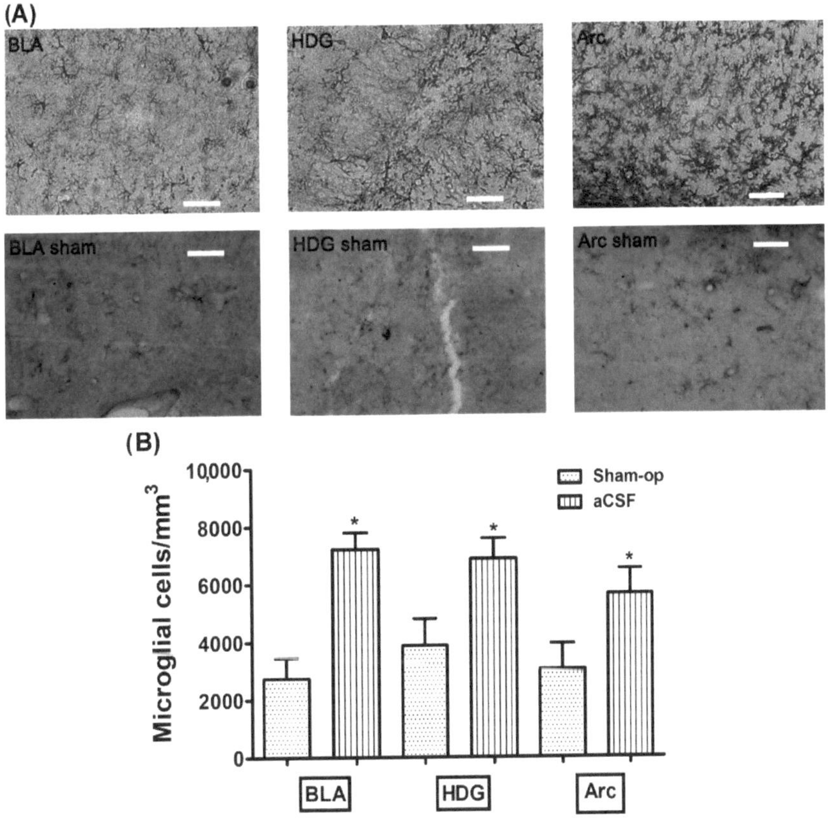

Figure 1. Stimulation of CBCs with NaCN increases the density of microglia in brain regions involved in the carotid chemoreflex. (**A**) The upper panel shows glial reactivity in the basolateral amygdala (BLA), the hilus of the dentate gyrus (HDG), and the arcuate nucleus (Arc), produced by aCSF infusion into the left LV prior to stimulation of CBCs with NaCN. (**B**) Shows the density of microglia cells in the analyzed regions. These values are presented as means ± S.E.M. The asterisk sign (*) indicates a significant difference with *Sham-operated* group (ANOVA, Tukey's post hoc test). Scale—50 μm.

Table 3. Morphology of microglia type in each area after NaCN infusion in CBCs for each treatment.

	Area		
Drug	**BLA**	**HDG**	**Arc**
aCSF	II (58%)	III (36%)	II (54%)
BDNF	I, II, III (29% each)	III (42%)	II (46%)
K252a	II (58%)	II, III (38% each)	II (67%)
Sham-operated	I (89%)	I (57%)	I (57%)

3.2. The Infusion of K252a in the Left LV after the Injection of NaCN in the CBC Prevented Microglial Reactivity in the Basolateral Amygdala

When the microglial reactivity in the basolateral amygdala (BLA) was analyzed, it was found that the density of microglia cells increased in the aCSF + NaCN group (7209 ± 488 microglia cells/mm^3) and in the BDNF + NaCN group (6095 ± 408 microglia cells/mm^3) compared to the *Sham-operated* group (p = 0.0005 and p = 0.0009). On the other hand, when the inhibitor K252a was administered, no reactions to the anoxic stimulus were observed (3556 ± 536 microglia cells/mm^3), resulting in a similar number to the *Sham-operated* group (2743 ± 711 microglia cells/mm^3 p = 0.459) (Table 2, Figure 2A,B). In relation to the morphological analysis, a diversity of microglial cells was present in this region. The K252a + NaCN group showed predominant type II cells (58%). The rats that received the icv infusion of BDNF + NaCN showed type I cells (a few cytoplasmic arborizations), type II cells (showing three to five branches), and type III cells, with small soma and numerous and long cytoplasmic branches (29% for each type) (Table 3, Figure 3).

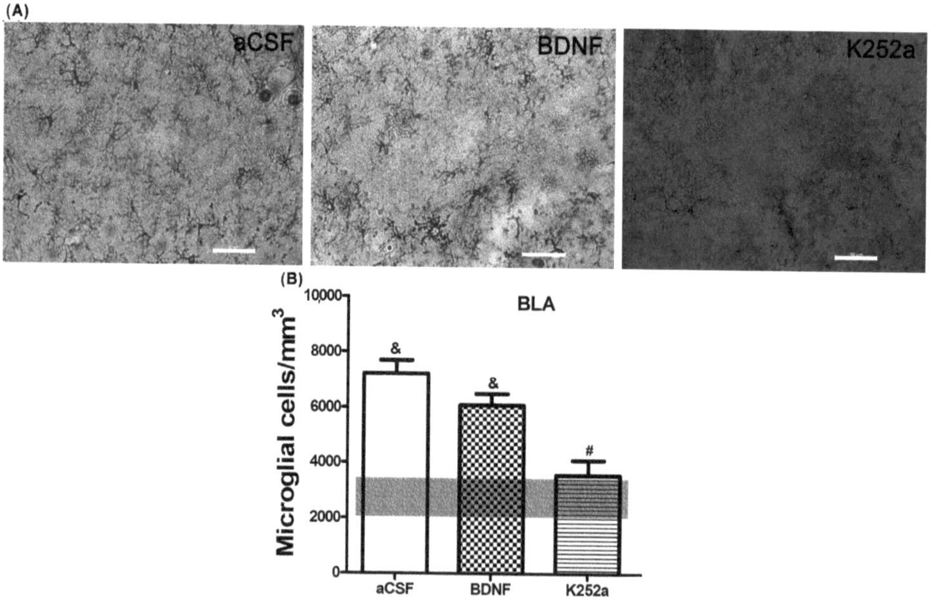

Figure 2. The K252a inhibitor does not increase microglial reactivity in the basolateral amygdala. (**A**) Decreased number density and morphological complexity were observed by K252a infusion into the left LV after NaCN injection into the CBC. (**B**) The transparent bar corresponds to the number of microglia cells in the *Sham-operated* group (2743 ± 711) that are statistically different to the aCSF and BDNF groups. These values are presented as means ± S.E.M. The ampersand sign (&) indicates a significant difference with the *Sham-operated* group, and the number sign (#) indicates a significant difference with the BDNF-infused group (ANOVA, Tukey's post hoc test). Scale—50 µm.

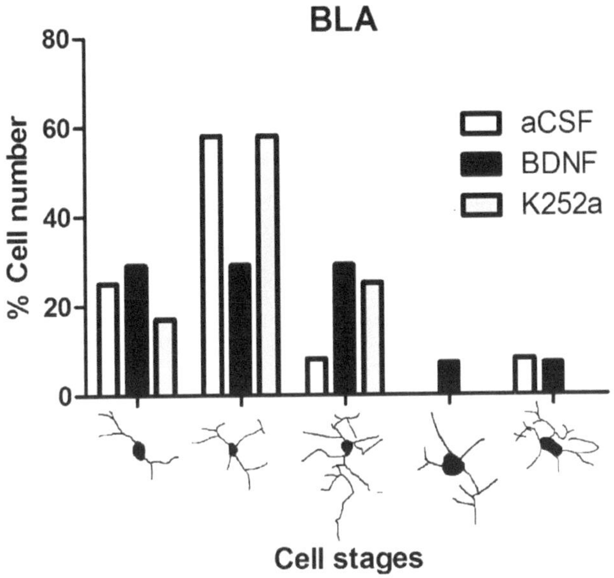

Figure 3. Morphological analysis of microglia cells in the basolateral amygdala. The graph shows the proportion of each morphological type (I–V) in the groups infused with aCSF, BDNF, or K252a. The main form observed was II, corresponding to the quiescence stage in the aCSF and K252a groups, and BDNF is probably a modulator of the activation state.

3.3. *Infusion of BDNF or K252a in the Left LV Prior to the CBC Stimulation Prevented the Increase in Microglial Cell Density in the Hilus of the Dentate Gyrus*

In the case of the dentate gyrus of the hippocampus, we found that the rats infused with BDNF + NaCN or K252a + NaCN had less microglial reactivity (3694 ± 491 microglia cells/mm^3 and 2241 ± 433 microglia cells/mm^3) compared to the aCSF + NaCN group (p = 0.0004 and p = 0.0000003, respectively) and that their numbers were similar to the *Sham-operated* group (p = 0.855 and p = 0.083, respectively) (Table 2, Figure 4A,B). There were no significant differences between these groups in relation to the microglia population. In this area, the microglial morphology in the BDNF group corresponded to type III (42%). The group infused with the inhibitor K252a exhibited types II and III (38% both) (Table 3, Figure 5).

3.4. *Blocking the TrkB Receptor Prior to CBC Stimulation Prevented the Increase in Microglia Density in the Arcuate Nucleus of the Hypothalamus*

Finally, upon analyzing the arcuate nucleus of the hypothalamus, we found that the rats infused with BDNF 4 min before the injection of NaCN had a similar amount of microglial cells (5012 ± 621 microglia cells/mm^3 p = 0.774) to the group infused with aCSF. On the other hand, the group of rats infused with the Trk inhibitor for the BDNF receptor, K252a, showed microglial reactivity (3288 ± 450 microglia cells/mm^3) similar to the *Sham-operated* group (p = 0.811) (Table 2, Figure 6A,B). The predominant microglia morphology in the three different NaCN treatments consisted of small cellular bodies with three to five cytoplasmic processes, corresponding to type II cells (46% BDNF + NaCN; 67% K252a + NaCN) (Table 3, Figure 7).

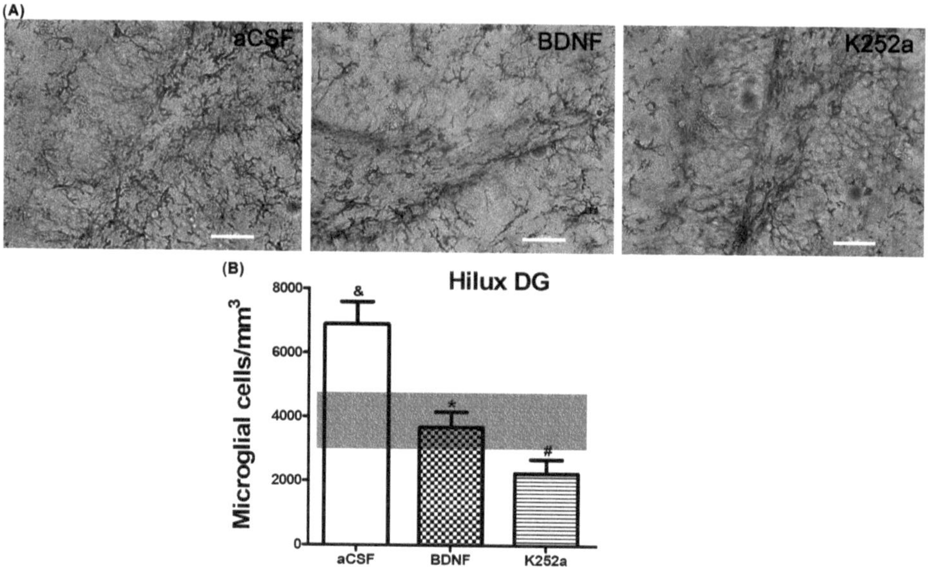

Figure 4. BDNF or K252a does not increase microglial cell density in the hilus of the dentate gyrus of the hippocampus. (**A**) BDNF or K252a infusion into the left LV prior to stimulation of CBCs with NaC has a down-regulating effect on active microglial cells. (**B**) The transparent bar corresponds to the number of microglia cells in the *Sham-operated* group (3879 ± 927) that are statistically different to the aCSF group. These values are presented as means ± S.E.M. The ampersand sign (&) indicates a significant difference with the *Sham-operated* group, the asterisk sign (*) indicates a significant difference with aCSF, and the number sign (#) indicates a significant difference with BDNF (ANOVA, post hoc Tukey test). Scale—50 μm.

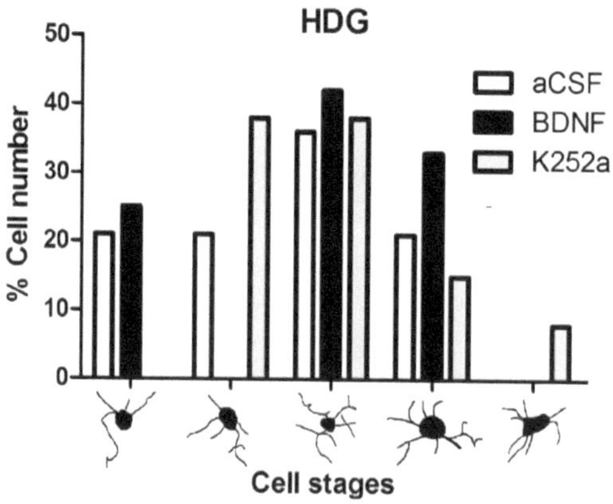

Figure 5. Morphological analysis of microglia cells in the hilus of the dentate gyrus. The graph shows the proportion of each morphological type in the groups infused with aCSF, BDNF, or K252a. Although some forms correspond to phagocytic microglia acting as intracerebral macrophages, these numbers are due to the normal variability of resident, nonproliferative microglia.

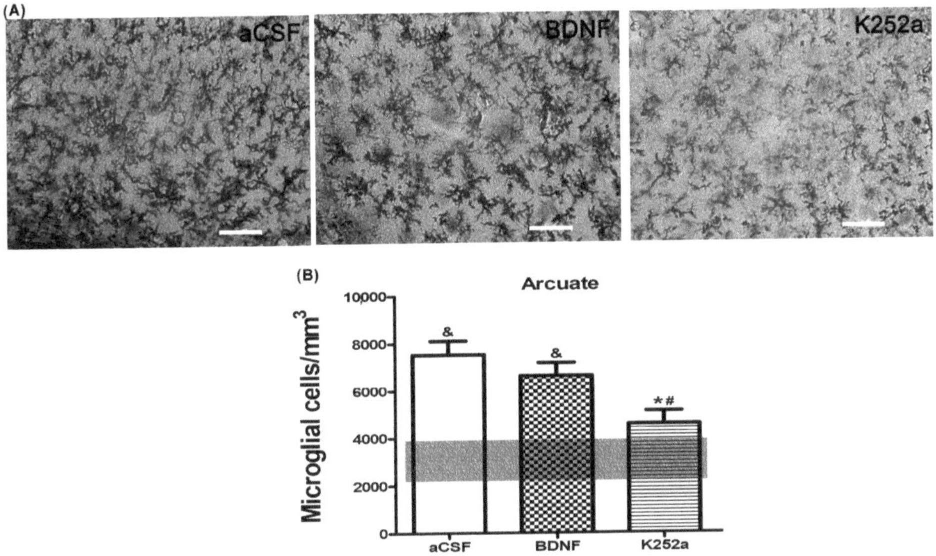

Figure 6. TrkB receptor blockade does not increase the density of microglia in the arcuate nucleus of the hypothalamus. (**A**) K252a infusion showed microglial reactivity (3288 ± 450 microglia cells/mm^3) similar to the *Sham-operated* group (p = 0.811), but the decrease was not significant compared to BDNF infusion. (**B**) The transparent bar corresponds to the number of microglia cells in the *Sham-operated* group (3048 ± 880) that are statistically different to the aCSF and BDNF groups. These values are presented as means ± S.E.M. The ampersand sign (&) indicates a significant difference with the *Sham-operated* group, the asterisk sign (*) indicates a significant difference with aCSF and the number sign (#) indicates a significant difference with BDNF (ANOVA, Tukey's post hoc test). Scale—50 µm.

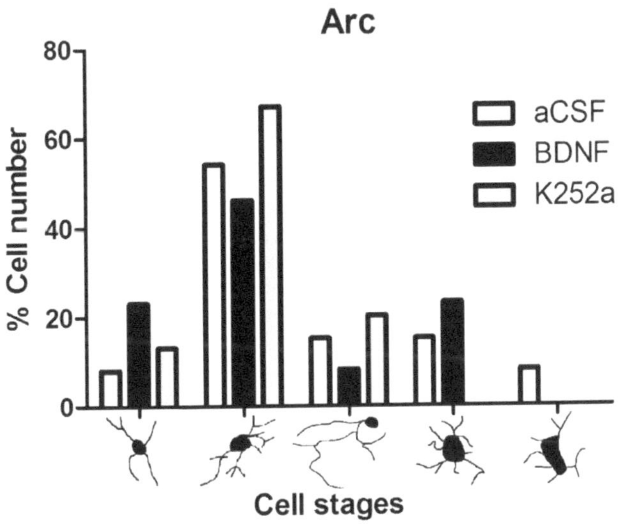

Figure 7. Morphological analysis of microglia cells in the arcuate nucleus of the hypothalamus. The graph shows the proportion of each morphological type in the groups infused with aCSF, BDNF, and K252a. The highest proportion found in the arcuate nucleus presents type II, which corresponds to a resting phenotype; stimulation of CBCs with NaCN was not sufficient to achieve microglial activation.

4. Discussion

In this study, we first evidenced microglial changes in regions related to the reflex chemosensory response after exposure to sodium cyanide (NaCN). Subsequently, we investigated the role of brain-derived neurotrophic factor (BDNF) and its receptor tyrosine kinase B (TrkB) in microglial activity induced by NaCN stimulation. Our results demonstrated that the manipulation of the BDNF–TrkB system was able to suppress the inflammatory response associated with microglia in the hypothalamus, hippocampus, and amygdala.

We briefly stimulated the carotid body with a sodium cyanide solution to induce a localized hypoxic response [5]. Then, we examined the potential relationship between NaCN stimulation and microglia response, which may involve BDNF [33,49,50]. To this end, one group of animals received an icv infusion of aCSF, while another group was administered BDNF, and a similar group was subjected to the TrkB inhibitor [51–53]. We compared the results to a control group of rats. The animals were sacrificed 48 h after the NaCN stimulus to ensure microglial reactivity. This model has been previously proven to induce cyanide toxicity by binding to the mitochondria, blocking electron transport, and inhibiting respiration [54], so it can cause severe CNS depression and death [55]. Our group has previously used this model to analyze c-fos activity and GluR2/3 receptor expression [56,57]. Since our previous studies evidenced that activity changes were not restricted to neurons, we proposed that astrocytes and microglia should be involved in the cerebrum's response to NaCN CBC stimulation. Furthermore, previous research validated our model, evidencing that after icv infusion, exogenous BDNF is dispersed throughout areas near the ventricles. Its diffusion depends on the distribution of its receptor TrkB [58,59]. However, incomplete forms of TrkB without the kinase domain may limit its diffusion [60].

Regional susceptibility was evaluated through measuring microglial reactivity in regions such as the hippocampus, amygdala, and hypothalamus. The hippocampus is an area related to memory and spatial learning that is particularly affected in animals exposed to intermittent hypoxia [61–63] or in patients with obstructive sleep apnea [64]. It has been demonstrated that hypoxic events may induce a loss of neurons in the hippocampus [65,66]. In addition, studies on ischemia–reperfusion evidenced that all layers of the dentate gyrus showed an activated form of microglia and greater cell numbers compared to a *Sham-operated* group [67]. Evidence also exists showing that the BDNF synthesized by microglia is present in various regions of the CNS during the course of various neurological disorders such as traumatic injury [68] and ischemia [69]. In our study, aCSF caused an increase in the density of microglia in the hilus region of the dentate gyrus of the hippocampus compared to the *Sham-operated* group, while exogenous BDNF and the inhibitor K252a caused a decrease in microglial reactivity after the NaCN-induced chemotoxic hypoxia. These effects were not observed in the other regions already analyzed. This was probably due to the fact that the dentate gyrus is the least excitable region since it has a strong inhibitory GABAergic innervation with very few excitatory synapses [70]. BDNF released by microglia acts on different types of neighboring cells, such as neurons, astrocytes, and possibly the same microglia, through a paracrine pathway. Yet, our results revealed that the regional reactivity of microglia in the dentate gyrus was lower compared to the other studied regions. In this line of evidence, other experiments demonstrated that microglial reactions may occur sooner in the dentate gyrus than in the CA1 zone, at least under hypothermic ischemia [71]. Then, it seems that this region is more resistant to an ischemic/hipoxic injury, while the CA1 region could be more vulnerable [72,73]. Therefore, this suggests that the effects of cyanide stimulation and BDNF/TrkB must vary according to regional parameters.

The basolateral amygdala (BLA) is also a region involved in the chemosensory reflex pathway that receives stimuli through the NTS from the brainstem [74]. We noted here that groups infused with aCSF and BDNF showed a greater number of microglia cells compared to the *Sham-operated* group, while tissue from the rats that received the inhibitor K252a showed a low expression of microglia compared to the same groups. Again, blocking the TrkB receptor in the BLA decreased the number of microglia after the hypoxic stimulation. This effect suggests that the stimulating action of BDNF on microglial reactivity caused by the hypoxic stimulus was inhibited. The rats that received the inhibitor K252a or aCSF showed microglia morphology type II (resting), while in the group with infusion of BDNF, the three predominant forms were observed: I, II, and III (activated). In other experiments, microglia have been detected 14 days after a cardiac infarction; these microglial cells show hypertrophy of the cytoplasm and their processes are retracted, while the *Sham-operated* group had small and thin cytoplasmic processes, indicating that the microglia were in their resting form [75]. An elevated synthesis of BDNF has been confirmed in different regions of the CNS after an ischemic event [76]. With this, the results of our work could indicate that the neurotrophin BDNF promotes the transformation of microglia acting in cells that have a different amount of TrkB receptors, so the effects can differentially be detected in these cells. Since the BLA region is an important area for emotional and cognitive processing, it follows that stimulation with NaCN could cause cognitive function disorders. More studies are needed to evaluate the effects of this stimulation on cognition and emotion.

Finally, we studied the regional susceptibility of the arcuate nucleus of the hypothalamus to NaCN stimulation since the chemo-reflex response to hypoxia is coordinated by a brain stem pathway, which includes the solitary tract nucleus (NTS), the rostral ventrolateral medulla (RVLM), and the medullary pathways projecting to the paraventricular nucleus of the hypothalamus [77]. The arcuate nucleus, which has abundant TrkB receptors, also regulates behavior, sleep, energy expenditure, and feeding [78,79]. We evidenced arcuate nucleus susceptibility to NaCN stimulation since more microglial cells were counted in the groups with aCSF and BDNF infusions compared to the *Sham-operated* group. However, we also evidenced that blocking the TrkB receptor with the inhibitor K252a prevented the increase in microglia, making the number of microglia similar to the amount in the *Sham-operated* group. As stated before, microglia have a primary function of responding to changes in the extracellular environment, protecting neurons, and promoting the repair of nervous tissue in pathological situations [21,80,81]. When faced with damaging signals, microglia change morphologically and functionally, resulting in migration, proliferation, and increased production and secretion of proinflammatory cytokines and neurotrophic factors. Such changes are known as "microglial activation" or microgliosis [82–87]. The activated microglia can modulate the inflammatory response to produce mediators that work to eliminate the source of the inflammatory stimulus. However, an excessive activation of microglia, as can occur in hypoxia, may result in chronic neuroinflammation and promote neurodegenerative diseases. Then, our findings suggest that the BDNF–TrkB system may play a critical role in regulating microglial activity and protecting against neurodegeneration. Studies from other groups shed light on the potential therapeutic benefits of targeting microglial activity in response to hypoxic stimuli. One study found that melatonin may be a viable pharmacological target for protecting against hypoxic-derived neuronal damage, as it attenuated the toxicity of chemical hypoxia on microglia through the involvement of SIRT1 and AMPK [87]. Another study demonstrated that therapeutic hypothermia could be effective in attenuating microglial activation and migration following brain damage caused by an ischemic stroke or a traumatic brain injury [88]. In addition, another study evidenced that microglia have the ability to secrete constitutive BDNF in a non-stimulated phase, which has anti-inflammatory effects that inhibit microglia activation [89,90]. Together, these findings suggest that targeting microglial activity may be a promising approach for treating neuroinflammatory conditions and protecting against hypoxic-induced neuronal damage. However, further research is needed to determine the optimal time window for therapeutic interventions and to better understand the mechanisms underlying these effects.

5. Conclusions

We demonstrated that the stimulation of CBCs with NaCN induces a strong inflammatory response mediated by microglial cells in the arcuate nucleus, basolateral amygdala, and dentate gyrus hilus of the hippocampus. Furthermore, we evidenced that brain-derived neurotrophic factor (BDNF) and its receptor TrkB are involved in NaCN stimulation. The TrkB receptor inhibitor, K252a, significantly inhibited microglial reactivity, indicating a role for the tyrosine-kinase receptor in these changes. Thus, TrkB receptor inhibition could diminished chronic neuroinflammation after anoxic stimulation.

Author Contributions: All authors have contributed to this article as follows: conceptualization, J.G.-E. and E.R.d.Á.-B.; data curation, R.C.-P. and F.J.-H.; writing—original draft preparation, R.C.-P., F.J.-H., J.G.-E. and S.L.; methodology in vivo and study materials, R.C.-P., S.M. and M.L.; performing and analysis of IHC experiments and visualizing data, R.C.-P., Y.R.-D. and S.L. All authors have read and agreed to the published version of the manuscript.

Funding: This research was funded by the Consejo Nacional de Ciencia y Tecnología (CONACyT México), grant number 177047, and by fondo fortalecimiento UdeG grant number REC/CINV/335/2022.

Institutional Review Board Statement: The animal study protocol was approved by the Bioethics and Biosecurity Committee of the University of Colima (No. 2013-02) for studies involving animals.

Informed Consent Statement: Not applicable.

Data Availability Statement: The data presented in this study are available on request from the corresponding author.

Conflicts of Interest: The authors declare no conflict of interest.

References

1. Joyner, M.J.; Limberg, J.K.; Wehrwein, E.A.; Johnson, B.D. Role of the carotid body chemoreceptors in glucose homeostasis and thermoregulation in humans. *J. Physiol.* **2018**, *596*, 3079–3085. [CrossRef]
2. Limberg, J.K.; Taylor, J.L.; Mozer, M.T.; Dube, S.; Basu, A.; Basu, R.; Rizza, R.A.; Curry, T.B.; Joyner, M.J.; Wehrwein, E.A. Effect of bilateral carotid body resection on cardiac baroreflex control of blood pressure during hypoglycemia. *Hypertension* **2015**, *65*, 1365–1371. [CrossRef] [PubMed]
3. Iturriaga, R.; Alcayaga, J.; Chapleau, M.W.; Somers, V.K. Carotid body chemoreceptors: Physiology, pathology, and implications for health and disease. *Physiol. Rev.* **2021**, *101*, 1177–1235. [CrossRef]
4. Prabhakar, N.R.; Peng, Y.J.; Kumar, G.K.; Nanduri, J. Peripheral chemoreception and arterial pressure responses to intermittent hypoxia. *Comprehe. Physiol.* **2015**, *5*, 561–577.
5. Alvarez-Buylla, R.; de Alvarez-Buylla, E.R. Carotid sinus receptors participate in glucose homeostasis. *Respir. Physiol.* **1988**, *72*, 347–359. [CrossRef]
6. Pardal, R.; López-Barneo, J. Low glucose-sensing cells in the carotid body. *Nat. Neurosci.* **2002**, *5*, 197–198. [CrossRef] [PubMed]
7. Isom, G.E.; Way, J.L. Effects of oxygen on the antagonism of cyanide intoxication: Cytochrome oxidase, in vitro. *Toxicol. Appl. Pharmacol.* **1984**, *74*, 57–62. [CrossRef] [PubMed]
8. Housley, G.D.; Sinclair, J.D. Localization by kainic acid lesions of neurones transmitting the carotid chemoreceptor stimulus for respiration in rat. *J. Physiol.* **1988**, *406*, 99–114. [CrossRef] [PubMed]
9. Ruyle, B.C.; Klutho, P.J.; Baines, C.P.; Heesch, C.M.; Hasser, E.M. Hypoxia activates a neuropeptidergic pathway from the paraventricular nucleus of the hypothalamus to the nucleus tractus solitarii. *Am. J. Physiol. Regul. Integr. Comp. Physiol.* **2018**, *315*, R1167–R1182. [CrossRef]
10. Luo, Y.; Kaur, C.; Ling, E.A. Hypobaric hypoxia induces fos and neuronal nitric oxide synthase expression in the paraventricular and supraoptic nucleus in rats. *Neurosci. Lett.* **2000**, *296*, 145–148. [CrossRef]
11. Bабović, S.S.; Žigić, S.; Šakić, B. c-Fos protein expression in the anterior amygdaloid area and nc. accumbens in the hypoxic rat brain. *Med. Pregl.* **2014**, *67*, 379–384. [CrossRef]
12. Tadmouri, A.; Champagnat, J.; Morin-Surun, M.P. Activation of microglia and astrocytes in the nucleus tractus solitarius during ventilatory acclimatization to 10% hypoxia in unanesthetized mice. *J. Neurosci. Res.* **2014**, *92*, 627–633. [CrossRef] [PubMed]
13. Qiao, Y.; Liu, Z.; Yan, X.; Luo, C. Effect of intermittent hypoxia on neuro-functional recovery post brain ischemia in mice. *J. Mol. Neurosci.* **2015**, *55*, 923–930. [CrossRef] [PubMed]
14. Streit, W.J. The role of microglia in neurotoxicity. In *The Role of Glia in Neurotoxicity*; Aschner, M., Costa, L.G., Eds.; CRC Press: Boca Raton, FL, USA, 1996; pp. 29–40.

15. Graeber, M.B.; Li, W.; Rodriguez, M.L. Role of microglia in CNS inflammation. *FEBS Lett.* **2011**, *585*, 3798–3805. [CrossRef] [PubMed]
16. Hong, S.; Beja-Glasser, V.F.; Nfonoyim, B.M.; Frouin, A.; Li, S.; Ramakrishnan, S.; Merry, K.M.; Shi, Q.; Rosenthal, A.; Barres, B.A.; et al. Complement and microglia mediate early synapse loss in Alzheimer mouse models. *Science* **2016**, *352*, 712–716. [CrossRef]
17. Rodriguez, J.I.; Kern, J.K. Evidence of microglial activation in autism and its possible role in brain underconnectivity. *Neuron. Glia. Biol.* **2011**, *7*, 205–213. [CrossRef]
18. Lyman, M.; Lloyd, D.G.; Ji, X.; Vizcaychipi, M.P.; Ma, D. Neuroinflammation: The role and consequences. *Neurosci. Res.* **2014**, *79*, 1–12. [CrossRef]
19. Bachstetter, A.D.; Xing, B.; de Almeida, L.; Dimayuga, E.R.; Watterson, D.M.; Van Eldik, L.J. Microglial p38α MAPK is a key regulator of proinflammatory cytokine up-regulation induced by toll-like receptor (TLR) ligands or beta-amyloid (Aβ). *J. Neuroinflammation* **2011**, *8*, 79. [CrossRef]
20. Hirsch, E.C.; Vyas, S.; Hunot, S. Neuroinflammation in Parkinson's disease. *Parkinsonism Relat. Disord.* **2012**, *18* (Suppl. 1), S210–S212. [CrossRef]
21. Nimmerjahn, A.; Kirchhoff, F.; Helmchen, F. Resting microglial cells are highly dynamic surveillants of brain parenchyma in vivo. *Science* **2005**, *308*, 1314–8131. [CrossRef]
22. Colonna, M.; Butovsky, O. Microglia Function in the Central Nervous System During Health and Neurodegeneration. *Annu. Rev. Immunol.* **2017**, *35*, 441–468. [CrossRef] [PubMed]
23. Wu, G.; Chen, Z.; Wang, P.; Zhao, M.; Fujino, M.; Zhang, C.; Zhou, W.; Hirano, S.I.; Li, X.K.; Zhao, L. Hydrogen inhalation protects hypoxic-ischemic brain damage by attenuating inflammation and apoptosis in neonatal rats. *Exp. Biol. Med.* **2019**, *244*, 1017–1027. [CrossRef] [PubMed]
24. Zhang, J.; Rong, P.; Zhang, L.; He, H.; Zhou, T.; Fan, Y.; Mo, L.; Zhao, Q.; Han, Y.; Li, S.; et al. IL4-driven microglia modulate stress resilience through BDNF-dependent neurogenesis. *Sci. Adv.* **2021**, *7*, eabb9888. [CrossRef] [PubMed]
25. Sharma, D.; Barhwal, K.K.; Biswal, S.N.; Srivastava, A.K.; Bhardwaj, P.; Kumar, A.; Chaurasia, O.P.; Hota, S.K. Hypoxia-mediated alteration in cholesterol oxidation and raft dynamics regulates BDNF signalling and neurodegeneration in hippocampus. *J. Neurochem.* **2019**, *148*, 238–251. [CrossRef] [PubMed]
26. Kerschensteiner, M.; Stadelmann, C.; Dechant, G.; Wekerle, H.; Hohlfeld, R. Neurotrophic cross-talk between the nervous and immune systems: Implications for neurological diseases. *Ann. Neurol.* **2003**, *53*, 292–304. [CrossRef]
27. Madinier, A.; Bertrand, N.; Mossiat, C.; Prigent-Tessier, A.; Beley, A.; Marie, C.; Garnier, P. Microglial involvement in neuroplastic changes following focal brain ischemia in rats. *PLoS ONE* **2009**, *4*, e8101. [CrossRef]
28. Sun, P.; Li, X.; Chen, C.; Chen, Q.; Ouyang, Q.; Liu, F.; Xiang, Z.; Yuan, H. Activating transcription factor 4 modulates BDNF release from microglial cells. *J. Mol. Neurosci.* **2014**, *52*, 225–230. [CrossRef]
29. Huang, Y.; Cheung, L.; Rowe, D.; Halliday, G. Genetic contributions to Parkinson's disease. *Brain Res. Brain Res. Rev.* **2004**, *46*, 44–70. [CrossRef]
30. Imai, F.; Suzuki, H.; Oda, J.; Ninomiya, T.; Ono, K.; Sano, H.; Sawada, M. Neuroprotective effect of exogenous microglia in global brain ischemia. *J. Cereb. Blood. Flow. Metab.* **2007**, *27*, 488–500. [CrossRef]
31. Tansey, M.G.; Goldberg, M.S. Neuroinflammation in Parkinson's disease: Its role in neuronal death and implications for therapeutic intervention. *Neurobiol. Dis.* **2010**, *37*, 510–518. [CrossRef]
32. Chavez-Valdez, R.; Mason, A.; Nunes, A.R.; Northington, F.J.; Tankersley, C.; Ahlawat, R.; Johnson, S.M.; Gauda, E.B. Effect of hyperoxic exposure during early development on neurotrophin expression in the carotid body and nucleus tractus solitarii. *J. Appl. Physiol.* **2012**, *112*, 1762–1772. [CrossRef] [PubMed]
33. Montero, S.; Cuéllar, R.; Lemus, M.; Avalos, R.; Ramírez, G.; de Álvarez-Buylla, E.R. Brain-derived neurotrophic factor in the nucleus tractus solitarii modulates glucose homeostasis after carotid chemoreceptor stimulation in rats. *Adv. Exp. Med. Biol.* **2012**, *758*, 233–239. [PubMed]
34. Thibault, K.; Lin, W.K.; Rancillac, A.; Fan, M.; Snollaerts, T.; Sordoillet, V.; Hamon, M.; Smith, G.M.; Lenkei, Z.; Pezet, S. BDNF-dependent plasticity induced by peripheral inflammation in the primary sensory and the cingulate cortex triggers cold allodynia and reveals a major role for endogenous BDNF as a tuner of the affective aspect of pain. *J. Neurosci.* **2014**, *34*, 14739–14751. [CrossRef] [PubMed]
35. Zhou, L.J.; Yang, T.; Wei, X.; Liu, Y.; Xin, W.J.; Chen, Y.; Pang, R.P.; Zang, Y.; Li, Y.Y.; Liu, X.G. Brain-derived neurotrophic factor contributes to spinal long-term potentiation and mechanical hypersensitivity by activation of spinal microglia in rat. *Brain Behav. Immun.* **2011**, *25*, 322–334. [CrossRef] [PubMed]
36. Ding, H.; Chen, J.; Su, M.; Lin, Z.; Zhan, H.; Yang, F.; Li, W.; Xie, J.; Huang, Y.; Liu, X.; et al. BDNF promotes activation of astrocytes and microglia contributing to neuroinflammation and mechanical allodynia in cyclophosphamide-induced cystitis. *J. Neuroinflamm.* **2020**, *17*, 19. [CrossRef]
37. Infantino, R.; Schiano, C.; Luongo, L.; Paino, S.; Mansueto, G.; Boccella, S.; Guida, F.; Ricciardi, F.; Iannotta, M.; Belardo, C.; et al. MED1/BDNF/TrkB pathway is involved in thalamic hemorrhage-induced pain and depression by regulating microglia. *Neurobiol. Dis.* **2022**, *164*, 105611. [CrossRef]

38. Nakajima, K.; Kikuchi, Y.; Ikoma, E.; Honda, S.; Ishikawa, M.; Liu, Y.; Kohsaka, S. Neurotrophins regulate the function of cultured microglia. *Glia* **1998**, *24*, 272–289. [CrossRef]
39. Tanaka, K.; Okugawa, Y.; Toiyama, Y.; Inoue, Y.; Saigusa, S.; Kawamura, M.; Araki, T.; Uchida, K.; Mohri, Y.; Kusunoki, M. Brain-derived neurotrophic factor (BDNF)-induced tropomyosin-related kinase B (Trk B) signaling is a potential therapeutic target for peritoneal carcinomatosis arising from colorectal cancer. *PLoS ONE* **2014**, *9*, e96410. [CrossRef]
40. Thielen, R.J.; Engleman, E.A.; Rodd, Z.A.; Murphy, J.M.; Lumeng, L.; Li, T.K.; McBride, W.J. Ethanol drinking and deprivation alter dopaminergic and serotonergic function in the nucleus accumbens of alcohol-preferring rats. *J. Pharmacol. Exp. Ther.* **2004**, *309*, 216–225. [CrossRef]
41. Valvassori, S.S.; Arent, C.O.; Steckert, A.V.; Varela, R.B.; Jornada, L.K.; Tonin, P.T.; Budni, J.; Mariot, E.; Kapczinski, F.; Quevedo, J. Intracerebral Administration of BDNF Protects Rat Brain against Oxidative Stress Induced by Ouabain in an Animal Model of Mania. *Mol. Neurobiol.* **2015**, *52*, 353–362. [CrossRef]
42. Fukuhara, K.; Ishikawa, K.; Yasuda, S.; Kishishita, Y.; Kim, H.K.; Kakeda, T.; Yamamoto, M.; Norii, T.; Ishikawa, T. Intracerebroventricular 4-methylcatechol (4-MC) ameliorates chronic pain associated with depression-like behavior via induction of brain-derived neurotrophic factor (BDNF). *Cell. Mol. Neurobiol.* **2012**, *32*, 971–977. [CrossRef] [PubMed]
43. Paxinos, G.; Watson, C. *The Rat Brain in Stereotaxic Coordinates*; Academic Press: New York, NY, USA, 1986.
44. Streit, W.J.; Schulte, B.A.; Balentine, D.J.; Spicer, S.S. Histochemical localization of galactose-containing glycoconjugates in sensory neurons and their processes in the central and peripheral nervous system of the rat. *J. Histochem. Cytochem.* **1985**, *33*, 1042–1052. [CrossRef] [PubMed]
45. Streit, W.J.; Kreutzberg, G.W. Lectin binding by resting and reactive microglia. *J. Neurocytol.* **1987**, *16*, 249–260. [CrossRef] [PubMed]
46. Hatton, W.J.; von Bartheld, C.S. Analysis of cell death in the trochlear nucleus of the chick embryo: Calibration of the optical disector counting method reveals systematic bias. *J. Comp. Neurol.* **1999**, *409*, 169–186. [CrossRef]
47. Kaplan, S.; Odaci, E.; Canan, S.; Onger, M.E.; Aslan, H.; Unal, B. The Disector counting technique. *NeuroQuantology* **2012**, *10*, 44–53. [CrossRef]
48. Diz-Chaves, Y.; Pernía, O.; Carrero, P.; Garcia-Segura, L.M. Prenatal stress causes alterations in the morphology of microglia and the inflammatory response of the hippocampus of adult female mice. *J. Neuroinflamm.* **2012**, *9*, 71. [CrossRef]
49. Clark, C.G.; Hasser, E.M.; Kunze, D.L.; Katz, D.M.; Kline, D.D. Endogenous brain-derived neurotrophic factor in the nucleus tractus solitarius tonically regulates synaptic and autonomic function. *J. Neurosci.* **2011**, *31*, 12318–12329. [CrossRef]
50. Mizoguchi, Y.; Kato, T.A.; Seki, Y.; Ohgidani, M.; Sagata, N.; Horikawa, H.; Yamauchi, Y.; Sato-Kasai, M.; Hayakawa, K.; Inoue, R.; et al. Brain-derived neurotrophic factor (BDNF) induces sustained intracellular Ca^{2+} elevation through the up-regulation of surface transient receptor potential 3 (TRPC3) channels in rodent microglia. *J. Biol. Chem.* **2014**, *289*, 18549–18555. [CrossRef]
51. Barbacid, M. The Trk family of neurotrophin receptors. *J. Neurobiol.* **1994**, *25*, 1386–1403. [CrossRef]
52. Wang, Y.; Hagel, C.; Hamel, W.; Müller, S.; Kluwe, L.; Westphal, M. Trk A, B, and C are commonly expressed in human astrocytes and astrocytic gliomas but not by human oligodendrocytes and oligodendroglioma. *Acta. Neuropathol.* **1998**, *96*, 357–364. [CrossRef]
53. Spencer-Segal, J.L.; Waters, E.M.; Bath, K.G.; Chao, M.V.; McEwen, B.S.; Milner, T.A. Distribution of phosphorylated TrkB receptor in the mouse hippocampal formation depends on sex and estrous cycle stage. *J. Neurosci.* **2011**, *31*, 6780–6790. [CrossRef] [PubMed]
54. Geller, R.J.; Barthold, C.; Saiers, J.A.; Hall, A.H. Pediatric cyanide poisoning: Causes, manifestations, management, and unmet needs. *Pediatrics* **2006**, *118*, 2146–2158. [CrossRef] [PubMed]
55. Hawk, M.A.; Ritchie, G.D.; Henderson, K.A.; Knostman, K.A.; Roche, B.M.; Ma, Z.J.; Matthews, C.M.; Sabourin, C.L.; Wakayama, E.J.; Sabourin, P.J. Neurobehavioral and Cardiovascular Effects of Potassium Cyanide Administered Orally to Mice. *Int. J. Toxicol.* **2016**, *35*, 604–615. [CrossRef]
56. Cuéllar, R.; Montero, S.; Luquín, S.; García-Estrada, J.; Dobrovinskaya, O.; Melnikov, V.; Lemus, M.; de Álvarez-Buylla, E.R. Glutamatergic Receptor Activation in the Commisural Nucleus Tractus Solitarii (cNTS) Mediates Brain Glucose Retention (BGR) Response to Anoxic Carotid Chemoreceptor (CChr) Stimulation in Rats. *Adv. Exp. Med. Biol.* **2015**, *860*, 301–308. [PubMed]
57. Cuéllar, R.; Montero, S.; Luquín, S.; García-Estrada, J.; Melnikov, V.; Virgen-Ortiz, A.; Lemus, M.; Pineda-Lemus, M.; de Álvarez-Buylla, E. BDNF and AMPA receptors in the cNTS modulate the hyperglycemic reflex after local carotid body NaCN stimulation. *Auton. Neurosci.* **2017**, *205*, 12–20. [CrossRef] [PubMed]
58. Merlio, J.P.; Ernfors, P.; Jaber, M.; Persson, H. Molecular cloning of rat trkC and distribution of cells expressing messenger RNAs for members of the trk family in the rat central nervous system. *Neuroscience* **1992**, *51*, 513–532. [CrossRef]
59. Yan, Q.; Matheson, C.; Sun, J.; Radeke, M.J.; Feinstein, S.C.; Miller, J.A. Distribution of intracerebral ventricularly administered neurotrophins in rat brain and its correlation with trk receptor expression. *Exp. Neurol.* **1994**, *127*, 23–36. [CrossRef]
60. Croll, S.D.; Chesnutt, C.R.; Rudge, J.S.; Acheson, A.; Ryan, T.E.; Siuciak, J.A.; DiStefano, P.S.; Wiegand, S.J.; Lindsay, R.M. Co-infusion with a TrkB-Fc receptor body carrier enhances BDNF distribution in the adult rat brain. *Exp. Neurol.* **1998**, *152*, 20–33. [CrossRef]
61. Rudge, J.S.; Li, Y.; Pasnikowski, E.M.; Mattsson, K.; Pan, L.; Yancopoulos, G.D.; Wiegand, S.J.; Lindsay, R.M.; Ip, N.Y. Neurotrophic factor receptors and their signal transduction capabilities in rat astrocytes. *Eur. J. Neurosci.* **1994**, *6*, 693–705. [CrossRef]

62. Ju, J.; Chen, R.; Li, T.; Wang, J.; Xiao, Y.; Liu, C. Modulation of hippocampal synaptic transmission in a mouse model of chronic intermittent hypoxia: Implications for neurocognitive impairments. *Zhonghua Yi Xue Za Zhi.* **2016**, *96*, 610–614.
63. Fanselow, M.S.; Dong, H.W. Are the dorsal and ventral hippocampus functionally distinct structures? *Neuron* **2010**, *65*, 7–19. [CrossRef] [PubMed]
64. Gozal, D.; Daniel, J.M.; Dohanich, G.P. Behavioral and anatomical correlates of chronic episodic hypoxia during sleep in the rat. *J. Neurosci.* **2001**, *21*, 2442–2450. [CrossRef] [PubMed]
65. Bucks, R.S.; Olaithe, M.; Eastwood, P. Neurocognitive function in obstructive sleep apnoea: A meta-review. *Respirology* **2013**, *18*, 61–70. [CrossRef]
66. Torelli, F.; Moscufo, N.; Garreffa, G.; Placidi, F.; Romigi, A.; Zannino, S.; Bozzali, M.; Fasano, F.; Giulietti, G.; Djonlagic, I.; et al. Cognitive profile and brain morphological changes in obstructive sleep apnea. *Neuroimage* **2011**, *54*, 787–793. [CrossRef]
67. Ahn, J.H.; Shin, B.N.; Park, J.H.; Kim, I.H.; Cho, J.H.; Chen, B.; Lee, T.K.; Tae, H.J.; Lee, J.C.; Cho, J.H.; et al. Long-term observation of neuronal degeneration and microgliosis in the gerbil dentate gyrus after transient cerebral ischemia. *J. Neurol. Sci.* **2016**, *363*, 21–26. [CrossRef]
68. Dougherty, K.D.; Dreyfus, C.F.; Black, I.B. Brain-derived neurotrophic factor in astrocytes, oligodendrocytes, and microglia/macrophages after spinal cord injury. *Neurobiol. Dis.* **2000**, *7*, 574–585. [CrossRef]
69. Lee, Y.; Lee, S.R.; Choi, S.S.; Yeo, H.G.; Chang, K.T.; Lee, H.J. Therapeutically targeting neuroinflammation and microglia after acute ischemic stroke. *Biomed. Res. Int.* **2014**, *2014*, 297241. [CrossRef]
70. Chamberland, S.; Topolnik, L. Inhibitory control of hippocampal inhibitory neurons. *Front. Neurosci.* **2012**, *6*, 165. [CrossRef] [PubMed]
71. Yamashita, A.; Kunimatsu, T.; Yamada, K.; Kojo, A.; Yamamoto, T.; Sato, S.; Onozuka, M. Hypothermic and normothermic ischemia-reperfusion activate microglia differently in hippocampal formation. *Arch. Histol. Cytol.* **2010**, *73*, 73–80. [CrossRef] [PubMed]
72. Cervós-Navarro, J.; Diemer, N.H. Selective vulnerability in brain hypoxia. *Crit. Rev. Neurobiol.* **1991**, *6*, 149–182. [PubMed]
73. Blanquet, P.R.; Mariani, J.; Fournier, B. Identification of a biphasic signaling pathway involved in ischemic resistance of the hippocampal dentate gyrus. *Exp. Neurol.* **2006**, *202*, 357–372. [CrossRef]
74. Aston-Jones, G.; Waterhouse, B. Locus coeruleus: From global projection system to adaptive regulation of behavior. *Brain Res.* **2016**, *1645*, 75–78. [CrossRef]
75. Bae, E.; Hwang, I.K.; Yoo, K.Y.; Han, T.H.; Lee, C.H.; Choi, J.H.; Yi, S.S.; Lee, S.Y.; Ryu, P.D.; Yoon, Y.S.; et al. Gliosis in the amygdala following myocardial infarction in the rat. *J. Vet. Med. Sci.* **2010**, *72*, 1041–1045. [CrossRef]
76. Lee, T.H.; Kato, H.; Chen, S.T.; Kogure, K.; Itoyama, Y. Expression disparity of brain-derived neurotrophic factor immunoreactivity and mRNA in ischemic hippocampal neurons. *Neuroreport* **2002**, *13*, 2271–2275. [CrossRef]
77. Coldren, K.M.; Li, D.P.; Kline, D.D.; Hasser, E.M.; Heesch, C.M. Acute hypoxia activates neuroendocrine, but not presympathetic, neurons in the paraventricular nucleus of the hypothalamus: Differential role of nitric oxide. *Am. J. Physiol. Regul. Integr. Comp. Physiol.* **2017**, *312*, R982–R995. [CrossRef]
78. Wang, C.; Bomberg, E.; Billington, C.; Levine, A.; Kotz, C.M. Brain-derived neurotrophic factor in the hypothalamic paraventricular nucleus reduces energy intake. *Am. J. Physiol. Regul. Integr. Comp. Physiol.* **2007**, *293*, R1003–R1012. [CrossRef]
79. Teubner, B.J.; Leitner, C.; Thomas, M.A.; Ryu, V.; Bartness, T.J. An intact dorsomedial posterior arcuate nucleus is not necessary for photoperiodic responses in Siberian hamsters. *Horm. Behav.* **2015**, *70*, 22–29. [CrossRef]
80. Streit, W.J.; Mrak, R.E.; Griffin, W.S. Microglia and neuroinflammation: A pathological perspective. *J. Neuroinflammation* **2004**, *1*, 14. [CrossRef]
81. Kreutzberg, G.W. Microglia: A sensor for pathological events in the CNS. *Trends Neurosci.* **1996**, *19*, 312–318. [CrossRef]
82. Miwa, T.; Furukawa, S.; Nakajima, K.; Furukawa, Y.; Kohsaka, S. Lipopolysaccharide enhances synthesis of brain-derived neurotrophic factor in cultured rat microglia. *J. Neurosci. Res.* **1997**, *50*, 1023–1029. [CrossRef]
83. Ulmann, L.; Hatcher, J.P.; Hughes, J.P.; Chaumont, S.; Green, P.J.; Conquet, F.; Buell, G.N.; Reeve, A.J.; Chessell, I.P.; Rassendren, F. Up-regulation of P2X4 receptors in spinal microglia after peripheral nerve injury mediates BDNF release and neuropathic pain. *J. Neurosci.* **2008**, *28*, 11263–11268. [CrossRef]
84. Fujita, R.; Ma, Y.; Ueda, H. Lysophosphatidic acid-induced membrane ruffling and brain-derived neurotrophic factor gene expression are mediated by ATP release in primary microglia. *J. Neurochem.* **2008**, *107*, 152–160. [CrossRef]
85. Trang, T.; Beggs, S.; Wan, X.; Salter, M.W. P2X4-receptor-mediated synthesis and release of brain-derived neurotrophic factor in microglia is dependent on calcium and p38-mitogen-activated protein kinase activation. *J. Neurosci.* **2009**, *29*, 3518–3528. [CrossRef]
86. Sastre, M.; Richardson, J.C.; Gentleman, S.M.; Brooks, D.J. Inflammatory risk factors and pathologies associated with Alzheimer's disease. *Curr. Alzheimer. Res.* **2011**, *8*, 132–141. [CrossRef]
87. Merlo, S.; Luaces, J.P.; Spampinato, S.F.; Toro-Urrego, N.; Caruso, G.I.; D'Amico, F.; Capani, F.; Sortino, M.A. SIRT1 Mediates Melatonin's Effects on Microglial Activation in Hypoxia: In Vitro and In Vivo Evidence. *Biomolecules* **2020**, *10*, 364. [CrossRef]
88. Seo, J.W.; Kim, J.H.; Kim, J.H.; Seo, M.; Han, H.S.; Park, J.; Suk, K. Time-dependent effects of hypothermia on microglial activation and migration. *J. Neuroinflammation* **2012**, *9*, 164. [CrossRef]

89. Nakajima, K.; Honda, S.; Tohyama, Y.; Imai, Y.; Kohsaka, S.; Kurihara, T. Neurotrophin secretion from cultured microglia. *J. Neurosci. Res.* **2001**, *65*, 322–331. [CrossRef]
90. Mizoguchi, Y.; Monji, A.; Kato, T.; Seki, Y.; Gotoh, L.; Horikawa, H.; Suzuki, S.O.; Iwaki, T.; Yonaha, M.; Hashioka, S.; et al. Brain-derived neurotrophic factor induces sustained elevation of intracellular Ca^{2+} in rodent microglia. *J. Immunol.* **2009**, *183*, 7778–7786. [CrossRef]

toxics

Article

Biomonitoring of Serum Inorganic Element Concentrations in Morbidly Obese Patients: Impact of Bariatric Surgery

Álvaro Ramos-Luzardo [1,2], Pilar Fernández Valerón [1], Beatriz Vanessa Díaz-González [3], Manuel Zumbado [2,3], Katherine Simbaña-Rivera [2,4], Inmaculada Bautista-Castaño [5], Norberto Ruiz-Suárez [2], Elisabeth Hernández-García [3], Judith Cornejo-Torre [3], Octavio P. Luzardo [2,3,5], Lluis Serra-Majem [3,5] and Luis Alberto Henríquez-Hernández [2,3,5,*]

1 Department of Biochemistry and Molecular Biology, Physiology, Genetics, and Immunology, Universidad de Las Palmas de Gran Canaria, 35016 Las Palmas de Gran Canaria, Spain; alvaro.ramos@ulpgc.es (Á.R.-L.); pilarfdez.valeron@ulpgc.es (P.F.V.)
2 Toxicology Unit, Clinical Sciences Department, Universidad de Las Palmas de Gran Canaria, 35016 Las Palmas de Gran Canaria, Spain; manuel.zumbado@ulpgc.es (M.Z.); katherine.simbana@ulpgc.es (K.S.-R.); norberto.ruiz@ulpgc.es (N.R.-S.); octavio.perez@ulpgc.es (O.P.L.)
3 Research Institute of Biomedical and Health Sciences (IUIBS), Universidad de Las Palmas de Gran Canaria, 35016 Las Palmas de Gran Canaria, Spain; beatrizvanessa@gmail.com (B.V.D.-G.); elisabethernandez924@gmail.com (E.H.-G.); judith.cornejo.dn@gmail.com (J.C.-T.); lluis.serra@ulpgc.es (L.S.-M.)
4 Centro de Investigación para la Salud en América Latina (CISeAL), Facultad de Medicina, Pontificia Universidad Católica del Ecuador (PUCE), Quito 170525, Ecuador
5 Centro de Investigación Biomédica en Red Fisiopatología de la Obesidad y la Nutrición (CIBEROBN), Instituto de Salud Carlos III, 28029 Madrid, Spain; inmaculada.bautista@ulpgc.es
* Correspondence: luis.henriquez@ulpgc.es; Tel.: +34-928-451-461

Abstract: Adipose tissue, in addition to serving as energy storage, can store lipophilic substances, some of which may pose a health risk if massively mobilized during rapid weight loss. This study aimed to biomonitor inorganic elements in obese patients undergoing bariatric surgery and analyze the role of sociodemographic factors. ICP-MS was employed to quantify 55 elements, including elements identified from the ATSDR's Substance Priority List and rare-earth elements, in serum samples of 58 obese patients before and after bariatric surgery. A total of 39 out of 55 elements were detected, either before or after surgery, in at least one patient. Serum concentrations of gold, copper, mercury, platinum, and selenium significantly decreased after surgery. Serum concentrations of iron and zinc were significantly lower before surgery. Age, sex, diabetes status, arterial hypertension, and family history of obesity were demographic factors associated with the concentration of elements. Correlation analyses showed significant associations between elements and total lipid concentration or body mass index. Our findings indicate a complex interplay between inorganic elements and body fat and provide relevant information regarding the biomonitoring of these chemical elements in a specific and understudied population.

Keywords: obesity; bariatric surgery; inorganic elements; heavy metals; rare-earth elements; weight loss; obesogens

Academic Editor: Carina Ladeira

Received: 30 January 2025
Revised: 17 February 2025
Accepted: 20 February 2025
Published: 23 February 2025

Citation: Ramos-Luzardo, Á.; Valerón, P.F.; Díaz-González, B.V.; Zumbado, M.; Simbaña-Rivera, K.; Bautista-Castaño, I.; Ruiz-Suárez, N.; Hernández-García, E.; Cornejo-Torre, J.; Luzardo, O.P.; et al. Biomonitoring of Serum Inorganic Element Concentrations in Morbidly Obese Patients: Impact of Bariatric Surgery. *Toxics* **2025**, *13*, 152. https://doi.org/10.3390/toxics13030152

1. Introduction

Inorganic elements comprise the entirety of the 118 elements listed on the periodic table [1]. The biological significance of these elements varies widely; some are essential for human health in trace amounts, others have demonstrated deleterious effects on living organisms, and the effects of many remain undetermined. In any case, elevated levels of any of them, either acute or chronic, lead to physiological alterations and the possible development of diseases [2,3]. The group of heavy metals, defined as elements with high atomic weights and densities at least five times greater than that of water, is perhaps the most widely studied. The existing literature on the subject has shown adverse effects on the nervous system, cognition and behavior, skin, and the digestive system, among many others [4–6]. The increase in the production and use of electronic devices poses an environmental challenge. It is estimated that more than 50 million tons of electronic waste (e-waste) are produced each year, with recycling rates below 15% [7]. This means that many of the inorganic elements used in the manufacture of electronic devices are dumped into the environment, contaminating soils, water, and food. Among the materials used in the construction of electronic devices, rare-earth elements (REEs) hold particular importance. REEs is a collective term used to designate the 17 chemical elements in the periodic table (the 15 lanthanides, plus scandium and yttrium) characterized by their incompletely filled 4f atomic orbitals, which result in unique magnetic, optical, and chemical properties that are essential for many modern technologies [8]. Unlike other elements, the adverse effects of REEs on human health remain unclear, although it is known that they interact with a wide variety of biomolecules [9] and have been associated with the development of some diseases [10].

Recent research has transformed adipose tissue from a mere biological storage compartment into a metabolically active endocrine organ [11]. Although it plays a protective toxicological role by sequestering lipophilic xenobiotics, adipose tissue also acts as an internal source of chronic exposure due to lipolysis-induced release. Adipocytes express receptors that can be modulated by xenobiotics, reinforcing the obesogen theory [12,13]. Obesogens are substances that contribute to an increased risk of developing overweight and obesity conditions. Persistent organic pollutants are among the most prominent [14], although there is evidence suggesting that some inorganic elements may also be involved [13].

Obesity has doubled globally in the last three decades and has a severe socioeconomic impact [15]. According to the World Health Organization, 60% of European adults are overweight or obese [16]. Obesity is defined by a high body mass index (BMI) and classified as follows: overweight (25–30 kg/m^2), class I (30–35 kg/m^2), class II (35–40 kg/m^2), or class III ($\geq$40 kg/m^2) obesity [17]. It is caused by a combination of genetic, physiological, behavioral, lifestyle, dietary, and environmental factors. While once primarily linked to fat accumulation, obesity is now understood to involve systemic inflammation, gut microbiota dysregulation, adipocyte cell cycle changes, and oxidative stress [18]. Altogether, these factors destabilize energy homeostasis and increase the risk of obesity [18]. Beyond physical limitations, obesity significantly increases the risk of developing serious health conditions [19–21]. Significant advancements in obesity treatment have occurred [22]. Lifestyle interventions, pharmacological therapies (e.g., lipase inhibitors), oral hydrogel capsules, or endoscopic bariatric therapies are the most frequently applied to reduce adiposity [23]. However, while pharmacotherapy shows promising results, bariatric surgery remains the most effective to achieve long-term weight loss and reduce comorbidities [22,24,25]. This type of surgery is typically recommended for poor treatment responses, class II obesity with complications, or class III obesity.

In a recent study, we investigated the behavior of certain organic molecules in a population of morbidly obese individuals undergoing bariatric surgery [26]. The present study aimed to biomonitor serum concentrations of fifty inorganic elements, encompassing essential elements, heavy metals, and rare-earth elements (REEs) in patients subjected to bariatric surgery for the treatment of morbid obesity. This study sought to determine the variations in serum concentrations, elucidate their interactions with adipose tissue, and evaluate their potential contribution to obesogenesis.

2. Materials and Methods

2.1. Study Population and Sample Collection

A total of 58 adult patients (42 women; 16 men) aged 20–69, with BMIs ranging from 35 to 60 kg/m^2, were recruited. All participants underwent bariatric surgery following a clinical evaluation between 30 March and 7 July 2015. After surgery, they adhered to a strict dietary regimen for one year and participated in a second evaluation from 15 March to 23 December 2016. Weight, height, and body composition were measured at baseline and follow-up using the Tanita BC-420MA III device (Girod Medical, Madrid, Spain). Additionally, whole-blood samples were collected in 4 mL EDTA-containing tubes and centrifuged at 3000× *g* for 5 min to obtain serum, which was stored at −20 °C for subsequent chemical analysis. Weight loss outcomes were reported as total body weight loss (TBWL), percentage of total body weight loss (%TBWL), and percentage of excess body weight loss (%EWL). Changes in clinically relevant parameters, including lipidemia, smoking habits, comorbidities, and depressive state, were assessed longitudinally.

Ethical approval was obtained from the CHUIMI Ethics Committee (Complejo Hospitalario Universitario Insular Materno Infantil) on 26 June 2014, with the ethical approval code 2024-093-728.

2.2. Selection of Elements, Standards, and Sample Preparation

The analysis included 55 elements, comprising elements identified from the ATSDR's Substance Priority List of 2022 [27], which were selected due to their prevalence, toxicity, and potential for exposure in populations. Rare-earth elements (REEs) and other minor elements (MEs) were included. REEs and MEs are becoming increasingly prevalent in populations globally due to increasing use in technological, medical, and zootechnical sectors [8,28,29]. For the analysis of these elements, we utilized the compilation provided by Tansel (2017) [8] regarding elements used in computers and high-tech devices. The complete list of elements analyzed is detailed in Table S1.

The methodology for the study was adapted from protocols previously established by our research group, as detailed by Henríquez-Hernández in 2023 [30]. Briefly, 0.5 mL of serum samples were weighed using a high-precision analytical balance in a digestion vessel. To these, 2 mL of 65% high-purity nitric acid, distilled in a sub-boiling system equipped with a PTFE-TFM purification unit (SubCLEAN, Milestone SRL, Sorisole, Italy), and 7.5 mL of ultrapure water from a Milli-Q system were added to achieve complete mineralization of the samples. Microwave-assisted extraction (MAE) was conducted using a Milestone Ethos Up apparatus (Milestone, Bologna, Italy). The conditions applied were power (W)–temperature (°C)–time (min): step 1: 1800–100–5; step 2: 1800–150–5; step 3: 1800–200–8; and step 4: 1800–200–7. To verify that the MAE was effective, certified reference materials (Fish Muscle ERM®-BB422 (Institute for Reference Materials and Measurements; Geel, Belgium); Lobster Hepatopancreas NRC LUTS-1 (National Research Council; Ottawa, ON, Canada); Seronorm Trace Elements Whole Blood L-1 (SERO AS; Billingstad, Norway)) were added before the procedure. Following extraction, the samples were cooled and transferred into metal-free polypropylene tubes. Subsequently, aliquots of the samples were diluted in autosampler vials to a final concentration of 4% (*v*/*v*). After MAE, the internal standard (ISTD) was manually spiked to facilitate elemental analysis, as previously published [31].

The ISTD solution, containing scandium (Sc), germanium (Ge), rhodium (Rh), and iridium (Ir), each at 20 mg/mL, was introduced into the ICP-MS at a final concentration of 40 ppb. This procedure was carried out to evaluate reproducibility in terms of counts per second (CPS) and to determine the relative standard deviation (RSD) and recovery rates of four isotopes (45Sc, 72Ge, 103Rh, and 193Ir), which should fall within the 70–130%

range. Germanium was eliminated from analysis to avoid interferences caused by doubly charged ions of neodymium (Nd) and samarium (Sm). The standards for the purity of elements were acquired as certified reference materials from CPA Chem (Stara Zagora, Bulgaria) and consisted of a multi-element solution of 100 mg/L in 2% nitric acid (HNO_3), a single-element solution of mercury at 100 mg/L in 5% HNO_3, and individual solutions for the REEs and MEs at 1000 mg/L in diluted HNO_3 or HNO_3/HF for ICP-MS. Consequently, three different calibration curves were prepared using a 4% HNO_3 solution, except for the mercury calibration curve, which additionally required 1% HCl: (1) a multi-element curve including the main heavy metals, potentially toxic, and some minor elements (nine points ranging from 0.00 to 2000 ppb); (2) a mercury-specific curve (five points ranging from 0.00 to 50 ppb); and (3) a curve for REEs and the majority of MEs (six points from 0.00 to 25 ppb). These levels were chosen based on linearity, sensitivity, and precision to achieve superior analytical parameters and quantification quality. Furthermore, isotopes were selected to minimize interferences from doubly charged ions of REEs. To prevent memory effects, particularly with mercury, a cleaning solution of 2% HNO_3 and 1% HCl was employed between vials. Given the sample weight of 0.5 g and the subsequent dilution during digestion and analysis, the final matrix concentration in the analyzed sample accounts for 1.5%, suggesting a minimal matrix effect [31].

2.3. Analytical Procedure

The analysis was performed using an Agilent 7900 ICP-MS system (Agilent Technologies, Tokyo, Japan), equipped with standard nickel cones, a MicroMist glass concentric nebulizer, and an Ultra High Matrix Introduction (UHMI) system. The Integrated Sample Introduction System (ISIS) was configured for discrete sampling. To address isobaric interferences, specific isotopes were selected to ensure accurate measurements. The 4th generation Octopole Reaction System (ORS4) operated in helium (He) mode for all elements to minimize interferences from low-mass elements, thereby enhancing detection limits. Prior to analysis, a tuning solution containing a mixture of Cs, Co, Li, Mg, Th, and Y was used to optimize instrument performance concerning CPS, RSD values, and the recommended oxide ratio and double ion charges. Each sample was analyzed in triplicate using separate vials, with three complete readings taken automatically for each vial, providing nine individual measurements per sample. Quantification of elements was conducted using MassHunter v.4.2 ICP-MS Data Analysis software by Agilent Technologies.

The entire procedure underwent in-house validation using a certified reference material (Seronorm™ Trace Elements Whole Blood, Billingstad, Norway) at three concentration levels (L-1, L-2, and L-3). The reference material contained clinically relevant concentrations for all elements studied, except In and Ti which were assessed through fortified matrix assays. Five replicates from five independent vials at each level were digested following the sample protocol and analyzed under identical ICP-MS conditions. Linear calibration curves were appropriate for most elements, with regression coefficients (R^2) $\geq$ 0.995, and recoveries ranged from 79% to 121% for REEs and MEs, and from 76% to 124% for toxic elements listed by ATSDR. Ten reagent blanks, treated similarly to the samples, were also integrated into the work sequence during the analysis. Instrumental limits of detection (LODs) and quantification (LOQs) were, respectively, determined based on signals three and ten times higher than the average blank solution signal measured on the same day as the sample analysis (Table S1). This approach ensures accurate estimation of LODs and LOQs by accounting for daily instrumental condition variations.

Method accuracy and precision were evaluated through recovery studies using a 4% (v/v) nitric acid solution spiked with two mixtures at three concentration levels: one containing a multi-elemental mix, and one for the REEs and MEs. Recovery values were calculated by dividing the obtained concentration by the expected theoretical concentration based on the added levels of both solutions. LOQs for determining the sample concentration were calculated by multiplying the instrumental LOQ by the dilution factor applied during the procedure and adjusting for exact mass. All analytical values used for quantification were derived from measurements with RSD < 5%.

Quality control of blood sample analyses included independent vials every 24 sample replicates, each containing a digested blank, two customized mixes, and each level (L-1, L-2, and L-3) of the certified reference material. This practice ensured that analyte concentrations fell within 20% of the RSD, thus verifying the absence of contamination and ensuring accurate and reproducible results.

2.4. Statistical Analysis

Descriptive statistics were calculated. Continuous variables were summarized by mean, standard deviation, median, range, and interquartile range (25th and 75th percentiles). Categorical variables were summarized by frequency and proportion. Values below the limit of quantification (LOQ) were considered missing. The normality of continuous variables was assessed using the Shapiro–Wilk test. For variables with detection frequencies exceeding 30%, group comparisons were performed using either paired-sample parametric tests (Student's *t*-test) or non-parametric alternatives (Mann–Whitney U test and Wilcoxon's W test), following a Levene's test to assess the homogeneity of variances and ensure appropriate test selection. This approach was made considering both methodological robustness and public health relevance, and it does not lead to left censoring, as it does not arbitrarily remove low-concentration values and instead prioritizes elements with meaningful prevalence. McNemar's tests were employed for categorical variable comparisons. Spearman's or Pearson's correlation tests were used to assess correlations between inorganic elements and continuous variables.

Given the sample size and the number of variables analyzed, we conducted power calculations for key comparisons, particularly those involving elemental concentrations between exposure groups. The power analysis was based on effect sizes observed in previous studies assessing elemental bioaccumulation, assuming an alpha level of 0.05 and a two-tailed hypothesis. The achieved power ($1 - \beta$) for detecting medium-to-large effect sizes (Cohen's d or Rank Biserial Correlation (rrb) > 0.5) was above 80%, indicating that our study design was sufficiently robust to detect meaningful differences in quantitative variables. For qualitative variables, we applied Cramér's V to measure the strength of association between categorical variables, ensuring that relationships were statistically meaningful. This approach allowed us to assess the effect size of categorical comparisons while accounting for sample distribution across different groups.

Statistical analyses were performed using Jamovi v2.6.25. Graphs were created using BioRender software (version 4) (Toronto, ON, Canada).

3. Results

3.1. Influence of Bariatric Surgery on Sociodemographic Variables

After the follow-up (534 ± 24.1 days), a remarkable total body weight loss (TBWL) of 45.3 kg was observed, with mean body weight decreasing from 129.8 to 83.6 kg (*p*-value < 0.001, Cohen's d = 2.18). Thus, mean BMI significantly decreased from 47.7 kg/m^2 (severe obesity) to 30.9 kg/m^2 (overweight category) (*p*-value < 0.001, Cohen's d = 2.35). This translated into a mean %EWL of 63.9%. At the beginning of the study, 15.5% of the patients had obesity class II and 84.5% had obesity class III. After surgery, 20.7% of the cohort achieved normal weight (Table 1). Student's *t*-tests demonstrated a statistically significant reduction in both circulating total lipids (TLs) (from 6.6 to 5.1 mg/mL;

$p < 0.001$, Cohen's d = 1.12) and fasting glycemia (from 119.0 to 92.3 mg/dL; $p < 0.001$, Cohen's d = 0.93). A significant decrease in the frequency of artery hypertension (AHT) was also observed (p-value = 0.007).

Table 1. Physical condition and habits of the patients before and after bariatric surgery (mean follow-up = 534 days (SD = 24.1 days)).

Variables	Before Intervention (n = 58)	After Intervention (n = 56)	p-Value
Weight (kg)			
Mean ± SD	129.8 ± 24.4	83.6 ± 19.2	<0.001 †
Median (range)	127.8 (93.2–189.1)	83.7 (54.0–134.3)	<0.001 ††
Body mass index (kg/m^2)			
Mean ± SD	47.7 ± 6.5	30.9 ± 6.0	<0.001 †
Median (range)	47.9 (35.5–59.4)	30.7 (20.5–44.0)	<0.001 ††
Normal weight (n, (%))	0 (0)	12 (20.7)	<0.001 *
Overweight (n, (%))	0 (0)	17 (29.3)	<0.001 *
Obesity class I (n, (%))	0 (0)	13 (22.4)	<0.001 *
Obesity class II (n, (%))	9 (15.5)	11 (19.0)	0.617 *
Obesity class III (n, (%))	49 (84.5)	5 (8.6)	<0.001 *
Total body weight loss (kg)			
Mean ± SD		45.3 ± 20.8	NA
Median (range)		43.4 (−0.6–92.1)	NA
% Excess weight loss			
Mean ± SD		63.9 ± 24.3	NA
Median (range)		65.5 (−1.4–102.8)	NA
Smoking habit (n, (%))			
Yes	10 (18.2)	14 (25.5)	0.133 *
No	33 (60.0)	36 (65.5)	
Former smoker	12 (21.8)	5 (9.1)	
N° of cigarettes			
Mean ± SD	11.6 ± 6.3	15 ± 7.6	0.830 †
Median (range)	11 (4–20)	12 (4–30)	0.855 ††
Total lipids (mg/mL)			
Mean ± SD	6.6 ± 0.9	5.1 ± 1.1	<0.001 †
Median (range)	6.5 (4.9–10.0)	4.8 (2.0–8.2)	<0.001 ††
Artery hypertension (n, (%))			
Yes	25 (43.1)	13 (24.5)	0.007 *
No	33 (56.9)	40 (75.5)	
Glucose (mg/dL)			
Mean ± SD	119.0 ± 46.4	92.3 ± 17.6	<0.001 †
Median (range)	103.0 (85.0–321.0)	92 (63.0–155.0)	<0.001 ††

Abbreviations: NA, not applicable; SD, standard deviation. † Student's t-test. †† Wilcoxon's W test. * McNemar's test.

The cohort was segmented by age, sex, diabetes mellitus condition, AHT condition, and family history of obesity (obesity in first-degree relatives). Differences between groups before surgery are shown in Table S2. Briefly, total lipids were significantly higher in men (p-value = 0.02, rrb = 0.33) and diabetic patients (p-value = 0.003, rrb = 0.43), indicating moderate effect sizes.

Differences between groups after surgery are shown in Table S3. Younger patients exhibited significantly lower blood glucose ($p = 0.001$, rrb = 0.62) and TL levels ($p = 0.013$, rrb = 0.34), indicating strong and moderate effects, respectively. BMI was also significantly reduced in this group ($p = 0.001$, Cohen's d = −0.86), with a notably higher %EWL ($p = 0.004$), suggesting greater weight loss outcomes. BMI was significantly higher among patients with a family history of obesity (p-value = 0.03, Cohen's d = −0.59) and those with AHT (p-value = 0.008). Patients showing AHT had higher levels of blood glucose (p-value = 0.004, rrb = 0.65) and lower %EWL (p-value = 0.008, rrb = 0.44), displaying greater weight loss. No sex differences were found after surgery.

3.2. Effect of Bariatric Surgery on Serum Levels of Inorganic Contaminants

A total of 39 out of 55 analyzed elements were detected, either before or after surgery, in at least one patient (Table 2). Detection rates significantly decreased after surgery in the case of barium, chromium, gallium, palladium, platinum, and silicon (p-value < 0.05). Serum concentrations of gold (p-value = 0.030, Cohen's d = −0.123), copper (p-value < 0.001, rrb = 0.91), mercury (p-value < 0.001, Cohen's d = 0.85), platinum (p-value = 0.034, Cohen's d = −3.06), and selenium (p-value < 0.001, Cohen's d = 1.12) significantly decreased after surgery. Conversely, serum concentrations of iron and zinc significantly increased after surgery (p-value = 0.044 and 0.001, respectively).

Differences between groups, in relation to the concentration of organic elements before and after surgery, were also analyzed (Tables S2 and S3). Before surgery, iron serum levels were higher in men (p-value = 0.016) while copper serum levels were higher in women (p-value = 0.002, rrb = 0.53). In patients younger than 43 (n = 30, 51.7%), analyses revealed lower strontium concentrations compared with older patients (p-value = 0.015). Patients with diabetes showed higher serum levels of platinum and zinc (p-values = 0.007 and 0.012, respectively; Table S2). Finally, patients with a family history of obesity showed higher serum levels of selenium (p-value < 0.001).

After surgery, women exhibited significantly higher copper serum levels (p-value < 0.001, rrb = −0.62). Regarding age, patients older than 43 years old showed higher serum levels of bromine, selenium, and zinc (p-value = 0.012, 0.023, and < 0.001, respectively; Table S3). Serum levels of bromine and rubidium were higher among patients with diabetes (p-value = 0.002 and 0.021, respectively), while serum levels of strontium were lower (p-value = 0.009, rrb = 0.41). Serum levels of rubidium were also higher among patients with AHT (p-value = 0.010, Cohen's d = −0.79). Finally, maternal obesity was linked to significantly higher cobalt and zinc serum levels (p-value = 0.018, rrb = 0.59, and p-value = 0.004, Cohen's d = −0.89, respectively), reflecting strong to very strong effects.

Correlation tests were performed to identify possible associations between the inorganic elements whose detection rates were above 30%, as well as correlations with demographic and clinical variables. Before surgery, age positively correlated with bromine and strontium (Pearson's r = 0.483 and Spearman's rho = 0.396; p-value = 0.001 and 0.006, respectively) (Figure 1A; Table S4). Serum copper negatively correlated with TL (Spearman's rho = −0.316; p-value = 0.030), while rubidium showed a positive correlation with TL (Spearman's rho = 0.325; p-value = 0.026). Zinc concentration positively correlated with iron and mercury (Spearman's rho = 0.341 and 0.348; p-value = 0.020 and 0.026, respectively), and rubidium positively correlated with platinum (Spearman's rho = 0.525; p-value = 0.047).

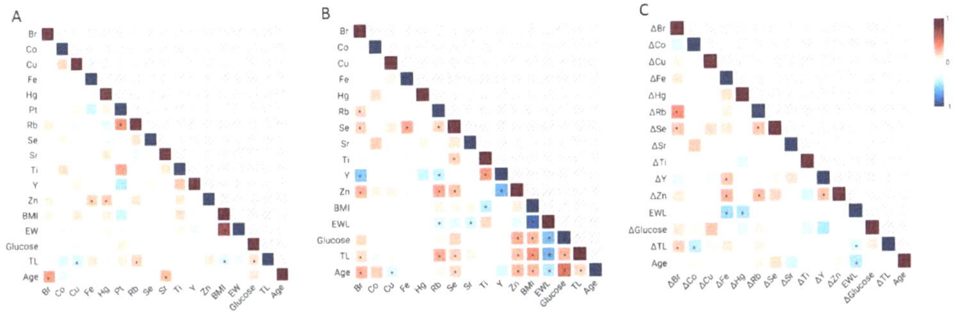

Figure 1. Correlations between serum levels of inorganic elements and sociodemographic variables before bariatric surgery (**A**), after bariatric surgery (**B**), and those observed in relation to the variation in serum concentrations of inorganic elements, defined as the concentration after surgery minus the concentration before surgery (**C**). * highlights significant correlations.

Table 2. Detection rates and serum concentrations of inorganic elements before and after surgery. Concentrations are expressed as median (p^{25th}–p^{75th}).

Element	Before Surgery (n = 47)		After Surgery (n = 50)		ΔConcentration (ng/mL)	*p*-Value #
	Detection Rate [N, (%)]	Concentration (ng/mL)	Detection Rate [N, (%)] +++	Concentration (ng/mL)		
Al (Aluminum)	4 (8.5)	106.8 (33.0–735.9)	7 (14.0)	0.32 (0.32–39.1)	0.21 (−24.3–7.86)	ND
As (Arsenic)	0 (0.0)	ND	1 (2.0)	22.1 (22.1–22.1)	ND	ND
Au (Gold)	14 (29.8)	2.71 (1.69–3.43)	11 (22.0)	2.67 (1.82–3.61)	0.87 (−2.99–1.80)	0.030
Ba (Barium)	7 (14.9)	0.64 (0.45–2.22)	0 (0.0) *	ND	−0.55 (−1.93–(−0.40))	ND
Be (Beryllium)	13 (27.7)	3.39 (3.31–3.45)	7 (14.0)	2.35 (1.25–2.40)	−3.27 (−3.41–0.12)	ND
Br (Bromine)	47 (100.0)	18.4 (16.2–20.6) θ	50 (100.0)	18.7 (15.4–21.8) θ	−0.90 (−2.76–1.81) θ	0.318
Cd (Cadmium)	2 (4.3)	0.20 (0.16–0.24)	8 (16.0)	0.29 (0.17–0.62)	0.21 (0.04–0.59)	ND
Ce (Cerium)	1 (2.1)	0.17 (0.17–0.17)	4 (8.0)	0.16 (0.07–0.27)	0.05 (0.04–0.23)	ND
Co (Cobalt)	28 (59.6)	0.27 (0.18–0.68)	25 (50.0)	0.18 (0.12–0.59)	−0.15 (−0.34–0.27)	0.842
Cr (Chromium)	9 (19.1)	2.18 (0.32–3.44)	1 (2.0) *	11.1 (11.1–11.1)	−2.15 (−3.42–(−0.30))	ND
Cs (Caesium)	1 (2.1)	0.34 (0.34–0.34)	7 (14.0)	0.24 (0.17–0.50)	0.26 (0.09–0.57)	ND
Cu (Copper)	47 (100.0)	1.30 (1.10–1.38) θ	50 (100.0)	1.03 (0.82–1.15) θ	−0.29 (−0.48–(−0.17)) θ	<0.001 †
Dy (Dysprosium)	1 (2.1)	0.06 (0.06–0.06)	1 (2.0)	0.06 (0.06–0.06)	0.00 (−0.02–0.02)	ND
Fe (Iron)	47 (100.0)	1.07 (0.88–1.42) θ	50 (100.0)	1.34 (0.94–1.75) θ	0.21 (−0.25–0.60) θ	0.044 †
Ga (Gallium)	8 (17)	0.30 (0.20–0.54)	0 (0.0) *	ND	−0.22 (−0.55–(−0.17))	ND
Gd (Gadolinium)	0 (0.0)	ND	3 (6.0)	0.07 (0.06–0.12)	0.05 (0.05–0.05)	ND
Hg (Mercury)	41 (87.2)	0.84 (0.65–1.52)	42 (84.0)	0.42 (0.16–0.75)	−0.57 (−0.68–(−0.01))	<0.001
In (Indium)	2 (4.3)	0.23 (0.18–0.29)	0 (0.0)	ND	−0.21 (−0.26–(−0.15))	ND
La (Lanthanum)	1 (2.1)	0.08 (0.08–0.08)	8 (16.0)	0.07 (0.05–0.14)	0.04 (0.03–0.09)	ND
Li (Lithium)	1 (2.1)	11.9 (11.9–11.9)	0 (0.0)	ND	ND	ND
Mn (Manganese)	4 (8.5)	1.70 (1.25–2.61)	14 (28.0)	1.87 (1.55–2.28)	1.57 (0.15–1.87)	0.760
Mo (Molybdenum)	0 (0.0)	ND	1 (2.0)	20.7 (20.7–20.7)	20.7 (20.7–20.7)	ND
Nd (Neodymium)	9 (19.1)	0.08 (0.06–0.10)	12 (24.0)	0.15 (0.13–0.16)	0.05 (−0.04–0.13)	0.287
Os (Osmium)	8 (17.0)	0.79 (0.74–0.90)	3 (6.0)	0.96 (0.85–1.27)	−0.72 (−0.76–0.16)	0.121
Pd (Palladium)	12 (25.5)	0.08 (0.07–0.10)	4 (8.0) *	0.05 (0.05–0.07)	−0.04 (−0.05–(−0.04))	0.577
Pr (Praseodymium)	0 (0.0)	ND	1 (2.0)	0.07 (0.07–0.07)	ND	ND
Pt (Platinum)	15 (31.9)	0.13 (0.08–0.16)	3 (6.0) *	0.13 (0.13–0.16)	−0.11 (−0.14–0.00)	0.034
Rb (Rubidium)	47 (100.0)	151.4 (132.2–168.3)	50 (100.0)	148.4 (129.6–167.7)	0.92 (−20.5–17.2)	0.720
Ru (Ruthenium)	0 (0.0)	ND	1 (2.0)	0.06 (0.06–0.06)	0.04 (0.04–0.04)	ND
Se (Selenium)	47 (100.0)	87.8 (75.4–93.5)	50 (100.0)	57.3 (45.8–73.9)	−25 (−42.6–(−11.5))	<0.001
Si (Silicon)	5 (10.6)	496.7 (329.4–785.4)	1 (2.0) *	353.4 (353.4–353.4)	−432 (−496.7–(−329.4))	ND
Sr (Strontium)	47 (100.0)	23.5 (17.8–29.8)	50 (100.0)	23.8 (19.6–27.9)	1.66 (−3.72–4.23)	0.694
Ta (Tantalum)	9 (19.1)	0.09 (0.09–0.13)	20 (40.0)	0.07 (0.06–0.10)	0.03 (−0.05–0.05)	0.601
Th (Thorium)	5 (10.6)	0.08 (0.08–0.22)	16 (32.0)	0.09 (0.06–0.14)	0.03 (−0.04–0.07)	ND
Ti (Titanium)	30 (63.8)	7.30 (5.84–9.69)	36 (72.0)	8.34 (6.50–11.2)	1.28 (−5.19–6.83)	0.060
U (Uranium)	3 (6.4)	0.25 (0.16–0.48)	5 (10.0)	0.25 (0.23–0.81)	0.08 (−0.14–0.5)	ND
Y (Yttrium)	22 (46.8)	0.12 (0.08–0.14)	34 (68.0)	0.10 (0.08–0.15)	0.05 (−0.05–0.08)	0.503
Yb (Ytterbium)	0 (0.0)	ND	1 (2.0)	0.06 (0.06–0.06)	ND	ND
Zn (Zinc)	47 (100.0)	469.6 (310.3–570.5)	50 (100.0)	556.8 (468.3–639.3)	99.4 (−39.0–277.9)	0.001

Student's *t*-test. † Wilcoxon's W test. Cu and Fe did not adjust to normal distribution (Shapiro–Wilk). +++ McNemar's test. * $p < 0.05$. θ Concentration expressed in μg/mL. ND: differences were not determined through paired-sample *t*-tests due to the lack of coincident measures before and after surgery.

Post-surgery, bromine levels correlated with TL (Spearman's rho = 0.289; *p*-value = 0.042) and age (Pearson's r = 0.424; *p*-value = 0.002). Tests also revealed significant correlations between copper and age (Spearman's rho = −0.285; p-value = 0.045). Rubidium correlated with %EWL (Pearson's r = −0.287; p-value = 0.043) and TL (Spearman's rho = 0.442; p-value = 0.001). Selenium correlated with TL (Spearman's rho = 0.327; p-value = 0.020) and age (Pearson's r = 0.303; p-value = 0.032) and titanium concentration correlated with BMI (Spearman's rho = −0.378; p-value = 0.024). Correlations between zinc levels and blood glucose (Spearman's rho = 0.444; p-value = 0.020), TL (Spearman's rho = 0.375; p-value = 0.007), and age (Spearman's rho = 0.413; p-value = 0.003) were observed (Figure 1B; Table S5). Levels of bromine showed positive correlations with rubidium (Pearson's r = 0.314; p-value = 0.026), selenium (Pearson's r = 0.337; p-value = 0.017), and zinc (Spearman's rho = 0.429; p-value = 0.002), as well as a negative correlation with yttrium (Spearman's rho = −0.534; p-value = 0.001). Tests also showed correlations between rubidium and selenium (Pearson's r = 0.320; p-value = 0.023), yttrium (Spearman's rho = −0.369; p-value = 0.032), and zinc (Spearman's rho = 0.462; p-value = 0.001). In parallel, selenium positively correlated with iron (Spearman's rho = 0.452; p-value = 0.001), titanium (Spearman's rho = 0.334; p-value = 0.047), and zinc (Spearman's rho = 0.390; p-value = 0.005). Finally, correlations between titanium and yttrium (Spearman's rho = 0.503; p-value = 0.011) and between yttrium and zinc (Spearman's rho = −0.577; p-value < 0.001) were found (Figure 1B; Table S5). Of note, correlations for platinum were not further analyzed due to its low detection rate after surgery.

Correlation analyses of the changes in concentration and sociodemographic variables displayed positive associations between Δ-bromine and Δ-TL (Spearman's rho = 0.326; p-value = 0.043). Δ-cobalt correlated with Δ-TL (Spearman's rho = −0.378; p-value = 0.040). Δ-iron correlated with %EWL (Spearman's rho = −0.448; p-value = 0.005). Lastly, a negative correlation between Δ-mercury and %EWL (Spearman's rho = −0.422; p-value = 0.009) was also found (Figure 1C; Table S6). Analyses of the changes in the serum concentration of inorganic elements revealed that Δ-bromine was positively correlated with Δ-rubidium (Pearson's r = 0.537; p-value < 0.001) and Δ-selenium (Pearson's r = 0.365; p-value = 0.022). Associations between Δ-iron and Δ-yttrium (Spearman's rho = 0.369; p-value = 0.030) and between Δ-iron and Δ-zinc (Spearman's rho = 0.417; p-value = 0.009) were also identified. Δ-rubidium positively correlated with Δ-selenium (Pearson's r = 0.369; p-value = 0.021) and Δ-zinc (Pearson's r = 0.400; p-value = 0.012) and Δ-yttrium was positively correlated with Δ-zinc (Pearson's r = 0.372; p-value = 0.028) (Figure 1C; Table S6).

4. Discussion

Of the 55 elements analyzed, 70.9% (39 elements) were detected in at least one individual before or after bariatric surgery. Of these, serum concentrations of seven elements (17.9%) underwent significant alterations following bariatric surgery, with gold, copper, mercury, platinum, and selenium exhibiting decreases, while iron and zinc levels increased.

Obesity leads to significant disruptions in the homeostasis of trace and other elements. Research has shown that obese individuals exhibit decreased levels of zinc, iron, selenium, and strontium, while levels of copper, chromium, and manganese are significantly increased [32,33]. This disruption seems to contribute to metabolic and oxidative stress [32,33]. While weight loss can ameliorate some of these disruptions, it may also lead to the release of stored compounds, potentially impacting metabolic function. Moreover, bariatric surgery has the potential to influence the serum concentrations of a wide range of inorganic elements, including heavy metals. Such surgical interventions may disrupt the homeostatic balance of essential metals by altering their absorption and metabolism. Specifically, while copper and zinc concentrations remain relatively stable, selenium shows a tendency to decline post-surgery [34,35]. However, while the observed changes in serum element concentrations are notable, further exploration reveals a more intricate interplay at the molecular level, including alterations in isotopic composition. Thus, bariatric surgery can alter the isotopic compositions of essential metals like copper, zinc, and iron, reflecting changes in their physiological processing rather than their concentrations [34,35]. In the present study, serum levels of copper, iron, and selenium were altered following surgery and rapid weight loss, reinforcing previous findings and suggesting a complex relationship with adipose tissue. The impact of bariatric surgery transcends weight loss and its immediate consequences on biochemical and clinical parameters such as total lipid levels and hypertension risk. By altering intestinal absorption mechanisms, this surgical procedure can significantly modify the body's elemental profile [36]. In routine clinical practice, patients who have undergone such surgery are recommended to take dietary supplements to maintain the homeostasis of trace elements and essential elements [35]. In this regard, iron and zinc, two elements commonly found in mineral supplements, were the only substances whose levels increased significantly after surgery. This increase may be attributed to physician-prescribed supplementation. However, this hypothesis cannot be confirmed as we do not have information regarding these prescriptions. On the contrary, the observed decreases in certain elements post-surgery can be attributed to the reduced absorption of these elements, particularly given that many are dietary in origin and, unlike lipophilic compounds [26], are not extensively stored in adipose tissue unless certain biotransformations increase the lipophilic characteristics of specific compounds (i.e., methylmercury). Due to its toxicological significance, the observed decrease in mercury concentration following surgery warrants special attention. A study on Norwegian obese patients undergoing bariatric surgery found that, 12 months post-surgery, the median whole-blood mercury concentration decreased by 31% [37]. In the present study, after a mean follow-up of 534 days, the median concentration decreased by 50%, highlighting the fact that the detection frequency did not change (87.2 and 84.0% before and after surgery, respectively). The underlying causes of this reduction remain unclear. Considering the relatively short half-life of mercury (≈50 days) [38], the observed reduction levels may be attributed to decreased fish and seafood consumption following surgery, as these are major contributors to mercury body burden [39].

We observed that age and sex were demographic factors associated with the concentration of certain elements. Specifically, bromide, selenium, and zinc were higher in older patients who underwent bariatric surgery. In general, serum levels of selenium and zinc show that these micronutrient levels tend to decrease with advancing age [40]. However, in a Spanish adult population, age was positively associated with serum selenium levels, suggesting that selenium levels might initially increase and then decline in very old age [41]. For analyses related to the influence of age on serum levels of the elements, the median of the age distribution, which was 43 years, was used as a cut-off point. This age is far from being considered an "advanced age". Thus, our results support previous observations. Interestingly, in a cohort of middle-aged and older Chinese adults, higher selenium levels were unexpectedly associated with increased risk factors for metabolic syndrome, suggesting that selenium status may not only decline with age but also play a more complex role in metabolic health [42]. This finding, along with the observed increased selenium concentration after bariatric surgery in the older population, focuses attention on this compound in relation to treatment failures, specifically weight regain post-surgery.

In relation to the influence that sex may exert on changes in element concentrations, our results have shown that, before surgery, iron levels were higher in men, and copper levels were higher in women. These findings are consistent with the existing literature, which reports higher copper levels in women [43]. Although women generally have higher serum iron levels, they also have a higher prevalence of anemia due to menstrual cycles, ultimately resulting in lower iron levels [44]. After surgery, only copper levels were significantly higher in women, supporting the hypothesis of mineral supplementation in bariatric surgery patients.

Associations and correlations were observed between some of the elements analyzed and medical parameters such as hypertension, diabetes, blood glucose, and total lipids. Prior to surgery, individuals with diabetes exhibited elevated serum zinc levels. After surgery, a significant positive correlation emerged between serum zinc concentrations and glucose levels. In line with the present results, previous studies have demonstrated a linear association between elevated serum zinc levels and an increased risk of both prediabetes and diabetes [45]. Moreover, lower serum zinc levels have been associated with poorer glycemic control in type 2 diabetes patients. Our findings suggest a potential link between zinc status and glucose homeostasis, highlighting the importance of zinc in glucose metabolism.

A post-surgical increase in serum rubidium levels was observed in hypertensive patients and patients with diabetes. The role of rubidium in diabetes remains unclear, as there are studies supporting this association [46] and others that have found no association [47]. Regarding hypertension, no previous association with serum rubidium levels has been described. However, associations between serum levels of other inorganic elements, such as lead, and the risk of hypertension and stroke have been previously reported [48].

Finally, a significant positive correlation was observed between total lipid levels and the serum concentrations of bromine, rubidium, selenium, and zinc, particularly postoperatively. Given that bromine is a constituent of widely used compounds such as polybrominated flame retardants, the metabolic fate of these compounds may influence serum bromine concentrations. While a positive correlation between total lipids and bromine levels was reported, the specific role of polybrominated biphenyls in this relationship remains complex [49,50]. Although the relationship between serum rubidium levels and total lipids remains unexplored, the literature consistently reports a positive correlation between selenium levels and lipid profiles [51–53]. Our findings corroborate these previous observations, further emphasizing the role of selenium in lipid metabolism and obesity. Regarding zinc, a positive correlation exists with the serum lipid profile [54,55]. Interestingly, zinc supplementation has been shown to significantly reduce serum cholesterol and triglyceride levels [56,57], positioning zinc as a central element in the context of obesity and highlighting the role of micronutrient supplementation in obesity and pre- or post-bariatric surgery interventions.

This study has several limitations that must be considered when interpreting the results. Firstly, there are limitations inherent to experimental design. The longitudinal design of this study is subjected to limitations related to sample selection and time-related factors. Secondly, there are limitations related to the influence of factors that determine the presence of these elements in the body. In this regard, the influence of the pre-intervention diet, as well as the post-surgical dietary intervention, will condition the intake of inorganic elements such as mercury. Additionally, certain toxic habits, particularly smoking, and the intensity and duration thereof, may exert a significant influence. Given the lack of detailed information regarding these variables, our results cannot be interpreted within these contexts. Certain medical prescriptions, specifically those related to dietary supplements post-surgery, may influence the serum levels of some of the elements considered, especially trace elements and micronutrients. Given the absence of this information, the observed variations in relation to bariatric surgery must be considered in light of this limitation. Finally, given that some of the analyzed elements are lipophilic and considered endocrine disruptors, biomonitoring these elements in adipose tissue is necessary. Having the inorganic element profile of adipose tissue could help to elucidate some of the alterations observed in serum levels after bariatric surgery. On the other hand, this study has several strengths. Longitudinal studies offer a unique perspective on dynamic and changing processes, allowing for the identification of patterns, trends, and causal relationships. Additionally, it provides relevant information about the behavior and influence of more than fifty inorganic elements in the context of obesity and rapid weight loss, a poorly explored area of research. Although the sample size is relatively small, rigorous statistical analyses were conducted, considering, for example, elements detected in more than 30% of individuals for correlation analyses. This strengthens the reported results despite the limitations.

5. Conclusions

This study demonstrates that bariatric surgery induces notable alterations in serum inorganic element levels, highlighting the intricate relationship between these elements and lipid metabolism. The results underscore the significance of zinc and selenium in this context.

Supplementary Materials: The following supporting information can be downloaded at https://www.mdpi.com/article/10.3390/toxics13030152/s1: Table S1: Limit of detection (LOD) and limit of quantification (LOQ) of the analyzed elements (n = 55) expressed in ppb; Table S2: Significant differences between groups before surgery; Table S3: Significant differences between groups after surgery; Table S4: Correlations between serum inorganic elements and sociodemographic variables before surgery; Table S5: Correlations between serum inorganic elements and sociodemographic variables after surgery; Table S6: Correlations between the variation in serum inorganic elements and sociodemographic variables.

Author Contributions: Conceptualization, L.S.-M. and L.A.H.-H.; methodology, Á.R.-L., B.V.D.-G., M.Z. and N.R.-S.; validation, O.P.L.; formal analysis, Á.R.-L., K.S.-R. and L.A.H.-H.; investigation, Á.R.-L. and L.A.H.-H.; resources, O.P.L.; data curation, Á.R.-L., B.V.D.-G., M.Z, J.C.-T. and E.H.-G.; writing—original draft preparation, Á.R.-L. and L.A.H.-H.; writing—review and editing, P.F.V., K.S-R, M.Z., N.R.-S., I.B.-C., L.S.-M., J.C.-T., E.H.-G., O.P.L. and B.V.D.-G.; supervision, P.F.V., L.S.-M. and L.A.H.-H.; funding acquisition, P.F.V., L.S.-M. and O.P.L. All authors have read and agreed to the published version of the manuscript.

Funding: This research was partially funded by a predoctoral grant to ÁR-L from the University of Las Palmas de Gran Canaria (ULPGC-2021-2022).

Institutional Review Board Statement: The study was conducted in accordance with the Declaration of Helsinki and approved by the Clinical Research Ethics Committee of the Complejo Hospitalario Universitario Insular Materno Infantil (protocol code 2024-093-728, on 26 June 2014).

Informed Consent Statement: Informed consent was obtained from all subjects involved in the study.

Data Availability Statement: All data will be available upon request.

Acknowledgments: The authors of this study express their sincere gratitude to Juan Ramón Hernández-Hernández, the digestive surgeon at the Complejo Hospitalario Universitario Insular Materno Infantil (Las Palmas de Gran Canaria, Spain) in charge of the bariatric surgeons.

Conflicts of Interest: The authors declare no conflicts of interest. The funder had no role in the design of the study; in the collection, analyses, or interpretation of the data; in the writing of the manuscript; or in the decision to publish the results.

References

1. Scerri, E.R. Happy 150th Birthday to the Periodic Table. *Chemistry* **2019**, *25*, 7410–7415. [CrossRef] [PubMed]
2. Cabrera-Rodriguez, R.; Luzardo, O.P.; Gonzalez-Antuna, A.; Boada, L.D.; Almeida-Gonzalez, M.; Camacho, M.; Zumbado, M.; Acosta-Dacal, A.C.; Rial-Berriel, C.; Henriquez-Hernandez, L.A. Occurrence of 44 elements in human cord blood and their association with growth indicators in newborns. *Environ. Int.* **2018**, *116*, 43–51. [CrossRef]
3. Henriquez-Hernandez, L.A.; Boada, L.D.; Carranza, C.; Perez-Arellano, J.L.; Gonzalez-Antuna, A.; Camacho, M.; Almeida-Gonzalez, M.; Zumbado, M.; Luzardo, O.P. Blood levels of toxic metals and rare earth elements commonly found in e-waste may exert subtle effects on hemoglobin concentration in sub-Saharan immigrants. *Environ. Int.* **2017**, *109*, 20–28. [CrossRef] [PubMed]
4. Budi, H.S.; Catalan Opulencia, M.J.; Afra, A.; Abdelbasset, W.K.; Abdullaev, D.; Majdi, A.; Taherian, M.; Ekrami, H.A.; Mohammadi, M.J. Source, toxicity and carcinogenic health risk assessment of heavy metals. *Rev. Environ. Health* **2024**, *39*, 77–90. [CrossRef] [PubMed]
5. Al Osman, M.; Yang, F.; Massey, I.Y. Exposure routes and health effects of heavy metals on children. *Biometals* **2019**, *32*, 563–573. [CrossRef] [PubMed]
6. Rehman, K.; Fatima, F.; Waheed, I.; Akash, M.S.H. Prevalence of exposure of heavy metals and their impact on health consequences. *J. Cell Biochem.* **2018**, *119*, 157–184. [CrossRef] [PubMed]
7. Andeobu, L.; Wibowo, S.; Grandhi, S. An assessment of e-waste generation and environmental management of selected countries in Africa, Europe and North America: A systematic review. *Sci. Total Environ.* **2021**, *792*, 148078. [CrossRef] [PubMed]
8. Tansel, B. From electronic consumer products to e-wastes: Global outlook, waste quantities, recycling challenges. *Environ. Int.* **2017**, *98*, 35–45. [CrossRef]
9. Samal, R.R.; Subudhi, U. Biochemical and biophysical interaction of rare earth elements with biomacromolecules: A comprehensive review. *Chemosphere* **2024**, *357*, 142090. [CrossRef] [PubMed]
10. Brouziotis, A.A.; Giarra, A.; Libralato, G.; Pagano, G.; Guida, M.; Trifuoggi, M. Toxicity of rare earth elements: An overview on human health impact. *Front. Environ. Sci.* **2022**, *10*, 948041. [CrossRef]
11. Lafontan, M. Historical perspectives in fat cell biology: The fat cell as a model for the investigation of hormonal and metabolic pathways. *Am. J. Physiol. Cell Physiol.* **2012**, *302*, C327–C359. [CrossRef] [PubMed]
12. Papalou, O.; Kandaraki, E.A.; Papadakis, G.; Diamanti-Kandarakis, E. Endocrine Disrupting Chemicals: An Occult Mediator of Metabolic Disease. *Front. Endocrinol.* **2019**, *10*, 112. [CrossRef] [PubMed]
13. Zangiabadian, M.; Jolfayi, A.G.; Nejadghaderi, S.A.; Amirkhosravi, L.; Sanjari, M. The association between heavy metal exposure and obesity: A systematic review and meta-analysis. *J. Diabetes Metab. Disord.* **2024**, *23*, 11–26. [CrossRef] [PubMed]
14. Henriquez-Hernandez, L.A.; Luzardo, O.P.; Valeron, P.F.; Zumbado, M.; Serra-Majem, L.; Camacho, M.; Gonzalez-Antuna, A.; Boada, L.D. Persistent organic pollutants and risk of diabetes and obesity on healthy adults: Results from a cross-sectional study in Spain. *Sci. Total Environ.* **2017**, *607–608*, 1096–1102. [CrossRef]
15. Ng, M.; Fleming, T.; Robinson, M.; Thomson, B.; Graetz, N.; Margono, C.; Mullany, E.C.; Biryukov, S.; Abbafati, C.; Abera, S.F.; et al. Global, regional, and national prevalence of overweight and obesity in children and adults during 1980-2013: A systematic analysis for the Global Burden of Disease Study 2013. *Lancet* **2014**, *384*, 766–781. [CrossRef] [PubMed]
16. Boutari, C.; Mantzoros, C.S. A 2022 update on the epidemiology of obesity and a call to action: As its twin COVID-19 pandemic appears to be receding, the obesity and dysmetabolism pandemic continues to rage on. *Metabolism* **2022**, *133*, 155217. [CrossRef] [PubMed]
17. Smith, K.B.; Smith, M.S. Obesity Statistics. *Prim. Care* **2016**, *43*, 121–135. [CrossRef] [PubMed]
18. Dhurandhar, N.V.; Petersen, K.S.; Webster, C. Key Causes and Contributors of Obesity: A Perspective. *Nurs. Clin. N. Am.* **2021**, *56*, 449–464. [CrossRef] [PubMed]
19. Vekic, J.; Zeljkovic, A.; Stefanovic, A.; Jelic-Ivanovic, Z.; Spasojevic-Kalimanovska, V. Obesity and dyslipidemia. *Metabolism* **2019**, *92*, 71–81. [CrossRef] [PubMed]
20. Santos, A.L.; Sinha, S. Obesity and aging: Molecular mechanisms and therapeutic approaches. *Ageing Res. Rev.* **2021**, *67*, 101268. [CrossRef]
21. Porta, M.; Pumarega, J.; Gasull, M.; Aguilar, R.; Henriquez-Hernandez, L.A.; Basagana, X.; Zumbado, M.; Villar-Garcia, J.; Rius, C.; Mehta, S.; et al. Individual blood concentrations of persistent organic pollutants and chemical elements, and COVID-19: A prospective cohort study in Barcelona. *Environ. Res.* **2023**, *223*, 115419. [CrossRef] [PubMed]
22. Ruban, A.; Stoenchev, K.; Ashrafian, H.; Teare, J. Current treatments for obesity. *Clin. Med.* **2019**, *19*, 205–212. [CrossRef] [PubMed]
23. Perdomo, C.M.; Cohen, R.V.; Sumithran, P.; Clement, K.; Fruhbeck, G. Contemporary medical, device, and surgical therapies for obesity in adults. *Lancet* **2023**, *401*, 1116–1130. [CrossRef]
24. Colquitt, J.L.; Pickett, K.; Loveman, E.; Frampton, G.K. Surgery for weight loss in adults. *Cochrane Database Syst. Rev.* **2014**, *2014*, CD003641. [CrossRef] [PubMed]

25. Sjostrom, L.; Gummesson, A.; Sjostrom, C.D.; Narbro, K.; Peltonen, M.; Wedel, H.; Bengtsson, C.; Bouchard, C.; Carlsson, B.; Dahlgren, S.; et al. Effects of bariatric surgery on cancer incidence in obese patients in Sweden (Swedish Obese Subjects Study): A prospective, controlled intervention trial. *Lancet Oncol.* **2009**, *10*, 653–662. [CrossRef]
26. Diaz-Gonzalez, B.V.; Ramos-Luzardo, A.; Henriquez-Hernandez, L.A.; Serra-Majem, L.; Bautista-Castano, I.; Acosta-Dacal, A.; Luzardo, O.P.; Hernandez-Garcia, E.; Cornejo-Torre, J.; Hernandez-Hernandez, J.R.; et al. Effect of bariatric surgery in the body burden of persistent and non-persistent pollutants: Longitudinal study in a cohort of morbidly obese patients. *Front. Endocrinol.* **2024**, *15*, 1412261. [CrossRef] [PubMed]
27. ATSDR. *The ATSDR 2022 Substance Priority List*; ATSDR: Atlanta, GA, USA, 2022.
28. Tansel, B. Increasing gaps between materials demand and materials recycling rates: A historical perspective for evolution of consumer products and waste quantities. *J. Environ. Manag.* **2020**, *276*, 111196. [CrossRef] [PubMed]
29. Srivastav, A.L.; Markandeya; Patel, N.; Pandey, M.; Pandey, A.K.; Dubey, A.K.; Kumar, A.; Bhardwaj, A.K.; Chaudhary, V.K. Concepts of circular economy for sustainable management of electronic wastes: Challenges and management options. *Environ. Sci. Pollut. Res. Int.* **2023**, *30*, 48654–48675. [CrossRef] [PubMed]
30. Henriquez-Hernandez, L.A.; Zumbado, M.; Rodriguez-Hernandez, A.; Duarte-Lopes, E.; Lopes-Ribeiro, A.L.; Alfama, P.M.; Livramento, M.; Diaz-Diaz, R.; Bernal-Suarez, M.D.M.; Boada, L.D.; et al. Human biomonitoring of inorganic elements in a representative sample of the general population from Cape Verde: Results from the PERVEMAC-II study. *Chemosphere* **2023**, *339*, 139594. [CrossRef] [PubMed]
31. Gonzalez-Antuna, A.; Camacho, M.; Henriquez-Hernandez, L.A.; Boada, L.D.; Almeida-Gonzalez, M.; Zumbado, M.; Luzardo, O.P. Simultaneous quantification of 49 elements associated to e-waste in human blood by ICP-MS for routine analysis. *MethodsX* **2017**, *4*, 328–334. [CrossRef] [PubMed]
32. Adnan, M.T.; Amin, M.N.; Uddin, M.G.; Hussain, M.S.; Sarwar, M.S.; Hossain, M.K.; Uddin, S.M.N.; Islam, M.S. Increased concentration of serum MDA, decreased antioxidants and altered trace elements and macro-minerals are linked to obesity among Bangladeshi population. *Diabetes Metab. Syndr.* **2019**, *13*, 933–938. [CrossRef] [PubMed]
33. Luciano-Mateo, F.; Cabre, N.; Nadal, M.; Garcia-Heredia, A.; Baiges-Gaya, G.; Hernandez-Aguilera, A.; Camps, J.; Joven, J.; Domingo, J.L. Serum concentrations of trace elements and their relationships with paraoxonase-1 in morbidly obese women. *J. Trace Elem. Med. Biol.* **2018**, *48*, 8–15. [CrossRef] [PubMed]
34. Hastuti, A.; Costas-Rodriguez, M.; Anoshkina, Y.; Parnall, T.; Madura, J.A., 2nd; Vanhaecke, F. High-precision isotopic analysis of serum and whole blood Cu, Fe and Zn to assess possible homeostasis alterations due to bariatric surgery. *Anal. Bioanal. Chem.* **2020**, *412*, 727–738. [CrossRef] [PubMed]
35. Papamargaritis, D.; Aasheim, E.T.; Sampson, B.; le Roux, C.W. Copper, selenium and zinc levels after bariatric surgery in patients recommended to take multivitamin-mineral supplementation. *J. Trace Elem. Med. Biol.* **2015**, *31*, 167–172. [CrossRef] [PubMed]
36. Miedziaszczyk, M.; Ciabach, P.; Szalek, E. The Effects of Bariatric Surgery and Gastrectomy on the Absorption of Drugs, Vitamins, and Mineral Elements. *Pharmaceutics* **2021**, *13*, 2111. [CrossRef] [PubMed]
37. Mikalsen, S.M.; Bjorke-Monsen, A.L.; Flaten, T.P.; Whist, J.E.; Aaseth, J. Cadmium, lead and mercury in Norwegian obese patients before and 12 months after bariatric surgery. *J. Trace Elem. Med. Biol.* **2019**, *54*, 150–155. [CrossRef]
38. Smith, J.C.; Farris, F.F. Methyl mercury pharmacokinetics in man: A reevaluation. *Toxicol. Appl. Pharmacol.* **1996**, *137*, 245–252. [CrossRef] [PubMed]
39. Abass, K.; Koiranen, M.; Mazej, D.; Tratnik, J.S.; Horvat, M.; Hakkola, J.; Jarvelin, M.R.; Rautio, A. Arsenic, cadmium, lead and mercury levels in blood of Finnish adults and their relation to diet, lifestyle habits and sociodemographic variables. *Environ. Sci. Pollut. Res. Int.* **2017**, *24*, 1347–1362. [CrossRef] [PubMed]
40. Ravaglia, G.; Forti, P.; Maioli, F.; Nesi, B.; Pratelli, L.; Savarino, L.; Cucinotta, D.; Cavalli, G. Blood micronutrient and thyroid hormone concentrations in the oldest-old. *J. Clin. Endocrinol. Metab.* **2000**, *85*, 2260–2265. [CrossRef] [PubMed]
41. Gonzalez-Estecha, M.; Palazon-Bru, I.; Bodas-Pinedo, A.; Trasobares, E.; Palazon-Bru, A.; Fuentes, M.; Cuadrado-Cenzual, M.A.; Calvo-Manuel, E. Relationship between serum selenium, sociodemographic variables, other trace elements and lipid profile in an adult Spanish population. *J. Trace Elem. Med. Biol.* **2017**, *43*, 93–105. [CrossRef]
42. Fang, C.; Wu, W.; Gu, X.; Dai, S.; Zhou, Q.; Deng, H.; Shen, F.; Chen, J. Association of serum copper, zinc and selenium levels with risk of metabolic syndrome: A nested case-control study of middle-aged and older Chinese adults. *J. Trace Elem. Med. Biol.* **2019**, *52*, 209–215. [CrossRef] [PubMed]
43. Milne, D.B.; Johnson, P.E. Assessment of copper status: Effect of age and gender on reference ranges in healthy adults. *Clin. Chem.* **1993**, *39*, 883–887. [CrossRef] [PubMed]
44. Blydt-Hansen, I.; Swarup, V.; Kishibe, T.; Bekdache, C.; Crowther, M.A.; Weyand, A.C.; Tang, G.; Sholzberg, M. Evaluating Hemoglobin Thresholds on the Basis of Sex: Preliminary Insights from a Systematic Review. *Blood* **2023**, *142*, 7200. [CrossRef]
45. Zhang, J.; Hu, J.; Zhao, J.; Li, J.; Cai, X. Serum zinc concentrations and prediabetes and diabetes in the general population. *Biol. Trace Elem. Res.* **2022**, *200*, 1071–1077. [CrossRef] [PubMed]

46. Feng, W.; Cui, X.; Liu, B.; Liu, C.; Xiao, Y.; Lu, W.; Guo, H.; He, M.; Zhang, X.; Yuan, J.; et al. Association of urinary metal profiles with altered glucose levels and diabetes risk: A population-based study in China. *PLoS ONE* **2015**, *10*, e0123742. [CrossRef]
47. Schlienger, J.L.; Grunenberger, F.; Maier, E.A.; Simon, C.; Chabrier, G.; Leroy, M.J. Disorders of plasma trace elements in diabetes. Relation to blood glucose equilibrium. *Presse Med.* **1988**, *17*, 1076–1079. [PubMed]
48. Medina-Estevez, F.; Zumbado, M.; Luzardo, O.P.; Rodriguez-Hernandez, A.; Boada, L.D.; Fernandez-Fuertes, F.; Santandreu-Jimenez, M.E.; Henriquez-Hernandez, L.A. Association between Heavy Metals and Rare Earth Elements with Acute Ischemic Stroke: A Case-Control Study Conducted in the Canary Islands (Spain). *Toxics* **2020**, *8*, 66. [CrossRef] [PubMed]
49. Bjermo, H.; Aune, M.; Cantillana, T.; Glynn, A.; Lind, P.M.; Ridefelt, P.; Darnerud, P.O. Serum levels of brominated flame retardants (BFRs: PBDE, HBCD) and influence of dietary factors in a population-based study on Swedish adults. *Chemosphere* **2017**, *167*, 485–491. [CrossRef]
50. Zhao, X.; Yang, X.; Du, Y.; Li, R.; Zhou, T.; Wang, Y.; Chen, T.; Wang, D.; Shi, Z. Polybrominated diphenyl ethers in serum from residents living in a brominated flame retardant production area: Occurrence, influencing factors, and relationships with thyroid and liver function. *Environ. Pollut.* **2021**, *270*, 116046. [CrossRef] [PubMed]
51. Ju, W.; Ji, M.; Li, X.; Li, Z.; Wu, G.; Fu, X.; Yang, X.; Gao, X. Relationship between higher serum selenium level and adverse blood lipid profile. *Clinical Nutrition* **2018**, *37*, 1512–1517. [CrossRef] [PubMed]
52. Nie, Q.; Wang, C.; Zhou, L. Association between Serum Selenium Levels and Lipids among People with and without Diabetes. *Nutrients* **2023**, *15*, 3190. [CrossRef] [PubMed]
53. Stranges, S.; Laclaustra, M.; Ji, C.; Cappuccio, F.P.; Navas-Acien, A.; Ordovas, J.M.; Rayman, M.; Guallar, E. Higher selenium status is associated with adverse blood lipid profile in British adults. *J. Nutr.* **2010**, *140*, 81–87. [CrossRef] [PubMed]
54. Hiller, R.; Seigel, D.; Sperduto, R.D.; Blair, N.; Burton, T.C.; Farber, M.D.; Gragoudas, E.S.; Gunter, E.W.; Haller, J.; Seddon, J.M.; et al. Serum zinc and serum lipid profiles in 778 adults. *Ann. Epidemiol.* **1995**, *5*, 490–496. [CrossRef] [PubMed]
55. Yeung, D.C.; Lam, K.S.; Wang, Y.; Tso, A.W.; Xu, A. Serum zinc-alpha2-glycoprotein correlates with adiposity, triglycerides, and the key components of the metabolic syndrome in Chinese subjects. *J. Clin. Endocrinol. Metab.* **2009**, *94*, 2531–2536. [CrossRef] [PubMed]
56. Banaszak, M.; Gorna, I.; Przyslawski, J. Zinc and the Innovative Zinc-alpha2-Glycoprotein Adipokine Play an Important Role in Lipid Metabolism: A Critical Review. *Nutrients* **2021**, *13*, 2023. [CrossRef] [PubMed]
57. Ranasinghe, P.; Wathurapatha, W.S.; Ishara, M.H.; Jayawardana, R.; Galappatthy, P.; Katulanda, P.; Constantine, G.R. Effects of Zinc supplementation on serum lipids: A systematic review and meta-analysis. *Nutr. Metab.* **2015**, *12*, 26. [CrossRef] [PubMed]

MDPI AG
Grosspeteranlage 5
4052 Basel
Switzerland
Tel.: +41 61 683 77 34

Toxics Editorial Office
E-mail: toxics@mdpi.com
www.mdpi.com/journal/toxics

www.ingramcontent.com/pod-product-compliance
Lightning Source LLC
Chambersburg PA
CBHW041416210926
51850CB00013B/3291